# John S. Nkoma

# INTRODUCTION TO BASIC CONCEPTS
# FOR
# ENGINEERS AND SCIENTISTS

## Electromagnetic, Quantum, Statistical and Relativistic Concepts

MKUKI NA NYOTA
DAR—ES—SALAAM

Professor John S Nkoma obtained his BSc in Physics and Mathematics from the University of Dar es Salaam, Tanzania, and later his MSc and PhD in Physics from the University of Essex, UK. Prof. Nkoma has over 30 years experience of university teaching and research in Botswana, Italy, Tanzania and UK, and has been external examiner to several universities. His research interests are in condensed matter physics, materials science and ICTs regulation in telecommunications, internet, broadcasting and postal communications. He is a Fellow of the African Academy of Sciences. He also served as the Director General/CEO of the Tanzania Communications Regulatory Authority (TCRA) during 2004 to 2015. His previous books include: 1. Introduction to Optics: Geometrical, Physical and Quantum, Bay Publishers, Gaborone ISBN 99912-511-6-2 (2003) by J S Nkoma and P K Jain, and 2. Introduction to Mechanics: Kinematics, Newtonian and Lagrangian, Bay Publishers, Gaborone ISBN 99912-561-4-8 (2004) by P K Jain and J S Nkoma.

PUBLISHED BY
Mkuki na Nyota Publishers Ltd
P. O. Box 4246
Dar es Salaam, Tanzania
www.mkukinanyota.com

©John S. Nkoma, 2018

ISBN 978-9987-08-328-2

Typesetting and layout by John S. Nkoma

Visit www.mkukinanyota.com to read more about and to purchase any of Mkuki na Nyota books.You will also find featured authors, interviews and news about other publisher/author events. Sign up for our e-newsletters for updates on new releases and other announcements.

Distributed world wide outside Africa by African Books Collective.
www.africanbookscollective.com

# Contents

CONTENTS

CONTENTS

# Preface

Science and Technology are ubiquitous in the modern world as evidenced by digital lifestyles through mobile phones, computers, digital financial services, digital music, digital television, online newspapers, digital medical equipment and services including e-services (e-commerce, e-learning, e-health, e-government) and the internet. This book, *"Introduction to Basic concepts for Engineers and Scientists: Electromagnetic, Quantum, Statistical and Relativistic Concepts."* is written with the objective of imparting basic concepts for engineering, physics, chemistry students or indeed other sciences, so that such students get an understanding as to what is behind all these modern advances in science and technology.

The basic concepts covered in this book include electromagnetic, quantum, statistical and relativistic concepts, and are covered in 20 chapters. The choice of these concepts is not accidental, but deliberate so as to highlight the importance of these basic science concepts in modern engineering and technology. Electromagnetic concepts, are covered in chapters 1 to 6 with chapters 1 (Maxwell's equations), 2 (Electromagnetic waves at boundaries), 3 (Diffraction and Interference), 4 (Optical fiber communications), 5 (Satellite communications) and 6 (Mobile cellular communications). Quantum concepts are covered in chapters 7 to 15 with chapters 7 (Wave-particle duality), 8 (The wave function and solutions of the Schrödinger equation in different systems), 9 (Introduction to the structure of the atom), Introduction to materials science I, II, III and IV, in four chapters: 10 (I: Crystal structure), 11 (II: Phonons), 12 (III: Electrons) and 13 (IV: Magnetic materials), 14 (Semiconductor devices), and 15 (Quantum Optics). Statistical concepts are covered in chapters 16 to 19, with chapters 16 (Introduction to statistical mechanics), 17 (Statistical mechanics distribution functions, covering Maxwell-Boltzmann statistics, Fermi-Dirac statistics and Bose-Einstein statistics), 18 (Transport theory) and 19 (Phase transitions). Finally, chapter 20 (Relativity) where Galilean, Special and General Relativity are discussed.

After the 20 chapters, there are detailed appendices, bibliography and an index. The appendices contain information on some useful constants and mathematical relations. The bibliography consists of books and journal articles, some are recent while others are old but still useful for the purposes of this book. Also, recognizing that the Internet is a rich source of information, there are several references to Internet resources, for which we acknowledge. Finally there is a handy detailed index to help the reader find topics of interest.

A word about notation and mathematics. With vectors, there are a lot of notations, for example, by hand, some use $\vec{A}$, or $\underline{A}$, and in print $\mathbf{A}$. In this book, we use $\mathbf{A}$. Generally, mathematics is a language of physics and engineering. Understanding mathematics for a particular concept makes that concept more clearer. Further, the language of mathematics is universal. It is for this reason that in this book, mathematics has been used freely to explain the many concepts of science and technology discussed herein.

The book is suitable for modules or courses in degree programs in engineering, physics or other science and technology areas. A particular course can find some chapters relevant. There are a large number of problems and exercises at the end of each chapter. Problem solving is an essential part in the training of scientists and engineers.

## Acknowledgments

I acknowledge all the institutions I have been associated with in different capacities over the years. The contents of this book reflects the experience of the author in working in several institutions, in particular the University of Dar es Salaam in Tanzania, the University of Botswana, the University of Essex in UK, the International Center for Theoretical Physics (ICTP) in Italy, the Tanzania Communications Regulatory Authority (TCRA).

I would like to express my sincere gratitude to a number of colleagues for carefully reading the manuscript and making many valuable suggestions, and these include, from the University of Dar es Salaam: Prof. J. Kondoro, Prof. H. Kundaeli, Dr I. Makundi and Dr. L. Massawe, and from the University of Botswana: Prof. P. K. Jain, Prof. P. V. C. Luhanga, Prof. M. Masale, and finally Eng. P. Ulanga and Dr. E. Manasseh, former colleagues at the Tanzania Communications Regulatory Authority (TCRA). I also thank the publisher for having produced the book in an excellent manner. Patience and understanding of my family and friends is gratefully acknowledged. Lastly, but not least, I thank my many students over the years who have attended my lectures from which this book has been developed.

**John S. Nkoma**
Professor of Physics and Ex-Director General/CEO, TCRA
Mbezi Beach
Dar es Salaam
February 2018

# Chapter 1

# Introduction to Maxwell's Equations

## 1.1 Introduction

In this first chapter of this book, we introduce Maxwell's equations, which are one of the most elegant ways of describing the electric and magnetic fields comprising the electromagnetic field. Maxwell introduced these equations in 1864. As we shall see, these equations predict correctly the speed of light, thus explaining the electromagnetic spectrum. Maxwell's equations underpin most of the modern information and communication technologies, including telecommunications, broadcasting, internet and so on.

## 1.2 Synthesis of Maxwell's Equations

### 1. Electric fields and charges

An electric field, $\mathbf{E}$ arises due to electric charges, $q$ as illustrated in Figure 1.1.

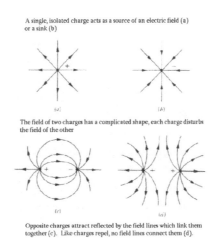

Figure 1.1: Electric fields and charges

$$\text{Electric flux, } \phi \;=\; \int_S \mathbf{E} \cdot \mathbf{dS}$$

where the right hand side represents a surface integral over $dS$

Gauss law of flux says

$$\text{Electric flux, } \phi \;=\; \frac{\sum q}{\epsilon_0} = \frac{q}{\epsilon_0}$$

$$\int_S \mathbf{E} \cdot \mathbf{dS} \;=\; \frac{q}{\epsilon_0} = \frac{1}{\epsilon_0} \int_V \rho dV \tag{1.1}$$

which is *Gauss law for electric fields in Integral form.*

But Divergence theorem from vector analysis says

$$\int_S \mathbf{E} \cdot \mathbf{dS} \;=\; \int_V \nabla \cdot \mathbf{E} dV$$

$$\int_V \nabla \cdot \mathbf{E} dV \;=\; \frac{1}{\epsilon_0} \int_V \rho dV$$

*or*

$$\nabla \cdot \mathbf{E} \;=\; \frac{\rho}{\epsilon_0} \tag{1.2}$$

which is *Gauss law for electric fields in differential form.*

## 2. Magnetic fields

Always, a magnet has a north pole and south pole, as illustrated in Figure 1.2 and Figure 1.3. A magnetic monopole does not exixt or shall we say has never been observed so far.

Magnets exist as dipoles. Magnetic field lines are continous.
Net magnetic flux is given by

$$\phi_B \;=\; \int_S \mathbf{B} \cdot \mathbf{dS} = 0$$

$$\text{Hence } \int_S \mathbf{B} \cdot \mathbf{dS} \;=\; 0 \tag{1.3}$$

which is *Gauss law for magnetic fields in Integral form.*

But Divergence theorem says

$$\int_S \mathbf{B} \cdot \mathbf{dS} \;=\; \int_V \nabla \cdot \mathbf{B} dV$$

*or*

$$\nabla \cdot \mathbf{B} \;=\; 0 \tag{1.4}$$

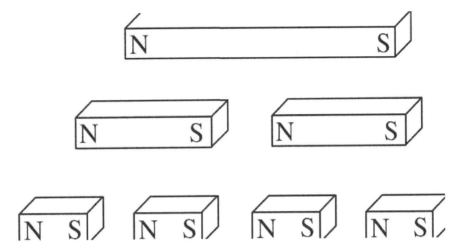

Figure 1.2: A magnet has always a N-pole and S-pole

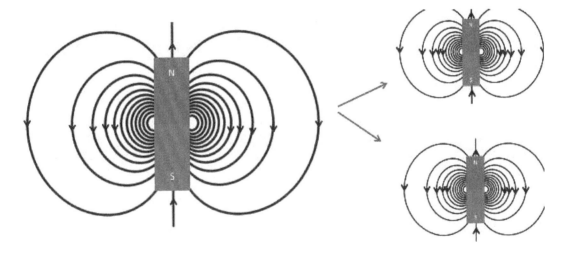

Figure 1.3: Magnetic-fields

which is *Gauss law for magnetic fields in differential form.*

### 3. Faraday's law of electromagnetic induction

Faraday's law states that the induced emf, $\varepsilon$, is proportional to the rate of change of flux. Experimentally, this is illustrated in Figure 1.4.

$$\varepsilon \ \propto \ \frac{d\phi}{dt} \qquad \text{Neumann's law}$$

$$\varepsilon \ = \ -\frac{d\phi}{dt} \qquad \text{Negative due to Lenz's law}$$

$$\varepsilon \ = \ \int \mathbf{E} \cdot \mathbf{dl}$$

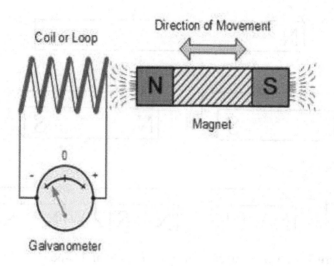

Figure 1.4: Faraday's law: A changing magnetic flux induces an emf.

$$\phi_B = \int_S \mathbf{B} \cdot \mathbf{dS}$$

$$\varepsilon = -\frac{d\phi}{dt}$$

$$= -\frac{d}{dt}\int_S \mathbf{B} \cdot \mathbf{dS}$$

$$= -\int_S \frac{d\mathbf{B}}{dt} \cdot dS$$

Hence

$$\varepsilon = \int \mathbf{E} \cdot \mathbf{dl} = -\frac{\mathbf{d}\phi}{\mathbf{dt}}$$

where the right hand side        represents a line integral over $dl$

$$\int \mathbf{E} \cdot \mathbf{dl} = -\int_S \frac{d\mathbf{B}}{dt} \cdot dS \qquad (1.5)$$

which is *Faraday's law in Integral form*.

But Stoke's theorem says

$$\int_S \nabla \wedge \mathbf{E} \cdot \mathbf{dS} = \int \mathbf{E} \cdot \mathbf{dl}$$

$$\varepsilon = \int_S \nabla \wedge \mathbf{E} \cdot \mathbf{dS}$$

$$\varepsilon = \int_S \nabla \wedge \mathbf{E} \cdot \mathbf{dS} = -\int_\mathbf{S} \frac{\partial \mathbf{B}}{\partial \mathbf{t}} \cdot \mathbf{dS}$$

$$\nabla \wedge \mathbf{E} = -\frac{\partial \mathbf{B}}{\partial t} \qquad (1.6)$$

which is *Faraday's law in differential form.*

### 4. Magnetic effect of an electric current

**Case (i)**

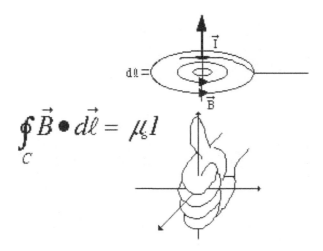

$$\oint_C \vec{B} \bullet \vec{d\ell} = \mu_0 I$$

Figure 1.5: Magnetic effect of an electric current.

**Case (ii)**

$$
\begin{aligned}
\int \mathbf{B} \cdot \mathbf{dl} &= \mu_0 I_d \\
&= \mu_0 \frac{dQ}{dt} \\
\text{But } \phi &= \frac{Q}{\epsilon_0} = \int_S \mathbf{E} \cdot \mathbf{dS} \\
\int \mathbf{B} \cdot \mathbf{dl} &= \mu_0 \frac{d}{dt}(\phi \epsilon_0) = \mu_0 \epsilon_0 \frac{d\phi}{dt} \\
&= \mu_0 \epsilon_0 \frac{d}{dt} \int_S \mathbf{E} \cdot \mathbf{dS}
\end{aligned}
$$

whence

$$I_d = \epsilon_0 \frac{d\phi}{dt} = \epsilon_0 \frac{d}{dt} \int_S \mathbf{E} \cdot \mathbf{dS}$$

Generally

$$\int \mathbf{B} \cdot \mathbf{dl} = \mu_0(I + I_d)$$

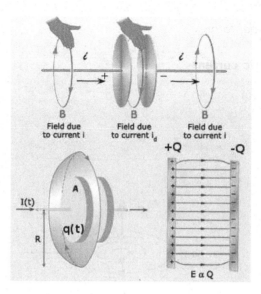

Figure 1.6: The displacement current.

$$= \mu_0(I + \epsilon_0 \frac{d}{dt}\int_S \mathbf{E}\cdot\mathbf{dS})$$

But $I = \int \mathbf{J}\cdot\mathbf{dS}$      where $\mathbf{J}$ is the current density

$$\int \mathbf{B}\cdot\mathbf{dl} = \mu_0\int \mathbf{J}\cdot\mathbf{dS} + \epsilon_0\frac{d}{dt}\int_{\mathbf{S}} \mathbf{E}\cdot\mathbf{dS}$$

$$= \mu_0\int_S \left(\mathbf{J} + \epsilon_0\frac{d}{dt}\mathbf{E}\right)\cdot\mathbf{dS} \tag{1.7}$$

which is *Ampere-Maxwell law in Integral form*.

Using Stoke's theorem says

$$\int_l \mathbf{B}\cdot\mathbf{dl} = \int_S \nabla\wedge\mathbf{B}\cdot\mathbf{dS}$$

$$\int_S \nabla\wedge\mathbf{B}\cdot\mathbf{dS} = \mu_0\int_S \left(\mathbf{J} + \epsilon_0\frac{d}{dt}\mathbf{E}\right)\cdot\mathbf{dS}$$

$$\nabla\wedge\mathbf{B} = \mu_0\mathbf{J} + \mu_0\epsilon_0\frac{d}{dt}\mathbf{E} \tag{1.8}$$

which is *Ampere-Maxwell law in differential form*.

Since $\mathbf{B} = \mu_0\mathbf{H}$ and $\mathbf{D} = \epsilon_0\mathbf{E}$, the above equation can be written in the form

$$\nabla\wedge\mathbf{H} = \mathbf{J} + \frac{d}{dt}\mathbf{D}$$

So, let us now collect the form of four equations in integral form (1.1, 1.3, 1.5 and 1.7) as well as the four equations in differential form (1.2, 1.4, 1.6 and 1.8).

## 1.3   Maxwell's Equations in Integral form

The electric field $\mathbf{E}$ and magnetic field $\mathbf{B}$ satisfy Maxwell's equations which are given below in the integral form.

$$
\begin{aligned}
\int_S \mathbf{E} \cdot \mathbf{dS} &= \frac{q}{\epsilon_0} = \frac{1}{\epsilon_0} \int_V \rho \, dv \\
\int_S \mathbf{B} \cdot \mathbf{dS} &= 0 \\
\int \mathbf{E} \cdot \mathbf{dl} &= -\frac{\partial \phi_B}{\partial t} = -\int_S \frac{\partial \mathbf{B}}{\partial t} \cdot \mathbf{dS} \\
\int \mathbf{B} \cdot \mathbf{dl} &= \mu_0 \int_S \left( \mathbf{J} + \epsilon_0 \frac{\partial \mathbf{E}}{\partial \mathbf{t}} \right) \cdot \mathbf{dS}
\end{aligned}
$$

where

$$
\begin{aligned}
\rho \quad &\text{is} \quad \text{the charge density,} \\
\epsilon_0 &= 8.854 \times 10^{-12} Fm^{-1} \text{is the permittivity of free space,} \\
\mu_0 &= 4\pi \times 10^{-7} Hm^{-1} \text{is the permeability of free space,} \\
\mathbf{J} \quad &\text{is} \quad \text{the current density vector, and}
\end{aligned}
$$

## 1.4   Maxwell's Equations in Differential form

The electric field $\mathbf{E}$ and magnetic field $\mathbf{B}$ satisfy Maxwell's equations which are given below in the differential form.

$$
\begin{aligned}
\nabla \cdot \mathbf{E} &= \frac{\rho}{\epsilon_0} \\
\nabla \cdot \mathbf{B} &= 0 \\
\nabla \wedge \mathbf{E} &= -\frac{\partial \mathbf{B}}{\partial t} \\
\nabla \wedge \mathbf{B} &= \mu_0 \mathbf{J} + \mu_0 \epsilon_0 \frac{\partial \mathbf{E}}{\partial \mathbf{t}}
\end{aligned}
$$

## 1.5   Maxwell's Equations and the Wave Equation in Free Space

Maxwell's equations in free space, where $\rho = 0, j = 0$, reduce to

$$
\begin{aligned}
\nabla \cdot \mathbf{E} &= 0 \\
\nabla \cdot \mathbf{B} &= 0 \\
\nabla \wedge \mathbf{E} &= -\frac{\partial \mathbf{B}}{\partial t} \\
\nabla \wedge \mathbf{B} &= \mu_0 \epsilon_0 \frac{\partial \mathbf{E}}{\partial t}
\end{aligned}
$$

Taking the curl on both sides of the third equation,

$$\nabla \wedge (\nabla \wedge \mathbf{E}) = \nabla \wedge \left(-\frac{\partial \mathbf{B}}{\partial t}\right)$$

Using $\quad \mathbf{A} \wedge (\mathbf{B} \wedge \mathbf{C}) = \mathbf{B}(\mathbf{A} \cdot \mathbf{C}) - \mathbf{C}(\mathbf{A} \cdot \mathbf{B})$

$$\nabla(\nabla \cdot \mathbf{E}) - \nabla^2 \mathbf{E} = -\frac{\partial}{\partial t}(\nabla \wedge \mathbf{B})$$

$$= -\frac{\partial}{\partial t}\left(\mu_0 \epsilon_0 \frac{\partial}{\partial t} \mathbf{E}\right)$$

$$\nabla(\nabla \cdot \mathbf{E}) - \nabla^2 \mathbf{E} = -\mu_0 \epsilon_0 \frac{\partial^2 \mathbf{E}}{\partial t^2}$$

$$-\nabla^2 \mathbf{E} = -\mu_0 \epsilon_0 \frac{\partial^2 \mathbf{E}}{\partial t^2}$$

$$\nabla^2 \mathbf{E} = \frac{1}{c^2}\frac{\partial^2 \mathbf{E}}{\partial t^2} \qquad (1.9)$$

which is known as the Wave Equation, and

$$c^2 = \frac{1}{\mu_0 \epsilon_0}$$

$$c = \frac{1}{\sqrt{\mu_0 \epsilon_0}} \qquad (1.10)$$

and

$$\mu_0 = 4\pi \times 10^{-7} Hm^{-1} \text{is the permeability of free space,}$$

$$\epsilon_0 = 8.854 \times 10^{-12} Fm^{-1} \text{is the permittivity of free space,}$$

Note the following,

$$\frac{1}{4\pi\epsilon_0} = 9 \times 10^9 Nm^2 C^{-2}$$

Hence

$$c = \frac{1}{\sqrt{\mu_0\epsilon_0}}$$

$$= \frac{1}{\left(4\pi \times 10^{-7} \times \frac{1}{4\pi \times 9 \times 10^9}\right)^{1/2}}$$

$$= \left(9 \times 10^7 \times 10^9\right)^{1/2}$$

$$= \left(9 \times 10^{16}\right)^{1/2}$$

$$c = \frac{1}{\sqrt{\mu_0\epsilon_0}} = 3 \times 10^8 \qquad ms^{-1}$$

which is the velocity of light in free space.

Consider a solution of the electric field of the form

$$E = E_0 e^{i(\mathbf{k}\cdot\mathbf{r} - \omega t)}$$

$$\frac{\partial \mathbf{E}}{\partial t} = -i\omega \mathbf{E}$$

$$\frac{\partial^2}{\partial t^2}\mathbf{E} = (-i\omega)(-i\omega \mathbf{E}) = -\omega^2 \mathbf{E}$$

It can also be shown that

$$\nabla^2 \mathbf{E} = -k^2 \mathbf{E}$$

Hence, from the Wave Equation, we obtain

$$(-k^2 \mathbf{E}) = \frac{1}{c^2}(-\omega^2)\mathbf{E}$$

$$k^2 = \frac{1}{c^2}\omega^2$$

$$\omega = ck$$

$$\text{or} \quad \text{using } \omega = 2\pi\nu \text{ and } k = \frac{2\pi}{\lambda}$$

$$2\pi\nu = c\frac{2\pi}{\lambda}$$

$$c = \lambda\nu \tag{1.11}$$

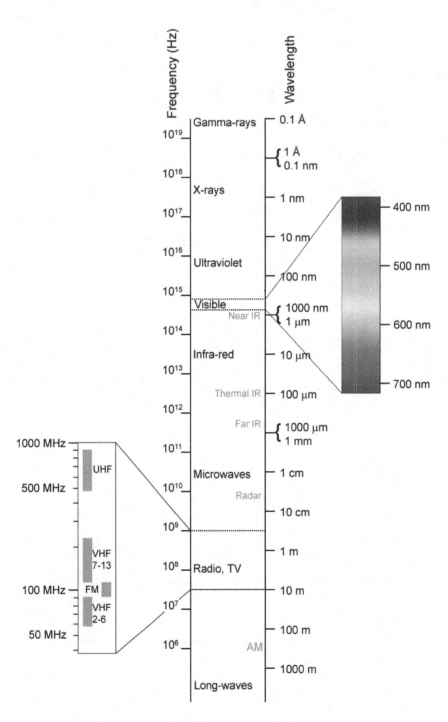

Figure 1.7: The Electromagnetic-Spectrum.

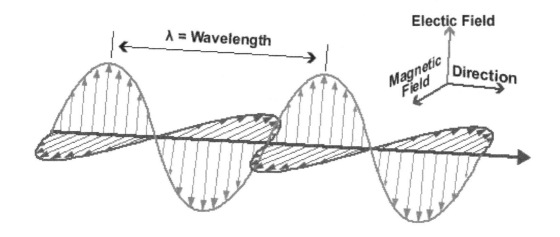

Figure 1.8: Electromagnetic waves.

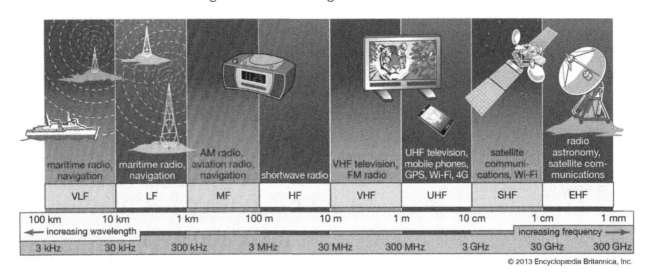

Figure 1.9: Radio Frequency Services (Source: www.britannica.com)

## 1.6   Maxwell's Equations in a Dielectric Medium

Consider a material medium which is described by a dielectric function $\epsilon(\omega)$. The dielectric function is related to the refractive index, $n(\omega)$, by the relation $\epsilon(\omega) = n^2(\omega)$, or $n(\omega) = \sqrt{\epsilon(\omega)}$. The electric field $\mathbf{E}$ and magnetic field $\mathbf{B}$ satisfy Maxwell's equations which are given below in the differential form.

$$
\begin{aligned}
\nabla \cdot \mathbf{E} &= 0 \\
\nabla \cdot \mathbf{B} &= 0 \\
\nabla \wedge \mathbf{E} &= -\frac{\partial \mathbf{B}}{\partial t} \\
\nabla \wedge \mathbf{B} &= \mu_0 \frac{\partial \mathbf{D}}{\partial t}
\end{aligned}
$$

where the displacement vector $\mathbf{D}$ is related to the electric field $\mathbf{E}$ and the polarization vector $\mathbf{P}$ by $\mathbf{D} = \epsilon_0 \mathbf{E} + \mathbf{P} = \epsilon_0 \epsilon(\omega)\mathbf{E}$, which for vacuum is only $\mathbf{D} = \epsilon_0 \mathbf{E}$.

It can be seen from Maxwell's equations that

$$
\begin{aligned}
\nabla \wedge (\nabla \wedge \mathbf{E}) &= \nabla \wedge \left( -\frac{\partial \mathbf{B}}{\partial t} \right) \\
&= -\frac{\partial}{\partial t} (\nabla \wedge \mathbf{B}) \\
&= -\frac{\partial}{\partial t} \left( \mu_0 \frac{\partial}{\partial t} \epsilon_0 \epsilon(\omega) \mathbf{E} \right) \\
\nabla(\nabla \cdot \mathbf{E}) - \nabla^2 \mathbf{E} &= -\mu_0 \epsilon_0 \epsilon(\omega) \frac{\partial^2 \mathbf{E}}{\partial t^2}
\end{aligned}
$$

Consider a solution of the electric field of the form

$$
\begin{aligned}
E &= E_0 e^{i(\mathbf{k} \cdot \mathbf{r} - \omega t)} \\
\frac{\partial \mathbf{E}}{\partial t} &= -i\omega \mathbf{E} \\
\frac{\partial^2}{\partial t^2} \mathbf{E} &= (-i\omega)(-i\omega \mathbf{E}) = -\omega^2 \mathbf{E}
\end{aligned}
$$

It can also be shown that

$$
\nabla^2 \mathbf{E} = -k^2 \mathbf{E}
$$

Hence, we obtain

$$
\begin{aligned}
0 - (-k^2 \mathbf{E}) &= -\mu_0 \epsilon_0 \epsilon(\omega)(-)\omega^2 \mathbf{E} \\
k^2 &= \frac{1}{c^2} \epsilon(\omega) \omega^2 \\
\epsilon(\omega) &= \frac{c^2 k^2}{\omega^2} \tag{1.12}
\end{aligned}
$$

The above equation relates the frequency $\omega$ and wavenumber $k$, and generally such equations are known as *dispersion relations*.

## 1.7   Maxwell's Equations in a Metal (Conducting Medium)

Consider a conducting medium which has a conductivity $\sigma$ and a current density $\mathbf{J} = \sigma \mathbf{E}$. The electric field $\mathbf{E}$ and magnetic field $\mathbf{B}$ satisfy Maxwell's equations which are given below in the differential form.

$$
\begin{aligned}
\nabla \cdot \mathbf{E} &= 0 \\
\nabla \cdot \mathbf{B} &= 0 \\
\nabla \wedge \mathbf{E} &= -\frac{\partial \mathbf{B}}{\partial t} \\
\nabla \wedge \mathbf{B} &= \mu_0 \mathbf{J} + \mu_0 \frac{\partial \mathbf{D}}{\partial t}
\end{aligned}
$$

where the displacement vector $\mathbf{D} = \epsilon_0 \mathbf{E} + \mathbf{P} = \epsilon_0 \epsilon(\omega) \mathbf{E}$, and $\mathbf{J} = \sigma \mathbf{E}$.
It can be seen from Maxwell's equations that

$$
\begin{aligned}
\nabla \wedge (\nabla \wedge \mathbf{E}) &= \nabla \wedge \left( -\frac{\partial \mathbf{B}}{\partial t} \right) \\
&= -\frac{\partial}{\partial t} (\nabla \wedge \mathbf{B}) \\
&= -\frac{\partial}{\partial t} \left( +\mu_0 \sigma \mathbf{E} + \mu_0 \frac{\partial}{\partial t} \epsilon_0 \epsilon(\omega) \mathbf{E} \right) \\
\nabla(\nabla \cdot \mathbf{E}) - \nabla^2 \mathbf{E} &= -\mu_0 \sigma \frac{\partial \mathbf{E}}{\partial t} - \mu_0 \epsilon_0 \epsilon(\omega) \frac{\partial^2 \mathbf{E}}{\partial t^2}
\end{aligned}
$$

Consider a solution of the electric field of the form

$$
\begin{aligned}
E &= E_0 e^{i(\mathbf{k} \cdot \mathbf{r} - \omega t)} \\
\frac{\partial \mathbf{E}}{\partial t} &= -i\omega \mathbf{E} \\
\frac{\partial^2}{\partial t^2} \mathbf{E} &= (-i\omega)(-i\omega \mathbf{E}) = -\omega^2 \mathbf{E}
\end{aligned}
$$

It can also be shown that

$$
\nabla^2 \mathbf{E} = -k^2 \mathbf{E}
$$

Hence, we obtain

$$
\begin{aligned}
0 - (-k^2 \mathbf{E}) &= -\mu_0 \sigma (-i\omega) \mathbf{E} - \mu_0 \epsilon_0 \epsilon(\omega)(-)\omega^2 \mathbf{E} \\
k^2 &= i\mu_0 \sigma \omega + \frac{1}{c^2} \epsilon(\omega) \omega^2 \\
\frac{c^2 k^2}{\omega^2} &= \epsilon(\omega) + i\frac{\sigma}{\epsilon_0 \omega}
\end{aligned}
$$

which makes the wavenumber $k$ in a conducting medium to be a complex quantity and hence the electric field and magnetic field will undergo a reduced amplitude due to absorption.

## 1.8   Maxwell's Equations in a Non-Linear Medium

Consider a non-linear material medium which is described by a polarization $P$ of the form

$$
\begin{aligned}
\mathbf{P} &= \epsilon_0 \left[ \chi^{(1)}\mathbf{E} + \chi^{(2)}\mathbf{E}^{(2)} + \chi^{(3)}\mathbf{E}^{(3)} + \cdots \right] \\
&= \epsilon_0\chi^{(1)}\mathbf{E} + \epsilon_0 \left[ \chi^{(2)}\mathbf{E}^{(2)} + \chi^{(3)}\mathbf{E}^{(3)} + \cdots \right] \\
\mathbf{P} &= \epsilon_0\chi^{(1)}\mathbf{E} + \mathbf{P}^{\mathbf{NL}}
\end{aligned}
\tag{1.13}
$$

Recall that the displacement vector $\mathbf{D}$ in a linear media is

$$
\begin{aligned}
\mathbf{D} &= \epsilon_0\mathbf{E} + \mathbf{P} \\
&= \epsilon_0\mathbf{E} + \epsilon_0\chi\mathbf{E} \\
\mathbf{D} &= \epsilon_0(1+\chi)\mathbf{E}
\end{aligned}
\tag{1.14}
$$

The displacement vector $\mathbf{D}$ in a non-linear media is

$$
\begin{aligned}
\mathbf{D} &= \epsilon_0\mathbf{E} + \mathbf{P} \\
&= \epsilon_0\mathbf{E} + \epsilon_0\chi^{(1)}\mathbf{E} + \mathbf{P}^{\mathbf{NL}} \\
&= \epsilon_0(1+\chi^{(1)})\mathbf{E} + \mathbf{P}^{\mathbf{NL}} \\
\mathbf{D} &= \epsilon_0\epsilon(\omega)\mathbf{E} + \mathbf{P}^{\mathbf{NL}}
\end{aligned}
\tag{1.15}
$$

Maxwell's equations in a non-linear media becomes

$$
\begin{aligned}
\nabla\cdot\mathbf{D} &= 0 \\
\nabla\cdot\mathbf{B} &= 0 \\
\nabla\wedge\mathbf{E} &= -\mu_0\frac{\partial\mathbf{H}}{\partial t} \\
\nabla\wedge\mathbf{H} &= \frac{\partial\mathbf{D}}{\partial t} = \epsilon_0\epsilon(\omega)\frac{\partial E}{\partial t} + \frac{\partial\mathbf{P}^{\mathbf{NL}}}{\partial t}
\end{aligned}
$$

It can be seen from Maxwell's equations in a non-linear medium, we obtain

$$
\begin{aligned}
\nabla\wedge(\nabla\wedge\mathbf{E}) &= \nabla\wedge\left(-\mu_0\frac{\partial\mathbf{H}}{\partial t}\right) \\
&= -\mu_0\frac{\partial}{\partial t}\left(\nabla\wedge\mathbf{H}\right) \\
&= -\mu_0\frac{\partial}{\partial t}\left[\epsilon_0\epsilon(\omega)\frac{\partial E}{\partial t} + \frac{\partial\mathbf{P}^{\mathbf{NL}}}{\partial t}\right] \\
\nabla\wedge(\nabla\wedge\mathbf{E}) &= -\mu_0\epsilon_0\epsilon(\omega)\frac{\partial^2\mathbf{E}}{\partial t^2} - \mu_0\frac{\partial^2\mathbf{P}^{\mathbf{NL}}}{\partial t^2} \\
\nabla\wedge(\nabla\wedge\mathbf{E}) + \mu_0\epsilon_0\epsilon(\omega)\frac{\partial^2\mathbf{E}}{\partial t^2} &= -\mu_0\frac{\partial^2\mathbf{P}^{\mathbf{NL}}}{\partial t^2} \\
\nabla\wedge(\nabla\wedge\mathbf{E}) + \frac{\epsilon(\omega)}{c^2}\frac{\partial^2\mathbf{E}}{\partial t^2} &= -\frac{1}{c^2\epsilon_0}\frac{\partial^2\mathbf{P}^{\mathbf{NL}}}{\partial t^2}
\end{aligned}
\tag{1.16}
$$

**Some examples of non linear optics phenomena**

**1. Raman Scattering**

Raman scattering is the inelastic scattering of light by excitations in matter. It is a nonlinear $\chi^{(2)}$ effect. Consider incident light of frequency $\omega_i$ interacting with matter in such a way that the scattered frequency is $\omega_s$ such that

$$\omega_i - \omega \;=\; \omega_s \qquad \text{Stokes Component}$$
$$\text{and}$$
$$\omega_i + \omega \;=\; \omega_s \qquad \text{Anti-Stokes Component}$$

There is also a Rayleigh component with no change in frequency. These processes are illustrated in Figure 1.10.

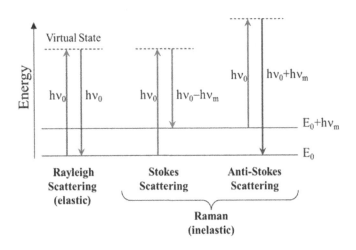

Figure 1.10: Raman Scattering

**2. Second Harmonic Generation**

Second Harmonic Generation (SHG), also known as frequency doubling, is a nonlinear optical effect in which photons with the same frequency interacting with a nonlinear optical material are effectively "combined" to generate new photons with twice the frequency and hence energy, and half the wavelength of the initial photons, as illustrated in Figure 1.11. Second harmonic generation, as an even-order nonlinear optical effect, is only allowed in media without inversion symmetry.

**3. Parametric Down Conversion**

Parametric Down Conversion(PDC), also known as parametric fluorescence, or parametric scattering

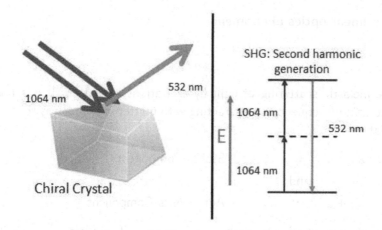

Figure 1.11: Second Harmonic Generation (SHG): Two photons of IR (1064 nm) interact with a chiral crystal to generate SHG (532 nm).

is a non linear optical effect which is an important process in quantum optics. This is as illustrated in Figure 1.12.

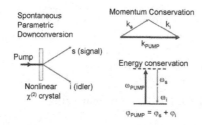

Figure 1.12: Parametric Down Conversion

## 4. Third Harmonic Generation (THG)

Third harmonic generation, is a nonlinear optical process. THG is a nonlinear $\chi^{(3)}$ effect, in which photons of frequency $\omega$ (wavelength $\lambda = 2\pi c/\omega$)) interacting with a nonlinear material are effectively combined to form new photons with three times the incident frequency i.e. $3\omega$ or a third of the wavelength ($\lambda/3$), as illustrated in Figure 1.13.

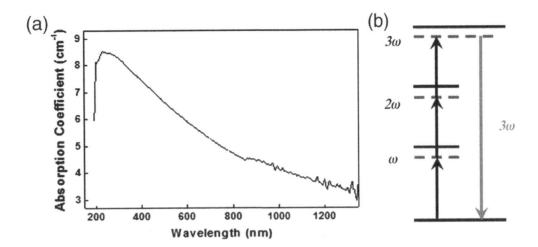

Figure 1.13: Third Harmonic Generation (Source: M. R. Tsai et.al, J. Biomedical Optics **18** (2), 026012 (2013)

## 1.9 The Poynting vector

The Poynting vector, **S**, of an electromagnetic wave of electric field **E** and magnetic field **H** is defined as

$$\mathbf{S} = \mathbf{E} \wedge \mathbf{H} \tag{1.17}$$

The Poynting vector represents the rate transfer of energy per area of an electromagnetic field. This can be understood by considering Maxwell's equations, say those of the conducting medium, where it can be seen that

$$
\begin{aligned}
\nabla \cdot (\mathbf{E} \wedge \mathbf{H}) &= \mathbf{H} \cdot \nabla \wedge \mathbf{E} - \mathbf{E} \cdot \nabla \wedge \mathbf{H} \\
&= \mathbf{H} \cdot \left(-\frac{\partial \mathbf{B}}{\partial t}\right) - \mathbf{E} \cdot \mathbf{J} - \mathbf{E} \cdot \frac{\partial \mathbf{D}}{\partial t} \\
&= -\mu_0 \mathbf{H} \cdot \frac{\partial \mathbf{H}}{\partial t} - \mathbf{E} \cdot \mathbf{J} - \epsilon_0 \epsilon(\omega) \mathbf{E} \cdot \frac{\partial \mathbf{E}}{\partial t}
\end{aligned}
$$

$$\text{But} \quad \mathbf{H} \cdot \frac{\partial \mathbf{H}}{\partial t} = \frac{1}{2}\frac{\partial H^2}{\partial t} \quad \text{and} \quad \mathbf{E} \cdot \frac{\partial \mathbf{E}}{\partial t} = \frac{1}{2}\frac{\partial E^2}{\partial t}$$

$$
\begin{aligned}
\nabla \cdot (\mathbf{E} \wedge \mathbf{H}) &= -\sigma E^2 - \frac{1}{2}\epsilon_0\epsilon(\omega)\frac{\partial E^2}{\partial t} - \frac{1}{2}\mu_0\frac{\partial H^2}{\partial t} \\
\nabla \cdot (\mathbf{E} \wedge \mathbf{H}) &= -\sigma E^2 - \frac{\partial}{\partial t}\left[\frac{1}{2}\epsilon_0\epsilon(\omega)E^2 + \frac{1}{2}\mu_0 H^2\right] \tag{1.18}
\end{aligned}
$$

which is the Poynting theorem. The integral form of Poynting theorem is given by

$$\int \nabla \cdot (\mathbf{E} \wedge \mathbf{H})dV = -\int \sigma E^2 dV - \frac{\partial}{\partial t}\int\left[\frac{1}{2}\epsilon_0\epsilon(\omega)E^2 + \frac{1}{2}\mu_0 H^2\right]dV \tag{1.19}$$

## 1.10    The Energy density

The Energy density, $U$, of an electromagnetic wave of electric field **E** and magnetic field **H** in a very general form is defined as

$$U = \frac{1}{2} \left\{ \epsilon_0 \frac{\partial}{\partial \omega} \left[ \omega \epsilon(\omega) E^2 \right] + \frac{\partial}{\partial \omega} \left[ \omega \mu(\omega) H^2 \right] \right\} \qquad (1.20)$$

The above form of the energy density, reduces to

$$U = \frac{1}{2} \left\{ \epsilon_0 \frac{\partial}{\partial \omega} \left[ \omega \epsilon(\omega) E^2 \right] + \mu_0 H^2 \right\} \quad \text{in a non-magnetic medium} \qquad (1.21)$$

$$U = \frac{1}{2} \left( \mathbf{E} \cdot \mathbf{D} + \mathbf{H} \cdot \mathbf{B} \right) \qquad (1.22)$$

$$U = \frac{1}{2} \epsilon_0 E^2 + \frac{1}{2} \mu_0 H^2 \quad \text{in vacuum} \qquad (1.23)$$

## 1.11    The Phase velocity, Group velocity and Energy velocity

The motion of waves is as illustrated in Figures 1.14 and 1.15. The motion of waves is identified with three velocities, the *phase velocity*, *group velocity* and *energy velocity* which are defined below.

### 1.11.1    The Phase velocity and Group velocity

$$\text{Phase velocity}, v_p = \frac{\omega}{k}$$

$$\text{Group velocity}, v_g = \frac{\partial \omega}{\partial k}$$

The phase velocity and group velocity can be obtained from a given dispersion relation of a particular system.

For example, from $\epsilon(\omega) = \frac{c^2 k^2}{\omega^2}$ can be found as

$$\text{Phase velocity}, v_p = \frac{c}{\sqrt{\epsilon(\omega)}}$$

$$\text{Group velocity}, v_g = \frac{c}{\epsilon^{1/2}(\omega) [1 + \frac{\omega}{2\epsilon(\omega)} \frac{\partial \epsilon(\omega)}{\partial \omega}]}$$

or in terms of the refractive index, we obtain

$$\text{Phase velocity}, v_p = \frac{c}{n(\omega)}$$

$$\text{Group velocity}, v_g = \frac{c}{n(\omega) + \omega \frac{\partial n(\omega)}{\partial \omega}}$$

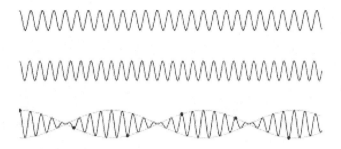

Figure 1.14: Waves move as a group

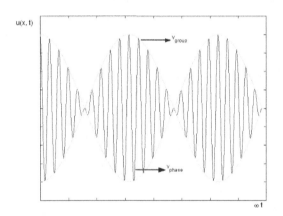

Figure 1.15: Phase velocity and group velocity

## 1.11.2 The Energy velocity

The energy velocity, $v_E$, is defined

$$v_E = \frac{S}{U} \tag{1.24}$$

where $S$ is the Poynting vector and $U$ is the energy density.

## 1.12   Exercises

**1.1**. (a) State Gauss's law for electric fields.
(b) A thin spherical shell of radius $r_0$ has a total net charge $Q$ that is uniformly distributed on it. Determine expressions for the electric field at points:
(i) outside the spherical shell,
(ii) inside the spherical shell,
(c) A solid spherical of radius $r_0$ has charge $Q$ uniformly distributed. Determine expressions for the electric field at points:
(i) outside the spherical shell,
(ii) inside the spherical shell,
(iii) Plot a schematic graph of how the electric field varies with distance from the center of the sphere $(r = 0)$ to outside the sphere $(r > r_0)$.

**1.2**. (a) State Maxwell's equations applicable to a conducting medium.
(b) Show that the Poynting vector $\mathbf{E} \wedge \mathbf{H}$ satifies the Poynting theorem

$$\int \nabla \cdot (\mathbf{E} \wedge \mathbf{H}) dV = -\int \sigma E^2 dV - \frac{\partial}{\partial t} \int \left[ \frac{1}{2} \epsilon_0 \epsilon(\omega) E^2 + \frac{1}{2} \mu_0 H^2 \right] dV$$

and explain the physical significance of this result.

**1.3**.(a) Give four examples of non-linear optical phenomena and describe each of them briefly.
(b) State Maxwell's equations applicable to a non-linear optical medium.
(c) Show that the following equation is applicable in a non-linear optical medium

$$\nabla \wedge (\nabla \wedge \mathbf{E}) + \frac{\epsilon(\omega)}{c^2} \frac{\partial^2 \mathbf{E}}{\partial t^2} = -\frac{1}{c^2 \epsilon_0} \frac{\partial^2 \mathbf{P}^{\mathbf{NL}}}{\partial t^2}$$

**1.4**. In Table 1.1, telecommunications operators and their allocated frequencies, are given. Calculate the respective wavelengths and complete Table 1.1.

**Table 1.1:** Telecommunications operators and their allocated frequencies.

| Operator | Allocated Frequency | Wavelength, $\lambda$ (Units ?) |
|---|---|---|
| LTE Operator | 800 MHz | ? |
| 2G GSM Operator/HSPA | 900 MHz | ? |
| GSM Operator/HSPA | 1800 MHz | ? |
| HSPA Operator | 2100 MHz | ? |
| WCS Operator | 2.3 GHz | ? |
| 4G Operator/LTE | 2.6 GHz | ? |

**1.5**. In Table 1.2, broadcasting operators and their allocated frequencies, are given. Calculate the respective wavelengths and complete Table 1.2.

**Table 1.2:** Broadcasting operators and their allocated frequencies.

| Operator | Allocated Frequency | Wavelength, $\lambda$ (Units ?) |
|---|---|---|
| FM Radio | 88.6 MHz | ? |
| FM Radio | 90.7 MHz | ? |
| MW Radio | 621 KHz | ? |
| MW Radio | 972 KHz | ? |
| DTT TV Station | 470 MHz | ? |
| C band downlink | 4.2 GHz | ? |

**1.6.** (a) How long does light from the sun take to reach the earth, given that light travels at a speed of $3 \times 10^8$ m/s and the earth-sun distance is $1.496 \times 10^{11}$ m.
(b) Calculate the distance travelled by light in one year(known as a *Light Year*) .

**1.7.** Consider the dispersion equation given by

$$\epsilon(\omega) = \frac{c^2 k^2}{\omega^2}$$

Show that the group velocity, $v_g = \frac{\partial \omega}{\partial k}$, for waves satisfying the above dispersion equation is given by

$$v_g = \frac{c}{\epsilon^{1/2}(\omega)[1 + \frac{\omega}{2\epsilon(\omega)} \frac{\partial \epsilon(\omega)}{\partial \omega}]}$$

**1.8.** Consider the dispersion equation given by

$$n(\omega) = \frac{ck}{\omega}$$

Show that the group velocity, $v_g = \frac{\partial \omega}{\partial k}$, for waves satisfying the above dispersion equation is given by

$$v_g = \frac{c}{n(\omega) + \omega \frac{\partial n(\omega)}{\partial \omega}}$$

**1.9.** (a) State Maxwell's equations in a non-metallic material.
(b) Consider medium 1 as having a positive dielectric constant $\epsilon_1$ occupying the half space $z > 0$ and medium 2 as having a frequency dependent dielectric function $\epsilon_2 = \epsilon(\omega)$ in the other the half space $z < 0$, such that the electric fields in the two media are $\mathbf{E_1} = E_{1x}\mathbf{i} + E_{1z}\mathbf{j}$ and $\mathbf{E_2} = E_{2x}\mathbf{i} + E_{2z}\mathbf{j}$, respectively, and the wavevectors in the two media are $\mathbf{q_1} = q_{1x}\mathbf{i} + q_{1z}\mathbf{j}$ and $\mathbf{q_2} = q_{2x}\mathbf{i} - q_{2z}\mathbf{j}$, respectively.
(i) Mention two boundary conditions at a surface of a solid, arising from electromagnetic theory.
(ii) What conditions do the tangential components of the wavevectors at the boundary satify?
(iii) From (i) and (ii) above show that

$$\frac{\epsilon_1}{\epsilon(\omega)} = \frac{q_{1z}}{q_{2z}}$$

(c) (i) Hence show that the dispersion relation for *surface polaritons* at a single interface is given by

$$\frac{c^2 q_{1x}^2}{\omega^2} = \frac{\epsilon_1 \epsilon(\omega)}{\epsilon_1 + \epsilon(\omega)}$$

where all symbols are in the usual notation, and you may assume that

$$q_{\lambda x}^2 + q_{\lambda z}^2 = \epsilon_\lambda \frac{\omega^2}{c^2}$$

and $\lambda = 1$ or 2 in a given region.

(ii) Plot schematically the dispersion curves for *phonon-type* surface polaritons.

**1.10**. Considering that the refractive index $n(\omega)$ is the square root of the dielectric function $\epsilon(\omega)$, that is, $n(\omega) = \sqrt{\epsilon(\omega)}$ or $n^2(\omega) = \epsilon(\omega)$. Suppose that $\epsilon(\omega)$ and $n(\omega)$ are complex, we have

$$\epsilon(\omega) = \epsilon'(\omega) + i\epsilon''(\omega) \text{ and } n(\omega) = \eta(\omega) + i\kappa(\omega)$$

where $\epsilon'(\omega)$ and $\epsilon''(\omega)$ are the real and imaginary parts of the dielectric function, and where $\eta(\omega)$ is the real part of the refractive index and $\kappa(\omega)$ is the imaginary part, usually called the extinction coefficient.

Show that the quantities $\eta(\omega)$, $\kappa(\omega)$, $\epsilon'(\omega)$, $\epsilon''(\omega)$ satisfy the following relations

$$\eta(\omega) = \{\frac{1}{2}[\epsilon'(\omega) + \sqrt{(\epsilon'(\omega)^2 + \epsilon''(\omega)^2)}]\}^{1/2}$$

$$\kappa(\omega) = \{\frac{1}{2}[-\epsilon'(\omega) + \sqrt{(\epsilon'(\omega)^2 + \epsilon''(\omega)^2)}]\}^{1/2}$$

# Chapter 2

# Electromagnetic Waves at Boundaries

## 2.1 Reflection and Refraction at a Single Interface

When electromagnetic waves are incident at a boundary, say at $z = 0$, part of the waves get reflected and part get refracted (transmitted), as illustrated in Figure 2.1.

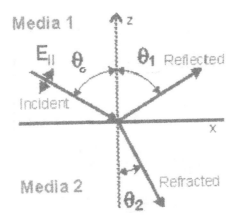

Figure 2.1: Incidence, Reflection and Refraction at a single interface.

According to Maxwell's equations, the incident field ($\lambda = 0$), the reflected field ($\lambda = 1$) and the refracted field ($\lambda = 2$), will satisfy

$$\nabla \cdot \mathbf{E}_\lambda = 0 \tag{2.1}$$

$$\nabla \cdot \mathbf{H}_\lambda = 0 \tag{2.2}$$

$$\nabla \wedge \mathbf{E}_\lambda = -\mu_0 \frac{\partial \mathbf{H}_\lambda}{\partial t} \tag{2.3}$$

$$\nabla \wedge \mathbf{H}_\lambda \;=\; \epsilon_0 \epsilon_\lambda(\omega) \frac{\partial \mathbf{E}_\lambda}{\partial t} \tag{2.4}$$

where

$$\lambda = \left\{ \begin{array}{c} 0 \\ 1 \\ 2 \end{array} \right.$$

Consider the *Incident fields* of the form

$$\begin{aligned} \mathbf{E_i} &= \mathbf{E_0} e^{i(\mathbf{k_0} \cdot \mathbf{r} - \omega t)} \\ &= \mathbf{E_0} e^{i(k_0 \mathbf{n_0} \cdot \mathbf{r} - \omega t)} \end{aligned}$$

where $\mathbf{k_0} = k_0 \mathbf{n_0}$ is the incident direction wavevector, with $\mathbf{n_0}$ being a unit vector in the incident direction.

Using the identity

$$\frac{\partial}{\partial t} = -i\omega t$$

we obtain

$$\mathbf{H_i} = \frac{1}{i\omega\mu_0} \nabla \wedge \mathbf{E_i}$$

Using the identity

$$\nabla = i\mathbf{k}_\lambda = ik_\lambda \mathbf{n}_\lambda k_0$$

we obtain

$$\mathbf{H_i} = \left( \frac{k_0}{\omega\mu_0} \right) \mathbf{n_0} \wedge \mathbf{E_i}$$

and hence

$$\mathbf{H_0} = \left( \frac{k_0}{\omega\mu_0} \right) \mathbf{n_0} \wedge \mathbf{E_0} \tag{2.5}$$

Similarly, the *Reflected fields* take the form

$$\begin{aligned} \mathbf{E_r} &= \mathbf{E_1} e^{i(\mathbf{k_1} \cdot \mathbf{r} - \omega t)} \\ &= \mathbf{E_1} e^{i(k_1 \mathbf{n_1} \cdot \mathbf{r} - \omega t)} \end{aligned}$$

where $\mathbf{k_1} = k_1 \mathbf{n_1}$ is the reflected direction wavevector, with $\mathbf{n_1}$ being a unit vector in the incident direction.

$$\mathbf{H_r} = \left( \frac{k_1}{\omega\mu_0} \right) \mathbf{n_1} \wedge \mathbf{E_r}$$

and hence

$$\mathbf{H_1} = \left( \frac{k_1}{\omega\mu_0} \right) \mathbf{n_1} \wedge \mathbf{E_1} \tag{2.6}$$

Similarly, the *Refracted (or Transmitted) fields* take the form

$$
\begin{aligned}
\mathbf{E_t} &= \mathbf{E_2} e^{i(\mathbf{k_2}\cdot\mathbf{r}-\omega t)} \\
&= \mathbf{E_2} e^{i(k_2\mathbf{n_2}\cdot\mathbf{r}-\omega t)}
\end{aligned}
$$

where $\mathbf{k_2} = k_2\mathbf{n_2}$ is the reflected direction wavevector, with $\mathbf{n_2}$ being a unit vector in the refracted direction.

$$
\mathbf{H_t} = \left(\frac{k_1}{\omega\mu_0}\right)\mathbf{n_2}\wedge\mathbf{E_t}
$$

and hence

$$
\mathbf{H_2} = \left(\frac{k_2}{\omega\mu_0}\right)\mathbf{n_2}\wedge\mathbf{E_2} \tag{2.7}
$$

**Boundary conditions**

The electric fields and magnetic fields satisfy the following *boundary conditions*:

1. The tangential component of the electric fields is continuous across the boundary at $z = 0$.

$$
\mathbf{n}\wedge(\mathbf{E_0}+\mathbf{E_1}) = \mathbf{n}\wedge\mathbf{E_2}
$$

2. The tangential component of the magnetic fields is continuous across the boundary at $z = 0$.

$$
\mathbf{n}\wedge(\mathbf{H_0}+\mathbf{H_1}) = \mathbf{n}\wedge\mathbf{H_2}
$$

or equivalently

$$
\mathbf{n}\wedge(\mathbf{n_0}\wedge\mathbf{E_0}+\mathbf{n_1}\wedge\mathbf{E_1})\left(\frac{k_1}{\omega\mu_0}\right) = \mathbf{n}\wedge(\mathbf{n_2}\wedge\mathbf{E_2})\left(\frac{k_2}{\omega\mu_0}\right)
$$

which after expanding the vector triple products, gives

$$
k_1\left\{(\mathbf{n}\cdot\mathbf{E_0})\mathbf{n_0}-(\mathbf{n}\cdot\mathbf{n_0})\mathbf{E_0}+(\mathbf{n}\cdot\mathbf{E_1})\mathbf{n_1}-(\mathbf{n}\cdot\mathbf{n_1})\mathbf{E_1}\right\} = k_2\left\{(\mathbf{n}\cdot\mathbf{E_2})\mathbf{n_2}-(\mathbf{n}\cdot\mathbf{n_2})\mathbf{E_2}\right\}
$$

There are two cases of interest that can be considered.

**Case 1:** *s-waves or Transverse Electric (TE) waves*

In the *s-waves* geometry, an incident electric field of amplitude $\mathbf{E_0}$, the reflected field $\mathbf{E_1}$ and the transmitted field $\mathbf{E_2}$ are *normal* to the plane of incidence.

Hence

$$
\mathbf{n}\cdot\mathbf{E_0} = \mathbf{n}\cdot\mathbf{E_1} = \mathbf{n}\cdot\mathbf{E_2}
$$

Note that

$$\mathbf{n} \cdot \mathbf{n_0} = \cos(\pi - \theta_0) = -\cos\theta_0$$
$$\mathbf{n} \cdot \mathbf{n_1} = \cos\theta_1$$
$$\mathbf{n} \cdot \mathbf{n_2} = \cos(\pi - \theta_2) = -\cos\theta_2$$

Hence we obtain the following equations for electric fields

$$E_0 + E_1 = E_2$$

$$E_0 \cos\theta_0 - E_1 \cos\theta_1 = \frac{k_2}{k_1} E_2 \cos\theta_2$$

These equations can be solved for $E_1$ and $E_2$ in terms of $E_0$. Multiply the first of the above equations by $\cos\theta_1$ and add to the second equation, we obtain

$$E_0 \cos\theta_1 + E_0 \cos\theta_0 = E_2 \cos\theta_1 + \frac{k_2}{k_1} E_2 \cos\theta_2$$
$$E_2 \{k_1 \cos\theta_1 + k_2 \cos\theta_2\} = k_1 (\cos\theta_0 + \cos\theta_1) E_0$$
$$E_2 = \frac{k_1 [\cos\theta_0 + \cos\theta_1]}{k_1 \cos\theta_1 + k_2 \cos\theta_2} E_0$$

Note that

$$E_1 = E_2 - E_0$$
$$= \left\{ \frac{k_1 [\cos\theta_0 + \cos\theta_1]}{k_1 \cos\theta_1 + k_2 \cos\theta_2} - 1 \right\} E_0$$
$$= \left[ \frac{k_1 \cos\theta_0 - k_2 \cos\theta_2}{k_1 \cos\theta_0 + k_2 \cos\theta_2} \right] E_0$$

Noting that $\theta_0 = \theta_1$, we obtain the Fresnel's equations given by

$$E_1 = \left[ \frac{k_1 \cos\theta_0 - k_2 \cos\theta_2}{k_1 \cos\theta_0 + k_2 \cos\theta_2} \right] E_0 \tag{2.8}$$

$$E_2 = \left[ \frac{2k_1 \cos\theta_0}{k_1 \cos\theta_0 + k_2 \cos\theta_2} \right] E_0 \tag{2.9}$$

**Case 2:** *p-waves or Transverse Magnetic (TM) waves*

In the *p-waves* geometry, an incident electric field of amplitude $\mathbf{E_0}$, the reflected field $\mathbf{E_1}$ and the transmitted field $\mathbf{E_2}$ are *parallel* to the plane of incidence, and hence the magnetic fields are *normal* to the plane of incidence. Hence

$$\mathbf{n} \cdot \mathbf{H_0} = \mathbf{n} \cdot \mathbf{H_1} = \mathbf{n} \cdot \mathbf{H_2}$$

Maxwell's equations give

$$\nabla \wedge \mathbf{H_\lambda} = \epsilon_0 \epsilon_\lambda(\omega) \frac{\partial \mathbf{E_\lambda}}{\partial t}$$

and hence

$$\mathbf{E}_\lambda = -\frac{\omega\mu_0}{k_\lambda}\mathbf{n}_\lambda \wedge \mathbf{H}_\lambda$$

where

$$\lambda = \left\{ \begin{array}{c} 0 \\ 1 \\ 2 \end{array} \right.$$

Using the boundary conditions
1. The tangential component of the electric fields is continuous across the boundary at $z = 0$.

$$\mathbf{n} \wedge (\mathbf{E_0} + \mathbf{E_1}) = \mathbf{n} \wedge \mathbf{E_2}$$

or equivalently

$$\mathbf{n} \wedge \left\{ \left(-\frac{\omega\mu_0}{k_0}\right)\mathbf{n_0} \wedge \mathbf{H_0} + \left(-\frac{\omega\mu_0}{k_1}\right)\mathbf{n_1} \wedge \mathbf{H_1} \right\} = \mathbf{n} \wedge \left(-\frac{\omega\mu_0}{k_2}\right)(\mathbf{n_2} \wedge \mathbf{H_2})$$

which after expanding the vector triple products, gives

$$\frac{1}{k_1}\left\{ (\mathbf{n} \cdot \mathbf{H_0})\mathbf{n_0} - (\mathbf{n} \cdot \mathbf{n_0})\mathbf{H_0} - (\mathbf{n} \cdot \mathbf{H_1})\mathbf{n_1} - (\mathbf{n} \cdot \mathbf{n_1})\mathbf{H_1} \right\} = \frac{1}{k_2}\left\{ (\mathbf{n} \cdot \mathbf{H_2})\mathbf{n_2} - (\mathbf{n} \cdot \mathbf{n_2})\mathbf{H_2} \right\}$$

As before, using

$$\begin{aligned}
\mathbf{n} \cdot \mathbf{n_0} &= -\cos\theta_0 \\
\mathbf{n} \cdot \mathbf{n_1} &= \cos\theta_1 \\
\mathbf{n} \cdot \mathbf{n_2} &= \cos(\pi - \theta_2) = -\cos\theta_2
\end{aligned}$$

we obtain

$$H_0 \cos\theta_0 - H_1 \cos\theta_1 = \frac{k_1}{k_2} H_2 \cos\theta_2$$

2. The tangential component of the magnetic fields is continuous across the boundary at $z = 0$.

$$\mathbf{n} \wedge (\mathbf{H_0} + \mathbf{H_1}) = \mathbf{n} \wedge \mathbf{H_2}$$

Hence we obtain the following equations for magnetic fields

$$H_0 \cos\theta_0 - H_1 \cos\theta_1 = \frac{k_1}{k_2} H_2 \cos\theta_2$$

$$H_0 + H_1 = H_2$$

These equations can be solved for $H_1$ and $H_2$ in terms of $H_0$. Multiply the second of the above equations by $\cos\theta_1$ and add to the first equation, we obtain

$$\begin{aligned}
H_0 \cos\theta_0 + H_0 \cos\theta_1 &= H_2 \left\{ \frac{k_1}{k_2}\cos\theta_2 + \cos\theta_1 \right\} \\
H_0 \left\{ \cos\theta_0 + \cos\theta_1 \right\} &= \frac{H_2}{k_2}(k_1 \cos\theta_2 + k_2 \cos\theta_1) \\
H_2 &= \frac{k_2 \left[ \cos\theta_0 + \cos\theta_1 \right]}{k_1 \cos\theta_2 + k_2 \cos\theta_1} H_0
\end{aligned}$$

Note that

$$
\begin{aligned}
H_1 &= H_2 - H_0 \\
&= \left\{ \frac{k_2 \left[\cos \theta_0 + \cos \theta_1\right]}{k_1 \cos \theta_2 + k_2 \cos \theta_1} - 1 \right\} H_0 \\
&= \left[ \frac{k_2 \cos \theta_0 - k_1 \cos \theta_2}{k_2 \cos \theta_1 + k_1 \cos \theta_2} \right] H_0
\end{aligned}
$$

Noting that $\theta_0 = \theta_1$, we obtain the Fresnel's equations given by

$$
H_1 = \left[ \frac{k_2 \cos \theta_0 - k_1 \cos \theta_2}{k_2 \cos \theta_0 + k_1 \cos \theta_2} \right] H_0 \tag{2.10}
$$

$$
H_2 = \left[ \frac{2k_2 \cos \theta_0}{k_2 \cos \theta_0 + k_1 \cos \theta_2} \right] H_0 \tag{2.11}
$$

## 2.2   Fresnel's Coefficients

In conclusion, we have derived **FRESNEL'S EQUATIONS**, which we summarise below

$$
\begin{aligned}
E_1 &= r_s E_0 \\
E_2 &= t_s E_0 \\
H_1 &= r_p H_0 \\
H_2 &= t_p H_0
\end{aligned}
$$

where we have introduced **FRESNEL'S COEFFICIENTS**, $r_s, t_s$ for s-waves and $r_p, t_p$ for p-waves.

$$
r_s = \left[ \frac{k_1 \cos \theta_0 - k_2 \cos \theta_2}{k_1 \cos \theta_0 + k_2 \cos \theta_2} \right] \tag{2.12}
$$

$$
t_s = \left[ \frac{2k_1 \cos \theta_0}{k_1 \cos \theta_0 + k_2 \cos \theta_2} \right] \tag{2.13}
$$

$$
r_p = \left[ \frac{k_2 \cos \theta_0 - k_1 \cos \theta_2}{k_2 \cos \theta_0 + k_1 \cos \theta_2} \right] \tag{2.14}
$$

$$
t_p = \left[ \frac{2k_2 \cos \theta_0}{k_2 \cos \theta_0 + k_1 \cos \theta_2} \right] \tag{2.15}
$$

We can define the Reflection Coefficients $R_p$ and $R_s$ for p-waves and s-waves respectively, as well as Transmission Coefficients $T_p$ and $T_s$ for p-waves and s-waves respectively, as below. These are illustrated graphically in Figure 2.2.

$$
R_p = |r_p|^2 = \left| \left[ \frac{k_1 \cos \theta_0 - k_2 \cos \theta_2}{k_1 \cos \theta_0 + k_2 \cos \theta_2} \right] \right|^2 \tag{2.16}
$$

$$
R_s = |r_s|^2 = \left| \left[ \frac{2k_1 \cos \theta_0}{k_1 \cos \theta_0 + k_2 \cos \theta_2} \right] \right|^2 \tag{2.17}
$$

$$
T_p = \frac{n_2}{n_1} |t_p|^2 = \frac{n_2}{n_1} \left| \left[ \frac{k_2 \cos \theta_0 - k_1 \cos \theta_2}{k_2 \cos \theta_0 + k_1 \cos \theta_2} \right] \right|^2 \tag{2.18}
$$

$$
T_s = \frac{n_2}{n_1} |t_s|^2 = \frac{n_2}{n_1} \left| \left[ \frac{2k_2 \cos \theta_0}{k_2 \cos \theta_0 + k_1 \cos \theta_2} \right] \right|^2 \tag{2.19}
$$

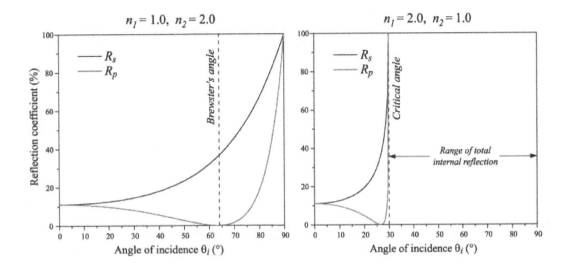

Figure 2.2: Reflection and Refraction Coefficients for p-waves and s-waves for $n_1 < n_2$ and $n_1 > n_2$. (Source: www.wikipedia.org)

Examining the graphs for $R_p$, $R_s$, $T_p$ and $T_s$, it can be noted that there are two angles of physical interest:

**1. Brewster Angle** is the angle when $R_p = 0$ which occurs when incidence is from a less dense medium to a more dense medium ($n_1 < n_2$). At this angle, there is no reflected ray.

**2. Critical Angle** is the angle when $T_p = 0$ which occurs when incidence is from a more dense medium to a less dense medium ($n_1 > n_2$). At this angle, there is no refracted ray, and there is TOTAL INTERNAL REFLECTION.

## 2.3  Brewster Angle

Brewster angle occurs when $R_p = 0$ or $r_p = 0$, noting that

$$r_p = \left[ \frac{k_2 \cos\theta_0 - k_1 \cos\theta_2}{k_2 \cos\theta_0 + k_1 \cos\theta_2} \right]$$

which is equal to zero if

$$k_2 \cos\theta_0 - k_1 \cos\theta_2 = 0$$

which implies

$$
\begin{aligned}
k_2 \cos\theta_0 &= k_1 \cos\theta_2 \\
\frac{ck_2}{\omega} \cos\theta_0 &= \frac{ck_1}{\omega} \cos\theta_2 \\
n_2 \cos\theta_0 &= n_1 \cos\theta_2
\end{aligned}
$$

$$\cos\theta_2 = \frac{n_2}{n_1}\cos\theta_0$$

But Snell's law says

$$n_1\sin\theta_0 = n_2\sin\theta_2$$

$$\sin\theta_2 = \frac{n_1}{n_2}\sin\theta_0$$

Hence

$$\sin^2\theta_2 + \cos^2\theta_2 = 1 = \frac{n_1^2}{n_2^2}\sin^2\theta_0 + \frac{n_2^2}{n_1^2}\cos^2\theta_0 = \sin^2\theta_0 + \cos^2\theta_0$$

$$\left(\frac{n_1^2}{n_2^2} - 1\right)\sin^2\theta_0 = \left(1 - \frac{n_2^2}{n_1^2}\right)\cos^2\theta_0$$

$$\left(\frac{n_1^2 - n_2^2}{n_2^2}\right)\sin^2\theta_0 = \left(\frac{n_1^2 - n_2^2}{n_1^2}\right)\cos^2\theta_0$$

$$\frac{\sin^2\theta_0}{\cos^2\theta_0} = \frac{n_2^2}{n_1^2}$$

$$\tan\theta_0 = \frac{n_2}{n_1}$$

$$\theta_0 = \tan^{-1}\left(\frac{n_2}{n_1}\right) = \theta_B \tag{2.20}$$

where $\theta_B$ is the Brewster angle, which only occurs when this condition is satisfied. This is very important in several applications, for example, polarisation effects where p-polarisation can be completely removed.

## 2.4  Critical Angle and Total Internal Refletion

Critical angle is the angle of incidence above which total internal reflection occurs.

But Snell's law says

$$n_1\sin\theta_0 = n_2\sin\theta_2$$

Critical angle is when $\theta_2 = 90°$

$$\sin\theta_0 = \left(\frac{n_2}{n_1}\sin 90°\right) = \frac{n_2}{n_1}$$

$$\theta_0 = \sin^{-1}\left(\frac{n_2}{n_1}\right) = \theta_c \tag{2.21}$$

where $\theta_c$ is the Critical angle, above which Total Internal Reflection occurs. This is very important in several applications, for example, optical fiber propagation.

## 2.5  Two interfaces:  Thin film

Thin films play an important part in modern technologies, such as thin film transistors (TFT) illustrated in Figure 2.3, solar cells illustrated in Figure 2.4, and others.

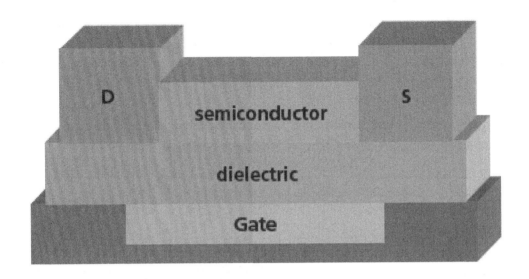

Figure 2.3: Thin Film Transistor (TFT)

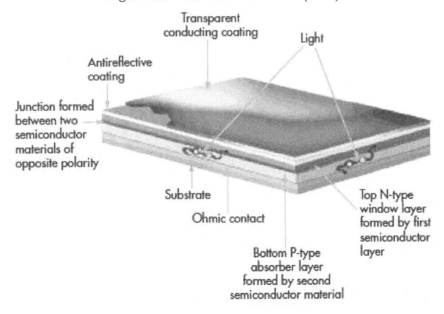

Figure 2.4: Thin Film Solar Cell (Source: www.circuitstoday.com)

## 2.6    Incidence, Reflection, Refraction and Transmission through a thin film

### 2.6.1    Path Difference and Phase Difference

Consider a ray incident from a medium of refractive index $n_0$, into a thin film of thickness $d$ and refractive index $n_1$. The ray is reflected as ray (1), and refracted into the thin film, where it is transmitted into a medium of refractive index $n_2$, and reflected within the thin film, and passes through the boundary back into the first medium as ray (2). We seek to find the path difference between ray (1) and ray (2).

$$\text{Distance travelled in a medium}, d_m = \text{Velocity in the medium}, v_m \times \text{time}, t$$
$$\text{Distance travelled in vacuum}, d_v = \text{Velocity in vacuum}, c \times \text{time}, t$$
$$\frac{d_v}{d_m} = \frac{c}{v_m} = n_1$$
$$d_v = n_1 d_m$$

where $n_1$ is the refractive index.

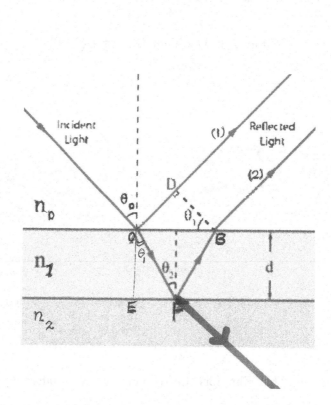

Figure 2.5: Phase Difference between Ray (1) and Ray (2).

Path difference between ray (1) and ray (2), $\Delta = n_1(OF + FB) - OD$

$$\text{But } OF = FB = \frac{d}{\cos\theta_1}$$

$$\Delta = n_1\frac{2d}{\cos\theta_1} - OB\sin\theta_0$$

$$\text{But } OB = 2d\tan\theta_1$$

$$\text{Hence } \Delta = n_1\frac{2d}{\cos\theta_1} - 2d\tan\theta_1\sin\theta_0$$

$$= \frac{2n_1d - 2d\sin\theta_1\sin\theta_0}{\cos\theta_1}$$

$$\text{But Snell's law gives, } \sin\theta_0 = n_1\sin\theta_1$$

$$\text{Hence } \Delta = \frac{2n_1d\left(1 - \sin^2\theta_1\right)}{\cos\theta_1}$$

$$\Delta = 2n_1d\cos\theta_1 \tag{2.22}$$

The path difference is related to the phase difference by the following relation,

$$\text{Phase Difference} = \frac{2\pi}{\lambda}\text{Path Difference}$$

$$= \frac{2\pi}{\lambda}\Delta$$

$$= \frac{2\pi}{\lambda}2n_1d\cos\theta_1$$

$$= 2\left(\frac{2\pi}{\lambda}n_1d\cos\theta_1\right)$$

$$= 2\delta_1 \tag{2.23}$$

$$\text{where } \delta_1 = \frac{2\pi}{\lambda}n_1d\cos\theta_1$$

The condition for constructive interference for a wavelength $\lambda$ is

$$\Delta = 2n_1d\cos\theta_1 = (m + \frac{1}{2})\lambda \tag{2.24}$$

and the condition for destructive interference for a wavelength $\lambda$ is

$$\Delta = 2n_1d\cos\theta_1 = m\lambda \tag{2.25}$$

## 2.6.2   Thin film Reflection, Refraction and Transmission

Consider electromagnetic waves incident onto a thin film of thickness $d$, where at the 0/1 the incident field will be partly reflected and refracted. We derive the Fresnel's coefficients for the thin film. At

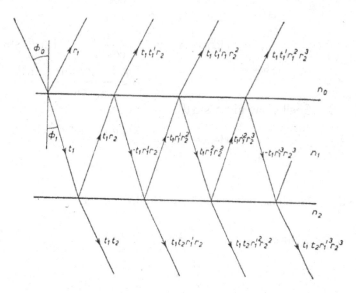

Figure 2.6: Thin Film: Incidence, Reflection and Transmission

the 0/1 interface, we have Fresnel's coefficients $r_1, t_1, r_1', t_1'$
By symmetry

$$r_1 = -r_1' \tag{2.26}$$

By conservation of energy

$$tt_1' + r_1^2 = 1 \tag{2.27}$$

At the 1/2 interface, we have Fresnel's coefficients $r_2$ and $t_2$.

**The reflected amplitude**, $R_A$ is given by

$$
\begin{aligned}
R_A &= r_1 + t_1 t_1' r_2 e^{-2i\delta_1} + t_1 t_1' r_1' r_2^2 e^{-4i\delta_1} + t_1 t_1' r_1'^2 r_2^3 e^{-6i\delta_1} + \cdots \\
&= r_1 + t_1 t_1' r_2 e^{-2i\delta_1} - t_1 t_1' r_1 r_2^2 e^{-4i\delta_1} + t_1 t_1' r_1^2 r_2^3 e^{-6i\delta_1} + \cdots
\end{aligned} \tag{2.28}
$$

Note that on the right hand side from the second term there is a geometrical progression (GP) with a common ratio

$$-r_1 r_2 e^{-2i\delta_1} \tag{2.29}$$

It is known that the sum to infinity of a geometrical progression is given by

$$S_{GP}^{\infty} = \frac{a}{1-r} \tag{2.30}$$

where $a$ is the first term of the GP and $r$ is a common ratio. Hence, we have

$$R_A = \left[ r_1 + \frac{t_1 t_1' r_2 e^{-2i\delta_1}}{1 + r_1 r_2 e^{-2i\delta_1}} \right]$$

$$\text{Using } t_1 t_1' = 1 - r_1^2$$

$$R_A = \left[ r_1 + \frac{[1 - r_1^2] r_2 e^{-2i\delta_1}}{1 + r_1 r_2 e^{-2i\delta_1}} \right]$$

$$= \left[ \frac{r_1 + r_1^2 r_2 e^{-2i\delta_1} + r_2 e^{-2i\delta_1} - r_1^2 r_2 e^{-2i\delta_1}}{1 + r_1 r_2 e^{-2i\delta_1}} \right]$$

$$R_A = \left[ \frac{r_1 + r_2 e^{-2i\delta_1}}{1 + r_1 r_2 e^{-2i\delta_1}} \right] \tag{2.31}$$

**The reflected intensity**, $I_R$ is given by

$$I_R = |R_A|^2 \tag{2.32}$$

$$= R_A R_A^\star$$

$$= \left[ \frac{r_1 + r_2 e^{-2i\delta_1}}{1 + r_1 r_2 e^{-2i\delta_1}} \right] \left[ \frac{r_1 + r_2 e^{+2i\delta_1}}{1 + r_1 r_2 e^{+2i\delta_1}} \right]$$

$$= \left[ \frac{r_1^2 + r_1 r_2 \left[ e^{+2i\delta_1} + e^{-2i\delta_1} \right] + r_2^2}{1 + r_1 r_2 \left[ e^{+2i\delta_1} + e^{-2i\delta_1} \right] + r_1^2 r_2^2} \right]$$

$$\text{But } e^{-2i\delta_1} + e^{-2i\delta_1} = 2 \cos 2\delta_1$$

$$\text{Hence } I_R = \left[ \frac{r_1^2 + 2 r_1 r_2 \cos 2\delta_1 + r_2^2}{1 + r_1^2 r_2^2 + 2 r_1 r_2 \cos 2\delta_1} \right]$$

$$= \left[ \frac{(r_1 + r_2)^2 - 2 r_1 r_2 \left[ 1 - \cos 2\delta_1 \right]}{(1 + r_1 r_2)^2 - 2 r_1 r_2 \left[ 1 - \cos 2\delta_1 \right]} \right]$$

$$I_R = \left[ \frac{(r_1 + r_2)^2 - 4 r_1 r_2 \sin^2 \delta_1}{(1 + r_1 r_2)^2 - 4 r_1 r_2 \sin^2 \delta_1} \right] \tag{2.33}$$

Note that the reflected intensity has a series of MAXIMA and MINIMA, which implies that an interference pattern is observed.

**The transmitted amplitude**, $T_A$ is given by

$$T_A = t_1 t_2' r_2 e^{-i\delta_1} + t_1 t_2 r_1' r_2 e^{-3i\delta_1} + t_1 t_2 r_1' r_1'^2 r_2^2 e^{-5i\delta_1} + \cdots$$

$$= t_1 t_2 e^{-i\delta_1} - t_1 t_2 r_1 r_2 e^{-3i\delta_1} + t_1 t_2 r_1^2 r_2^2 e^{-5i\delta_1} + \cdots \tag{2.34}$$

Note that on the right hand side above there is a geometrical progression (GP) with a common ratio

$$-r_1 r_2 e^{-2i\delta_1} \tag{2.35}$$

and hence the sum is

$$T_A = \left[ \frac{t_1 t_2 e^{-i\delta_1}}{1 + r_1 r_2 e^{-2i\delta_1}} \right] \tag{2.36}$$

**The transmitted intensity**, $I_T$ is given by

$$
\begin{aligned}
I_T &= \frac{n_3}{n_1}|T_A|^2 \tag{2.37}\\
&= \frac{n_3}{n_1}T_A T_A^\star \\
&= \frac{n_3}{n_1}\left[\frac{t_1 t_2 e^{-i\delta_1}}{1 + r_1 r_2 e^{-2i\delta_1}}\right]\left[\frac{t_1 t_2 e^{+2i\delta_1}}{1 + r_1 r_2 e^{+2i\delta_1}}\right] \\
&= \frac{n_3}{n_1}\frac{t_1^2 t_2^2}{1 + 2r_1 r_2 \cos 2\delta_1 + r_1^2 r_2^2} \\
I_T &= \frac{n_3}{n_1}\frac{t_1^2 t_2^2}{(1 + r_1 r_2)^2 - 4r_1 r_2 \sin^2 \delta_1} \tag{2.38}
\end{aligned}
$$

Note that the transmitted intensity has a series of MAXIMA and MINIMA, which implies that an interference pattern is observed.

What has been studied here finds applications in several devices, such as: 1. The Fabry-Perot Interferometer 2. Solar energy devices and other thin film devices.

## 2.7   Exercises

**2.1.** An electromagnetic wave is travelling from one medium of refractive index $n_1$ at an angle of incidence $\theta_0$, and enters another medium of refractive index $n_2$ at an angle of refraction $\theta_2$.
(a) Maxwell's equations give the following relations for the refractive indices, $n_1 = \frac{ck_1}{\omega}$ and $n_2 = \frac{ck_2}{\omega}$. Explain the meaning of all the symbols in these equations.
(b) State the boundary conditions for the electromagnetic fields at the boundary.
(c) State Snell's law.
(d) Show that Snell's law can also be expressed as the law of conservation of the tangential component of the wavevector.

**2.2.** (a) Explain what is meant by *p-waves* for electromagnetic waves.
(b) Show that Fresnel's equations for p-waves are given by

$$r_p = \left[ \frac{k_2 \cos \theta_0 - k_1 \cos \theta_2}{k_2 \cos \theta_0 + k_1 \cos \theta_2} \right]$$

$$t_p = \left[ \frac{2 k_2 \cos \theta_0}{k_2 \cos \theta_0 + k_1 \cos \theta_2} \right]$$

**2.3.** (a) Explain what is meant by *s-waves* for electromagnetic waves.
(b) Show that Fresnel's equations for s-waves are given by

$$r_s = \left[ \frac{k_1 \cos \theta_0 - k_2 \cos \theta_2}{k_1 \cos \theta_0 + k_2 \cos \theta_2} \right]$$

$$t_s = \left[ \frac{2 k_1 \cos \theta_0}{k_1 \cos \theta_0 + k_2 \cos \theta_2} \right]$$

**2.4.** (a) (i) Explain, briefly, what is meant by the "Brewster angle".
(ii) An electromagnetic wave is propagating in a medium of refractive index $n_1$ and enters another medium which is more dense whose refractive index is $n_2$. Show that the Brewster angle is given by

$$\theta_B = \tan^{-1} \left( \frac{n_2}{n_1} \right)$$

(b) An electromagnetic wave travelling in air (refractive index 1) is incident on a glass plate of refractive index 1.5. Calculate the Brewster angle.

**2.5.** (a) (i) Explain, briefly, what is meant by the "Critical angle".
(ii) An electromagnetic wave is propagating in a medium of refractive index $n_1$ and enters another medium which is more dense whose refractive index is $n_2$. Show that the critical angle is given by

$$\theta_c = \sin^{-1} \left( \frac{n_2}{n_1} \right)$$

(b) An electromagnetic wave travelling in a medium of refractive index 1.45 is incident on another medium of refractive index 1.43. Calculate the critical angle.

**2.6.** (a) Show that the reflection coefficient $R$ for s-waves in the case of normal incidence is given by

$$R = |\frac{n_1 - n_2}{n_1 + n_2}|^2$$

(b) If $n_1 = 1.0$ for vacuum, and $n_2 = \eta + i\kappa$, show that $R$ given above, can be written in the form

$$R = \frac{(1 - \eta)^2 + \kappa^2}{(1 + \eta)^2 + \kappa^2}$$

which reduces to

$$R = 1 - \frac{4\eta}{(\eta + 1)^2 + \kappa^2}$$

(c) Calculate the reflection coefficient at normal incidence for a specimen of InP with the following optical constants: $\eta = 3.549, \kappa = 0.302$ if it is bounded by vacuum.

**2.7.** (a) Consider a ray incident from a medium of refractive index $n_0$, incident at an angle $\theta_1$ into a thin film of thickness $d$ and refractive index $n_1$. Illustrate your answer by a clearly labelled diagram and show that path difference between neighbouring two rays is given by

$$\Delta = 2n_1 d \cos \theta_1$$

(b) State the conditions for constructive and destructive interference.
(c) Estimate, assuming very small $\theta_1$, the thickness of the film for one to see colours of wavelenghs (i) 455 nm (blue)(ii) 550 nm (green) and (iii) 700 nm (red).

**2.8.** Consider propagation of electromagnetic waves through a thin film of thickness $d$. Show that the reflected amplitude $R_A$ and transmitted amplitude, $T_A$ are given by

$$R_A = \left[ \frac{r_1 + r_2 e^{-2i\delta_1}}{1 + r_1 r_2 e^{-2i\delta_1}} \right]$$

$$T_A = \left[ \frac{t_1 t_2 e^{-i\delta_1}}{1 + r_1 r_2 e^{-2i\delta_1}} \right]$$

where $r_1, r_2$ are Fresnel's coefficients associated with the two interfaces of the thin film.

**2.9.** (a) Show that the reflected intensity, $I_R$, is given by

$$I_R = \left[ \frac{(r_1 + r_2)^2 - 4r_1 r_2 \sin^2 \delta_1}{(1 + r_1 r_2)^2 - 4r_1 r_2 \sin^2 \delta_1} \right]$$

(b) A thin film of soap is of thickness $d = 1\mu m$ and refractive index 1.4, is in air of refractive index 1.0. Plot the reflected intensity vs wavelength form $0.35\mu m$ to $0.75\ \mu m$. Comment on the graph.

(c) A thin film of $MgF_2$ is of thickness $d = 0.10\mu m$ and refractive index 1.38, is bounded by air of refractive index 1.0 and a substrate of refractive index 1.60. Plot the reflected intensity vs wavelength form $0.35\mu$m to $0.75~\mu$m. Comment on the graph.

**2.10**. Show that the transmitted intensity, $I_T$, is given by

$$I_T = \frac{n_3}{n_1} \frac{t_1^2 t_2^2}{(1+r_1 r_2)^2 - 4r_1 r_2 \sin^2 \delta_1}$$

# Chapter 3

# Diffraction and Interference

## 3.1   Introduction

[1]The observation that light travels in straight lines is well known leading to the observation of shadows cast when light illuminates objects. However, there is another observation whereby when light passes through some obstacles it appears to cast not an absolute dark shadow, but rather a pattern with dark and light bands. This is what is referred to as *diffraction*. Another effect, known as *interference* occurs under certain conditions when two or more waves overlap. Diffraction and Interference are wave properties experienced by electromagnetic waves and other waves. These phenomenon occur when the dimensions of the obstacle interacting with a wave is of the same order of magnitude as the wavelength.

In Part I of this chapter, *diffraction* is studied. Diffraction effects are conveniently classified into *Fraunhofer diffraction* and *Fresnel diffraction*. Fraunhofer diffraction occurs when the light source and the screen where the diffraction pattern is observed are effectively at infinite distances from the aperture or obstacle causing diffraction. Fresnel diffraction is such that either the source of light or the screen or both are at finite distances from the aperture causing diffraction. First, we shall consider Fresnel diffraction, for example, by a knife edge. Secondly, we consider several examples of Fraunhofer diffraction, namely Single slit diffraction, Rectangular aperture diffraction, Circular aperture diffraction, Double slits diffraction (Young's slits), N-slit diffraction. Part I of this chapter concludes with a discussion of X-ray diffraction, electron diffraction and neutron diffraction.

In Part II of this chapter, we discuss *interference*, a property which is exhibited by electromagnetic waves and other waves such as water waves, sound waves, under certain conditions, when two or more such waves travel and overlap in a region of space. The condition for interference to occur is that the overlapping beams must be coherent. Such interference is known to occur by two major ways: (1)Interference by Division of Wavefront and (2)Interference by Division of Amplitude.

---

[1]This chapter follows closely Chapters 6 (Interference) and 7 (Diffraction) of our previous book: J. S. Nkoma and P K Jain (2003), *Introduction to Optics: Geometrical, Physical and Quantum*, Bay Publishers, Gaborone, Botswana.

## 3.2   Part I: Diffraction

## 3.3   Fresnel Diffraction

### 3.3.1   Knife-edge diffraction geometry

In mobile communications, it is useful to develop models for diffraction of electromagnetic waves using knife edge to represent towers or mountainous or hilly terrain, as illustrated in Figure 3.1, where there is an obstruction of height $h$ at a distance $d_1$ from the transmitter and distance $d_2$ from the receiver.

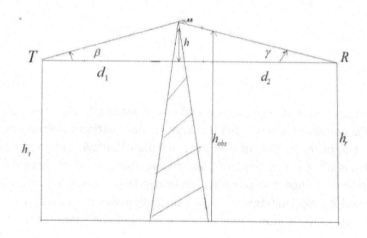

Figure 3.1: Kinife-edge difraction geometry.

The path difference betwee the direct path and the diffracted path is $\delta$ given

$$
\begin{aligned}
\delta &= \sqrt{d_1^2 + h^2} + \sqrt{d_2^2 + h^2} - (d_1 + d_2) \\
&\approx d_1(1 + \frac{h^2}{2d_1^2}) + d_2(1 + \frac{h^2}{2d_2^2}) - (d_1 + d_2) \\
&= \frac{h^2}{2d_1} + \frac{h^2}{2d_2} \\
&= \frac{h^2(d_1 + d_2)}{2d_1 d_2}
\end{aligned}
$$

Hence, Phase difference $\quad = \dfrac{2\pi}{\lambda}\delta = \dfrac{2\pi h^2(d_1 + d_2)}{\lambda 2 d_1 d_2}$

$$(3.1)$$

Hence, the Fresnel-Kirchoff diffraction parameter, $\nu_{FK}$, is given by

$$
\nu_{FK} = h\sqrt{\frac{2(d_1 + d_2)}{\lambda d_1 d_2}}
$$

$$\text{or} \quad \nu_{FK} = \alpha \sqrt{\frac{(2d_1 d_2)}{\lambda(d_1 + d_2)}} \qquad (3.2)$$

$$\text{and} \quad \phi = \frac{\pi \nu_{FK}^2}{2} \qquad (3.3)$$

Note that the phase difference is a function of the height of the obstruction, $h$, and the position of the obstruction from the transmitter and the receiver.

### 3.3.2 Fresnel diffraction by straight edges

Consider Fresnel diffraction by a system bound by straight edges, such as rectangular holes, slits, wires etc. The field at a point P is found by integrating all the differential contributions at the aperture.

From the geometry in Figure 3.2,

$$\begin{aligned}
\rho^2 &= \rho_0^2 + y^2 + z^2 \\
r^2 &= r_0^2 + y^2 + z^2 \\
\rho &\approx \rho_0 + \frac{y^2 + z^2}{2\rho_0} \\
r &\approx r_0 + \frac{y^2 + z^2}{2r_0}
\end{aligned}$$

and hence

$$\rho + r \approx \rho_0 + r_0 + (y^2 + z^2)\frac{\rho_0 + r_0}{2\rho_0 r_0}$$

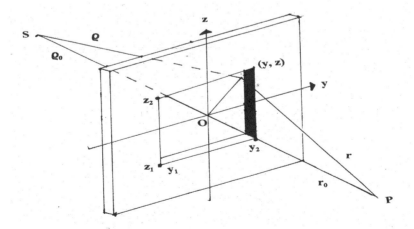

**Figure 3.2:** An illustration of Fresnel diffraction by straight edges.

The field at P is given by

$$E_P = \frac{E_0}{2(\rho_0 + r_0)} e^{i[k(\rho_0 + r_0) - \omega t]} \int_{u_1}^{u_2} e^{i\pi u^2/2} du \int_{v_1}^{v_2} e^{i\pi v^2/2} dv \qquad (3.4)$$

$$E_P = \frac{E_u}{2} \int_{u_1}^{u_2} e^{i\pi u^2/2} du \int_{v_1}^{v_2} e^{i\pi v^2/2} dv \qquad (3.5)$$

where

$$\frac{E_u}{2} = \frac{E_0}{2(\rho_0 + r_0)} e^{i[k(\rho_0 + r_0) - \omega t]} \tag{3.6}$$

$$u = y\left[\frac{2(\rho_0 + r_0)}{\lambda \rho_0 r_0}\right]^{1/2} \tag{3.7}$$

$$v = z\left[\frac{2(\rho_0 + r_0)}{\lambda \rho_0 r_0}\right]^{1/2} \tag{3.8}$$

The integral in equation (3.49) can be evaluated in terms of Fresnel integrals.

$$\int_{u_1}^{u_2} e^{i\pi u^2/2} du = \int_{u_1}^{u_2} \cos\left(\frac{\pi u^2}{2}\right) du + i \int_{u_1}^{u_2} \sin\left(\frac{\pi u^2}{2}\right) du \tag{3.9}$$

$$= C(u) + iS(u) \tag{3.10}$$

where

$$C(u) = \int_{u_1}^{u_2} \cos\left(\frac{\pi u^2}{2}\right) du \tag{3.11}$$

$$S(u) = \int_{u_1}^{u_2} \sin\left(\frac{\pi u^2}{2}\right) du \tag{3.12}$$

and similar expressions for the $v$ integral. The field $E_P$ can therefore be written in the form

$$E_P = \frac{E_u}{2}[C(u) + iS(u)]_{u_1}^{u_2}[C(v) + iS(v)]_{v_1}^{v_2} \tag{3.13}$$

Introducing $B(\omega)$ defined by

$$B(\omega) = C(\omega) + iS(\omega) \tag{3.14}$$

a curve of $S(\omega)$ against $C(\omega)$ is plotted on the complex plane for all values of $\omega$, and illustrated in Figure 3.3. This curve is known as the *Cornu spiral* .

The field $E_P$ for a rectangular aperture of straight edges is given by

$$E_P = \frac{E_u}{2}[B(u_2) - B(u_1)][B(v_2) - B(v_1)] \tag{3.15}$$

The diffracted intensity is the square of the electric field, given by

$$I_P = \frac{I_0}{2}|B_{12}(v)|^2$$

$$= \frac{I_0}{2}\left\{[C(v_2) - C(v_1)]^2 + [S(v_2) - S(v_1)]^2\right\} \tag{3.16}$$

A particular case of interest is the Fresnel diffraction by semi-infinite straight edge, whose diffracted intensity distribution is given by

$$I_P = \frac{I_0}{2}\left\{\left[\frac{1}{2} - C(v_1)\right]^2 + \left[\frac{1}{2} - S(v_1)\right]^2\right\} \tag{3.17}$$

and is illustrated in Figure 3.4.

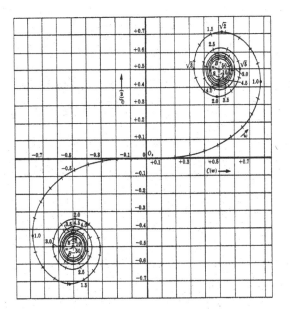

**Figure 3.3:** Cornu Spiral.

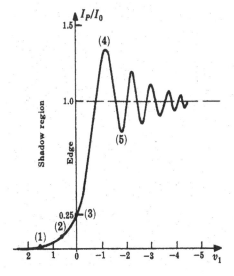

**Figure 3.4:** The intensity distribution in the diffraction pattern due to a straight edge. Points 1 and 2 are below the edge, 3 is at the edge, 4 and 5 are above the edge.

## 3.4   Fraunhofer Diffraction

### 3.4.1   Single slit diffraction

Consider light of angular frequency $\omega(= 2\pi\nu$, with $\nu$ being frequency) and wavevector $k(= 2\pi/\lambda$, with $\lambda$ being the wavelength) incident on a single slit $S_1$ of width $d$ as illustrated in Figure 3.5. The light is considered to be from a distant source, and can therefore be regarded as a plane wave. We wish to calculate the intensity distribution observed on a screen $S_2$ at a distance $D$ away.
The field of a wave emitted by a secondary source centred at O is given by

$$Re\left(A_0 e^{i(kr-\omega t)}\right) \tag{3.18}$$

where $Re$ represents the real part of the field.
Consider an element of length $dx$ at a distance $x$ from O. The field of a wave emitted by a secondary source centred at an element $dx$ is given by

$$Re A_0 e^{i(kr_1-\omega t)} = Re A_0 e^{i(kr-\omega t)} e^{-ikx\sin\theta} \tag{3.19}$$

where from Figure 3.5, we have used $r = r_1 + x\sin\theta$.

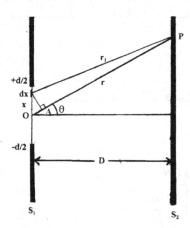

**Figure 3.5:** An illustration of Fraunhofer diffraction by a single slit of width $d$.

The total observable field will be the sum of all the elements from $-d/2$ to $+d/2$, given by $A(r,\theta)$

$$
\begin{aligned}
A(r,\theta) &= Re \int_{-d/2}^{+d/2} A_0 e^{i(kr-\omega t)} e^{-ikx\sin\theta} dx \\
&= A_0 d \frac{\sin\{\frac{1}{2}kd\sin\theta\}}{\{\frac{1}{2}kd\sin\theta\}} \cos(kr-\omega t) \\
&= A(0)\frac{\sin\beta}{\beta} \cos(kr-\omega t) \\
&= A(\theta)\cos(kr-\omega t) \tag{3.20}
\end{aligned}
$$

where $A(0) = A_0 d$, and the amplitude $A(\theta)$ of the resultant field is given by

$$A(\theta) = A(0)\frac{\sin\beta}{\beta} \qquad (3.21)$$

$$\beta = \frac{1}{2}kd\sin\theta \qquad (3.22)$$

The intensity distribution is the square of the amplitude distribution, given by

$$I(\theta) = I(0)\frac{\sin^2\beta}{\beta^2} \qquad (3.23)$$

where $I(0) = A^2(0)$. The intensity distribution given in equation 3.6 has a series of bright (maxima) and dark (minima) fringes. The graphical illustration of $I(\theta)$ against $\beta$ is shown in Figure 3.6.

The function $\sin\beta/\beta$ is sometimes referred to as $sinc\beta$, in terms of the $sinc$ function.

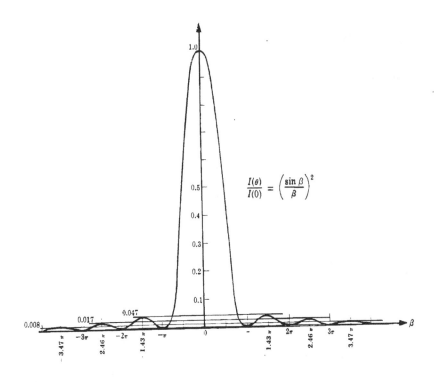

**Figure 3.6:** A graph of $I(\theta)/I(0)$ against $\beta$ for single slit Fraunhofer diffraction.

The maxima and minima in the graph of $I(\theta)$ against $\beta$ can be found by differentiating the expression for $I(\theta)$ with respect to $\beta$, as shown below.

$$\frac{dI}{d\beta} = \frac{d}{d\beta}\{I(0)\frac{\sin^2\beta}{\beta^2}\}$$

$$= I(0)\frac{2\sin\beta(\beta\cos\beta - \sin\beta)}{\beta^3}$$
$$= 0 \tag{3.24}$$

which implies that minima (or dark fringes) occur when $\sin\beta = 0$ and $\beta \neq 0$, which is when

$$\beta = \pm n\pi \text{ when n=1,2,3,} \cdots \tag{3.25}$$

The maximum central peak occurs at $\beta = 0$. The subsidiary maxima exist for nonzero $\beta$ satisfying

$$\beta\cos\beta - \sin\beta = 0 \tag{3.26}$$

or

$$\tan\beta = \beta \tag{3.27}$$

Equation (3.10) can be solved graphically as illustrated in Figure 3.7 by finding the points of intersection of two functions, $f_1$ and $f_2$, with $f_1 = \beta$ and $f_2 = \tan\beta$, and the values of $\beta$ are found as

$$\beta = \pm 1.4303\pi, \pm 2.4590\pi, \pm 3.4707\pi, \cdots.$$

*The secondary maxima are not quite midway the minima points.*

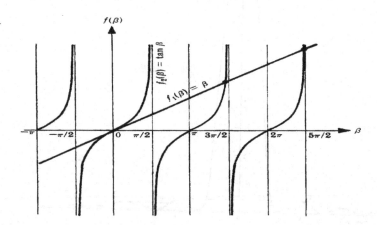

**Figure 3.7:** Graphs of $f_1 = \beta$ against $\beta$ and $f_2 = \tan\beta$ against $\beta$. Solutions of the equation $\tan\beta = \beta$ are the points of intersection of the two functions, $f_1$ and $f_2$.

Diffraction affects the ability of optical instruments, such as microscopes and telescopes to distinguish between closely spaced objects. This is what is referred to as *resolution* of images. To understand this, consider equation 3.22 and 3.25 from single slit diffraction. The equations imply

$$\sin\theta = n\frac{\lambda}{d}$$

or, considering the central bright fringe ($n = 1$), and noting that for small values of $\theta$, the approximation $\sin \theta \approx \theta$ can be used, one obtains

$$\theta = \frac{\lambda}{d}$$

where "$\theta$" is known as the *half angular width*. "$\theta$" sets the limit of resolution of an optical instrument. For example, a question arises, when are two images resolved? This is answered by considering what is referred to as *Rayleigh's criterion.*.

*Rayleigh's criterion states that two images or point sources are just resolved when the central maxima of one just coincides with the first maxima of the diffraction of the other.* This in turn implies that the central maxima due to the two sources must be separated by at least half the angular width "$\theta$" of the central maxima.

### 3.4.2 Diffraction by a rectangular aperture

Consider light of angular frequency $\omega (= 2\pi\nu$, with $\nu$ being frequency) and wavevector $k(= 2\pi/\lambda$, with $\lambda$ being the wavelength) incident from far on a rectangular aperture of width $a$ and height $b$ as illustrated in Figure 3.8

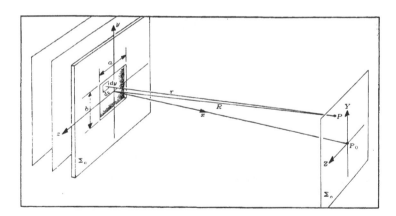

**Figure 3.8:** An illustration of Fraunhofer diffraction by a rectangular aperture of width $a$ and height $b$.

We wish to calculate the intensity distribution observed on a screen $S_2$ at a distance $x$ away. Each element of area $dS = dydz$ is a source of wavelets. An element at $(0, y, z)$ produces a field $dE_P$ at point $P(x, Y, Z)$ on screen $S_2$, of the form

$$dE_P = \frac{E_0}{r} e^{i(kr - \omega t)} \tag{3.28}$$

where from figure 3.8:

$$r = [x^2 + (Y - y)^2 + (Z - z)^2]^{1/2} \tag{3.29}$$
$$R = [x^2 + Y^2 + Z^2]^{1/2} \tag{3.30}$$

In view of $R$ being large, $y^2 \ll 2Yy$ and $z^2 \ll 2Zz$, the following approximation can be introduced

$$r \approx R\{1 - \frac{Yy + Zz}{R^2}\} \tag{3.31}$$

The total observable field will be the sum of all the elements from $-b/2$ to $+b/2$ along $y$ and from $-a/2$ to $+a/2$ along $z$, given by the real part of $E_P$.

$$
\begin{aligned}
E_P &= \frac{E_0}{R}e^{i(kR-\omega t)} \int_{-b/2}^{+b/2} \int_{-a/2}^{+a/2} e^{-ik(Yy+Zz)/R} dy dz \tag{3.32} \\
&= \frac{E_0}{R}e^{i(kR-\omega t)} \int_{-b/2}^{+b/2} e^{-ikYy/R} dy \int_{-a/2}^{+a/2} e^{-ikZz/R} dz \\
&= \frac{abE_0}{R}e^{i(kR-\omega t)} \frac{\sin(kbY/2R)}{(kbY/2R)} \frac{\sin(kaZ/2R)}{(kaZ/2R)} \tag{3.33}
\end{aligned}
$$

Taking the real part of $E_P$,

$$
\begin{aligned}
Re\,(E_P) &= \frac{abE_0}{R} \frac{\sin(kbY/2R)}{(kbY/2R)} \frac{\sin(kaZ/2R)}{(kaZ/2R)} \cos(kR - \omega t) \\
&= \frac{abE_0}{R} \frac{\sin \beta}{\beta} \frac{\sin \alpha}{\alpha} \cos(kR - \omega t) \tag{3.34} \\
\text{where } \beta &= kbY/2R \tag{3.35} \\
\alpha &= kaZ/2R \tag{3.36} \\
Re\,(E_P) &= E(Y, Z) \cos(kR - \omega t) \tag{3.37}
\end{aligned}
$$

where $E(Y, Z)$ is the amplitude. The intensity distribution due to diffraction by a rectangular aperture is the square of the amplitude distribution, given by

$$I(Y, Z) = E^2(Y, Z) \tag{3.38}$$

or

$$I(Y, Z) = I(0)\frac{\sin^2 \beta}{\beta^2} \frac{\sin^2 \alpha}{\alpha^2} \tag{3.39}$$

a result which could have been anticipated from the result of diffraction by a single slit given in equation 3.23. The photographic illustration of $I(Y, Z)$ is shown in Figure 3.9.

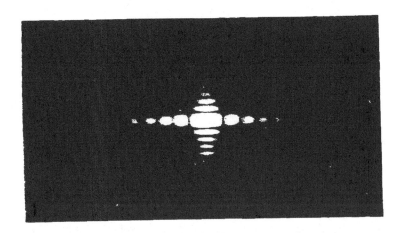

**Figure 3.9:** A photograph of the diffraction pattern produced by a rectangular aperture.

### 3.4.3  Diffraction by a circular aperture

Consider light of angular frequency $\omega(=2\pi\nu$, with $\nu$ being frequency) and wavevector $k(=2\pi/\lambda$, with $\lambda$ being the wavelength) incident from far on a circular aperture of diameter $d$ as illustrated in Figure 3.10.

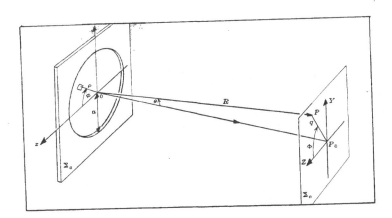

**Figure 3.10:** An illustration of Fraunhofer diffraction by a circular aperture of diameter $d$.

Without going into mathematical details, the intensity distribution observed on a screen at a distance away from the aperture, is given by

$$I(\theta) \;=\; I(0)\left[\frac{2J_1(\frac{1}{2}kd\sin\theta)}{\frac{1}{2}kd\sin\theta}\right]^2$$

$$=\; I(0)\left[\frac{2J_1(u)}{u}\right]^2 \tag{3.40}$$

$$\text{where } u \;=\; \frac{1}{2}kd\sin\theta \tag{3.41}$$

and $J_1(u)$ is the first order Bessel function defined by the series

$$J_1(u)=\frac{u}{2}\left[1-\frac{1}{1!2!}\left(\frac{u}{2}\right)^2+\frac{1}{2!3!}\left(\frac{u}{2}\right)^4-\frac{1}{3!4!}\left(\frac{u}{2}\right)^6+\cdots\right] \tag{3.42}$$

The intensity distribution of the diffraction pattern due to a circular aperture shows a series of maxima and minima as illustrated in the graph of $I(\theta)/I(0)$ against $\frac{1}{2}kd\sin\theta$ in Figure 3.11. A photograph of the diffraction pattern produced by a circular aperture is shown in Figure 3.12, where it can be noted that there is a central bright disc surrounded by alternating dark and bright rings, with progressively fainter rings.

The maxima of intensity occur at

$$\frac{1}{2}kd\sin\theta = 0, \pm5.14, \pm8.42, \cdots$$

and the minima of intensity occur at

$$\frac{1}{2}kd\sin\theta = 3.83, \pm7.02, \cdots$$

As in the case of single slit diffraction, the dominant maxima is at $\theta=0$, and the secondary maxima are not halfway between the minima.

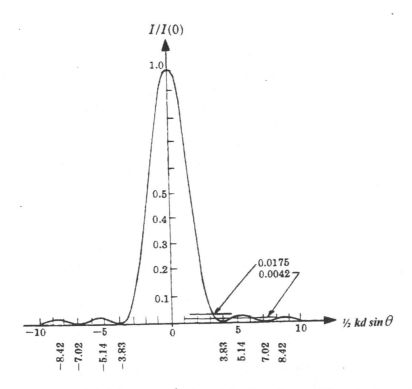

**Figure 3.11:** A graph of $I(\theta)/I(0)$ against $\frac{1}{2}kd\sin\theta$ for Fraunhofer diffraction by a circular aperture.

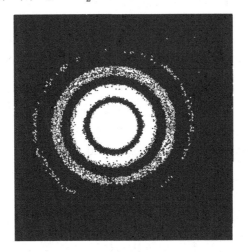

**Figure 3.12:** A photograph of the diffraction pattern produced by a circular aperture.

An application of the concept of *resolution* of images by a circular aperture shows that the half angular width is

$$\theta = 1.22\frac{\lambda}{d}$$

where $d$ is the diameter of the circular aperture and the factor of 1.22 arises from the Bessel function mentioned earlier, in equation 3.42.

### 3.4.4   Double slit diffraction (Young's slits)

Consider light of angular frequency $\omega(= 2\pi\nu$, with $\nu$ being frequency) and wavevector $k(= 2\pi/\lambda$, with $\lambda$ being the wavelength) incident from far on a system with double-slits (also known as Young's slits) , $S_1$ and $S_2$, with a slit-spacing $d$ as illustrated in Figure 3.13. We wish to calculate the intensity distribution observed at a point $P$ on a screen at a distance $D$ from the double slits.

It should be noted that each of the two slits produces diffraction effects as was discussed in section 3.4.1 when we considered single slit diffraction. Subsequently, the two diffracted intensities will superimpose to produce the resulting interference pattern.

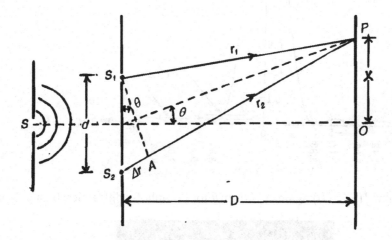

Figure 3.13: An illustration of Young's slits, $S_1$ and $S_2$, with a slit-spacing $d$.

The following observations can be made.
From the geometry, it can be noted that $D \gg d, D \gg x$ and $d \approx \lambda$, and
$r_2 = r_1 + d\sin\theta$
$2r_2 = (r_1 + r_2) + d\sin\theta$
An average distance, $r$, can be defined as

$$r = \frac{1}{2}(r_1 + r_2)$$

and hence $r_2 = r + \frac{1}{2}d\sin\theta$ and $r_1 = r - \frac{1}{2}d\sin\theta$
The path difference between waves through $S_1$ and $S_2$ is $AS_2$ given by

$$AS_2 = d\sin\theta$$

There is another quantity of physical interest, the phase difference, $\phi$, given by

$$
\begin{aligned}
\phi &= \frac{2\pi}{\lambda} \times \text{Path difference} \\
&= \frac{2\pi}{\lambda} d \sin\theta \\
&= kd \sin\theta \qquad\qquad (3.43)
\end{aligned}
$$

The total observable field at $P$ will be the sum of the fields through *each* of the double-slits, given by $y_P$ in the form

$$
y_P = ae^{i(kr_1 - \omega t)} + ae^{i(kr_2 - \omega t)} \qquad\qquad (3.44)
$$

where $a$ is the diffracted amplitude throught each of the slits, given by equation 3.21, that is

$$
a = A(0)\frac{\sin\beta}{\beta}
$$

and $\beta = \frac{1}{2}kb\sin\theta$, where we are introducing a slit width $b$ for each of the slits, which is different from the slit separation $d$ shown in Figure 3.13. We can express $kr_1$ and $kr_2$ in terms of $\phi$ as:

$$
\begin{aligned}
kr_2 &= kr + \frac{\phi}{2} \\
kr_1 &= kr - \frac{\phi}{2}, \text{ and hence} \\
y_P &= ae^{i(kr - \omega t - \phi/2)} + ae^{i(kr - \omega t + \phi/2)} \\
&= ae^{i(kr - \omega t)}\{e^{i\phi/2} + e^{-i\phi/2}\} \\
&= 2A(0)\frac{\sin\beta}{\beta}\cos\frac{\phi}{2}e^{i(kr - \omega t)} \qquad\qquad (3.45)
\end{aligned}
$$

Taking the real part, we obtain $Re\ y_P$,

$$
\begin{aligned}
Re\ y_P &= Re\ 2A(0)\frac{\sin\beta}{\beta}\cos\frac{\phi}{2}e^{i(kr - \omega t)} \\
&= 2A(0)\frac{\sin\beta}{\beta}\cos\frac{\phi}{2}\cos(kr - \omega t) \\
&= A(r,\theta)\cos(kr - \omega t) \qquad\qquad (3.46)
\end{aligned}
$$

where

$$
\begin{aligned}
A(r,\theta) &= 2A(0)\frac{\sin\beta}{\beta}\cos\left(\frac{1}{2}kd\sin\theta\right) \\
&= 2A(0)\frac{\sin\beta}{\beta}\cos\left(\frac{\phi}{2}\right) \\
&= 2A(0)\frac{\sin\beta}{\beta}\cos\gamma
\end{aligned}
$$

$$
(3.47)
$$

where we have introduced $\gamma = \phi/2$. The net intensity distribution observed on the screen is the square of the amplitude, given by

$$
\begin{aligned}
I(\theta) &= A^2(r,\theta) \\
&= 4A^2(0)\frac{\sin^2\beta}{\beta^2}\cos^2\frac{\phi}{2} \\
&= 4A^2(0)\frac{\sin^2\beta}{\beta^2}\cos^2\gamma
\end{aligned}
\tag{3.48}
$$

where one can identify

$$
\frac{\sin^2\beta}{\beta^2}
\tag{3.49}
$$

as the *diffraction term*, with $\beta = \frac{1}{2}kb\sin\theta$, and

$$
\cos^2\gamma
\tag{3.50}
$$

as the *interference term*, with $\gamma = \phi/2 = (1/2)kd\sin\theta$.

The resultant intensity will have a series of maxima (bright fringes) and minima (dark fringes). This can be understood as follows. The resultant intensity given in equation 3.48 will be zero if either the diffraction term is zero or the interference term is zero

The diffraction term is zero (*minima*) when

$$
\beta = \pm n\pi \text{ when n=1,2,3, } \cdots
$$

which implies

$$
\beta = \frac{1}{2}kb\sin\theta = \pm n\pi \text{ when n=1,2,3, } \cdots
$$

or

$$
b\sin\theta = \pm n\lambda = \pm\lambda, \pm 2\lambda, \pm 3\lambda, \cdots
$$

The interference term is zero (*minima*) when

$$
\gamma = (m + \frac{1}{2})\pi \text{ where } m = 0,1,2,\cdots
$$

that is, $(m + \frac{1}{2})$ is half odd integer, and thus

$$
\frac{1}{2}kd\sin\theta = \frac{\pi}{\lambda}d\sin\theta = (m + \frac{1}{2})\pi
$$

or

$$
d\sin\theta = (m + \frac{1}{2})\lambda = \frac{1}{2}\lambda, \frac{3}{2}\lambda, \frac{5}{2}\lambda, \cdots
$$

The position of the *maxima* are determined by $\beta = 0$ and also

$$
\gamma = m\pi \text{ where } m = 0,1,2,\cdots
$$

that is, $m$ is an integer, and thus

$$\frac{1}{2}kd\sin\theta = \frac{\pi}{\lambda}d\sin\theta = m\pi$$

or

$$d\sin\theta = m\lambda = 0, \lambda, 2\lambda, \cdots$$

A graph of the diffraction term is of illustrated in Figure 3.14 (a), while that of the interference term is illustrated in Figure 3,14 (b). The products of these two graphs gives the diffracted intensity through the Young's slits, as illustrated in Figure 3.14 (c).

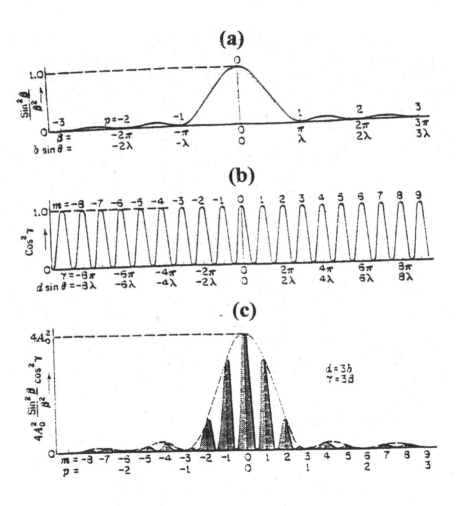

Figure 3.14: The intensity distribution of light through Young's slits, showing (a) the diffraction term (b) the interference term (c) the product of the diffraction term and the interference term which gives the resultant intensity distribution.

From Figure 3.14 one notes the following interesting features of the double slit diffraction pattern.

- The spacing between the maxima and minima of the fringes is determined by the interference effect, whereas the intensities of maxima are determined by the diffraction effect. Thus, while the fringes are equally spaced, they are not equally bright. The central fringe is the brightest, and the intensity diminishes on either side of it.

- The diffraction pattern of the single slit acts as the envelope of the interference fringes. As a result of the minima of the diffraction pattern some of the maxima of the interference pattern are suppressed. These are called the missing orders, and they occur where the condition of diffraction minimum coincides with the condition of the interference maximum.

### 3.4.5  N-slits diffraction or Diffraction Grating

Consider light of angular frequency $\omega(= 2\pi\nu$, with $\nu$ being frequency) and wavevector $k(= 2\pi/\lambda$, with $\lambda$ being the wavelength) incident from far on a system with $N$-slits, each of width $b$ and the slit-spacing is $d$ as illustrated in Figure 3.15. This system is referred to as a diffraction grating.

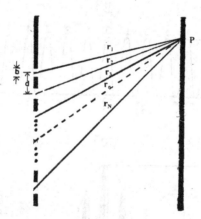

Figure 3.15:A diffraction grating with $N$-slits, each of width $b$ and slit-spacing $d$.

We wish to calculate the intensity distribution observed at a point $P$. The following observations can be made.

*First*, there will be single slit diffraction from each of the $N$-slits. *Second*, there will be interference of all the diffracted beams arising from each of the $N$-slits.

From geometry,
$$r_2 - r_1 = d\sin\theta.$$

$r_3 - r_1 = 2d \sin\theta$, and generally, we have
$r_n - r_1 = (n-1)d \sin\theta$.
An average distance, $r_{av}$, can be defined as

$$r_{av} = \frac{1}{2}(r_1 + r_N) = r_1 + (N-1)\frac{d}{2}\sin\theta$$

The total observable amplitude at $P$ will be the sum of all diffracted amplitudes through the $N$-slits, given by $E_P$ in the form

$$E_P = Ae^{-i\omega t}\{e^{ikr_1} + e^{ikr_2} + \cdots + e^{ikr_N}\} \tag{3.51}$$

where each of the slits contributes the amplitude $A$, and as discussed in the section dealing with single slit diffraction $A = A(0)\frac{\sin\beta}{\beta}$, as was shown in equation 3.21.
But note that

$$
\begin{aligned}
e^{ikr_n} &= e^{i[kr_1 + (n-1)kd\sin\theta]} \\
&= e^{i[kr_1 + (n-1)\phi]} \\
&= e^{ikr_1}e^{i(n-1)\phi} \tag{3.52} \\
\text{where } \phi &= kd\sin\theta \tag{3.53}
\end{aligned}
$$

and using this in equation (3.51), the total observable field at $P$ can be written in the form

$$
\begin{aligned}
E_P &= Ae^{-i\omega t}e^{ikr_1}\{1 + e^{i\phi} + e^{2i\phi} + \cdots + e^{i(N-1)\phi}\} \\
&= Ae^{-i\omega t}e^{ikr_1}\{1 + a + a^2 + \cdots + a^{(N-1)}\} \tag{3.54} \\
\text{where } a &= e^{i\phi} \tag{3.55}
\end{aligned}
$$

Note that the terms in the curly brackets in equation (3.54) form a geometrical progression (GP) of $N$ terms, with the first term in the curly brackets as 1, and the common ratio is $e^{i\phi}$, and thus the sum is given by

$$
\begin{aligned}
S_{GP}^N &= \frac{[a^N - 1]}{[a-1]} \\
&= \frac{[e^{iN\phi} - 1]}{[e^{i\phi} - 1]} \\
&= \frac{e^{iN\phi/2}[e^{iN\phi/2} - e^{-iN\phi/2}]}{e^{i\phi/2}[e^{i\phi/2} - e^{-i\phi/2}]} \\
&= e^{i(N-1)\phi/2}\frac{\sin N\phi/2}{\sin\phi/2} \tag{3.56}
\end{aligned}
$$

Using these results in equation (3.54), the total observable field at $P$ can be written in the form

$$
\begin{aligned}
E_P &= Ae^{-i\omega t}e^{[ikr_1 + i(N-1)\phi/2]}\frac{\sin N\phi/2}{\sin\phi/2} \\
&= Ae^{i(kr_{av} - \omega t)}\frac{\sin N\phi/2}{\sin\phi/2} \tag{3.57}
\end{aligned}
$$

Taking the real part, we obtain $ReE_P = E(r, \theta)$,

$$
\begin{aligned}
ReE_P &= ReAe^{i(kr_{av}-\omega t)}\frac{\sin N\phi/2}{\sin \phi/2}\\
E(r,\theta) &= A\cos(kr_{av}-\omega t)\frac{\sin N\phi/2}{\sin \phi/2}\\
&= A(\theta)\cos(kr_{av}-\omega t) \qquad\qquad\qquad (3.58)
\end{aligned}
$$

where

$$
A(\theta) = A\frac{\sin N\phi/2}{\sin \phi/2}
$$

and recalling that

$$
A = A(0)\frac{\sin \beta}{\beta}
$$

is the amplitude due to diffraction from a single slit.  Hence, $A(\theta)$ can explicitly be written in the form

$$
A(\theta) = A(0)\frac{\sin[\frac{1}{2}kb\sin\theta]}{[\frac{1}{2}kb\sin\theta]}\frac{\sin[\frac{1}{2}Nkd\sin\theta]}{[\sin \frac{1}{2}kd\sin\theta]} \qquad (3.59)
$$

The net intensity distribution observed on the screen is the square of the amplitude, given by

$$
\begin{aligned}
I(\theta) &= A^2(0)\frac{\sin^2[\frac{1}{2}kb\sin\theta]}{[\frac{1}{2}kb\sin\theta]^2}\frac{\sin^2[\frac{1}{2}Nkd\sin\theta]}{[\sin^2 \frac{1}{2}kd\sin\theta]}\\
&= I(0)\frac{\sin^2\beta}{\beta^2}\frac{\sin^2 N\gamma}{\sin^2\gamma} \qquad\qquad\qquad (3.60)
\end{aligned}
$$

$$
\text{where } I(0) = A^2(0) \qquad\qquad\qquad\qquad\qquad (3.61)
$$

and one can identify

$$
\frac{\sin^2\beta}{\beta^2} \qquad\qquad\qquad\qquad\qquad (3.62)
$$

as the *diffraction term*, with $\beta = \frac{1}{2}kb\sin\theta$, and

$$
\frac{\sin^2 N\gamma}{\sin^2\gamma} \qquad\qquad\qquad\qquad\qquad (3.63)
$$

as the *interference term*, with $\gamma = \frac{1}{2}kd\sin\theta$.

A graph of the diffraction term is of illustrated in Figure 3.16, while that of the interference term is illustrated in Figure 3.17 for $N = 6$. The products of these two graphs gives the diffracted intensity through the diffraction grating, and this is illustrated in Figure 3.17.

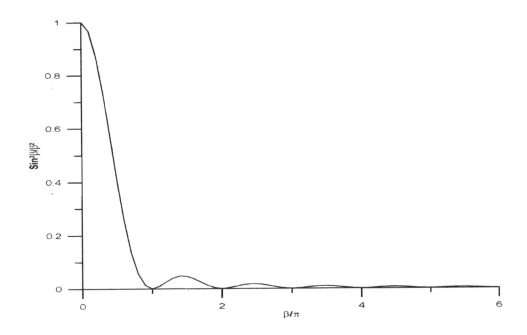

**Figure 3.16:** A graph of the diffraction term of a diffraction grating with $N = 6$.

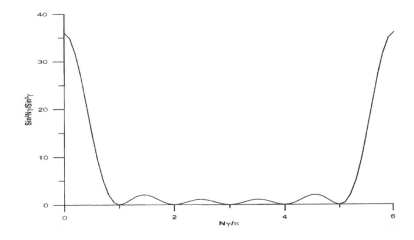

**Figure 3.17:** A graph of the interference term of a diffraction grating with $N = 6$.

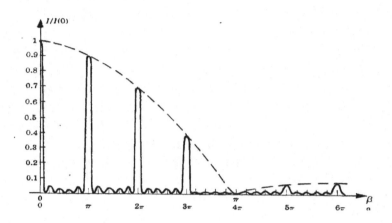

**Figure 3.18:** A graph of the intensity distribution of light diffracted by a diffraction grating with $N = 6$.

As in the case of the double slit diffraction, from Figures 3.16 to 3.18, we note that the diffraction pattern from the N-slits grating acts as an envelope of the interference pattern produces by the grating. Therefore, the locations of the maxima and minima in the resultant diffraction pattern are primarily determined by the interference term (equation 3.63).

The interference term is maximum, equal to one for $\gamma = m\pi$ where $m = 0, 1, 2, 3, \dots$ is an integer. From this, one otains the condition of $m^{th}$ order maximum as follows:

$$\frac{1}{2} k\, d\, sin\, \theta \;=\; m\, \pi = \frac{\pi}{\lambda}\, d\, sin\, \theta$$
$$or \quad d\, sin\, \theta \;=\; m\, \lambda$$

The zeroth order maximum is observed at $\theta = 0$, which is an obvious conclusion. d the spacing between the consequetive slits, also known as the *grating constant*, for a grating of length $L$ with $N$ number of slits is given by:

$$d = \frac{L}{N-1} \approx \frac{L}{N}, \quad for \;\; large \;\; N$$

The minimum of intensity are obtained when the diffraction term is zero, or when $sin\,(N\gamma) = 0$. This condition is satisfied for $N\gamma = n\pi$, where $n$ is an integer excluding $n = 0, N, 2N, 3N, \dots$. Explain why? The condition for minimum, thus reduces to:

$$N\, d\, sin\, \theta = n\, \lambda, \quad where \quad n \neq 0, N, 2N, 3N, \dots$$

Lastly, the effect of the diffraction term is such that at angular positions where the minima of diffraction pattern coincide with the maxima of interference, one encounters *missing orders* of the spectrum.

Diffraction gratings have important applications in physics, for example in determination of wavelengths. In this regard, the concept of *dispersive power, D* is introduced, defined as

$$D = \frac{d\theta}{d\lambda} \tag{3.64}$$

For example, for $d\sin\theta = m\lambda$, we obtain the dispersive power

$$D = \frac{d\theta}{d\lambda} = \frac{m}{d\cos\theta}$$

This equation has several factors of physical interest.

First, for a given small wavelength difference $d\lambda$, the angular separation $d\theta$ is *directly proportional* to the order $m$. Thus, the second order spectrum will be twice as wide as the first order spectrum.

Second, for a given small wavelength difference $d\lambda$, the angular separation $d\theta$ is *inversely proportional* to the slit separation $d$. Thus, the smaller the slit separation the wider the spectrum will be.

The effect of $\cos$ factor is that dispersion will be smallest when $\theta \approx 0$, that is close to the normal.

## 3.5 Diffraction by crystals

### 3.5.1 X-ray diffraction

X-rays can be diffracted by a crystal because the wavelength of the X-rays is of the same order of the magnitude as the interatomic spacing of the crystal. A schematic diagram of incident x-rays diffracted by crystal planes is illustrated in Figure 3.19. The fact that X-rays can be diffracted is a clear demonstration of their wave-like properties.

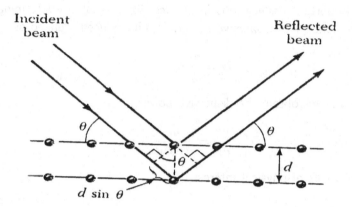

**Figure 3.19:** An illustration of X-ray diffraction from two parallel planes in a crystal. Note that the path difference between the two x-ray beams is $2d \sin \theta$.

The condition for the maxima of diffraction is given by Bragg's law, which states that the path difference between X-rays from two neighbouring planes must be an integer multiple of the wavelength.

$$2d \sin \theta = n\lambda \tag{3.65}$$

where

$n = 0, 1, 2, \cdots$
$d$ is the distance between neighbouring planes
$\theta$ is the diffraction angle
$\lambda$ is the wavelength

The distance $d$ is calculated for several angles of diffraction, $\theta$ using Bragg's law. The lattice spacing $a$ is calculated for several Miller indices $(hkl)$ , for example, for crystals with a cubic structure, one obtains.

$$a = \frac{d}{\sqrt{h^2 + k^2 + l^2}} \tag{3.66}$$

From such observations, lattice structure and lattice parameters can be obtained. There are three techniques for X-ray diffraction (XRD), namely Laue method, Rotating Crystal method and Powder method that are widely employed to study crystals. Modern XRD machines are computer controlled, and have made XRD studies highly automated.

Crystals are an ordered array of atoms of a solid. Crystals have what are known as a *direct lattice* and a *reciprocal lattice*, defined as below. X-rays map the reciprocal lattice.

If $\vec{a}_1, \vec{a}_2, \vec{a}_3$ are primitive vectors in the direct lattice, and $\vec{b}_1, \vec{b}_2, \vec{b}_3$ are primitive vectors in the reciprocal lattice, then

$$\vec{r}_j = x_j\vec{a}_1 + y_j\vec{a}_2 + z_j\vec{a}_3$$

is the direct lattice vector, with $x_j, y_j$ and $z_j$ being coordinates of atoms, and

$$\vec{G} = h\vec{b}_1 + k\vec{b}_2 + l\vec{b}_3$$

is the reciprocal lattice vector, with $(hkl)$ being the Miller indices.

The *X-ray Scattering Amplitude*, $A(\vec{G})$ is given by

$$A(\vec{G}) = N\int_{\text{cell}} n(r)e^{-i\vec{G}.\vec{r}}dV \tag{3.67}$$

$$= NS_G \tag{3.68}$$

where $S_G$ is the *Structure Factor*, $N$ is the number of unit cells in the crystal, and $n(r)$ is the concentration of atoms within an individual unit cell.

$$S_G = \int_{\text{cell}} n(\vec{r})e^{-i\vec{G}.\vec{r}}dV$$

$$= \sum_j \int n_j(\vec{r}-\vec{r}_j)e^{-i\vec{G}.\vec{r}}dV$$

$$= \sum_j e^{-i\vec{G}.\vec{r}_j}\int n_j(\vec{r}-\vec{r}_j)e^{-i\vec{G}.(\vec{r}-\vec{r}_j)}dV$$

$$= \sum_j f_j e^{-i\vec{G}.\vec{r}} \tag{3.69}$$

where

$$f_j = \int n_j(\vec{r}-\vec{r}_j)e^{-i\vec{G}.(\vec{r}-\vec{r}_j)}dV \tag{3.70}$$

is the *Atomic Form Factor*.

$$S_G \rightarrow F_{hkl}$$

where $F_{hkl}$ is the *Geometrical Structure Factor*.

$$F_{hkl} = \sum_j f_j e^{-i2\pi(hx_j+ky_j+lz_j)} \tag{3.71}$$

The diffracted intensity by a crystal is proportional to the square of the X-ray scattering amplitude given by equation (3.69), or equivalently to the square of the Geometrical Structure Factor given by equation (3.71). A typical diffraction pattern by crystals is given in Figure 3.20 (After Nkoma and Ekosse).

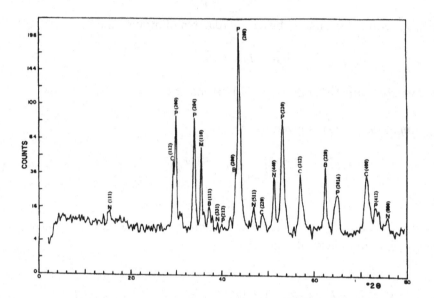

**Figure 3.20:** The X-ray Diffraction pattern of a sample of Cu-Ni orebody from Selebi-Phikwe, Botswana, showing peaks for pyrrhotite (P), pentlandite (N), chalcopyrite (C), magnetite (M) and bunsenite (B) (After Nkoma and Ekosse)

### 3.5.2   Electron diffraction

Electrons can be diffracted by a crystal because the de Broglie wavelength of the electrons is of the order of the magnitude of the interatomic spacing of the crystal. A pioneering study of electron diffraction was done by Davisson and Germer in 1927. The condition for electron diffraction is given by Bragg's law for *normal incidence*, as:

$$d \sin \theta = n\lambda \tag{3.72}$$

where $\lambda$ is the de Broglie wavelength of the electron, given as $\lambda = h/p$, with $p$ being the momentum of the electron, and $\theta$ is the diffrcation angle. If the electron is accelerated through a potential difference of $V$ volts, then

$$\frac{p^2}{2m} = eV \tag{3.73}$$

where $m$ is the mass of the electron, and hence

$$\lambda = \frac{h}{\sqrt{2meV}} = \frac{12}{(eV)^{\frac{1}{2}}} \mathring{A} \tag{3.74}$$

The fact that electrons can be diffracted is a demonstration of their wave-like properties. A typical electron diffraction pattern due to an aluminium film is illustrated in Figure 3.21, obtained using a Transmission Electron Microscope (TEM) (Courtesy: S H Coetzee, Electron Microscope Unit, Department of Physics, University of Botswana).

**Figure 3.21:** An electron diffraction pattern due to an aluminium film (Courtesy: S H Coetzee, Electron Microscope Unit, Department of Physics, University of Botswana).

### 3.5.3  Neutron diffraction

Neutrons can be diffracted by a crystal because the de Broglie wavelength of the neutrons is of the order of the magnitude of the interatomic spacing of the crystal. The condition for diffraction is given by Bragg's law for oblique incidence, as:

$$2d\sin\theta = n\lambda_n \tag{3.75}$$

where $\lambda_n$ is the de Broglie's wavelength of the neutron, given as

$$\lambda_n = \frac{h}{p} = \frac{h}{\sqrt{2m_n E}} = \frac{0.28}{[E(eV)]^{\frac{1}{2}}} \mathring{A} \tag{3.76}$$

where $p$ is the momentum of the neutron of mass $m_n$ and energy $E$. The fact that neutrons can be diffracted is a demonstration of their wave-like properties. A typical neutron diffraction pattern due to diamond is illustrated in Figure 3.22. The different peaks are from different crystal planes with Miller indices $(hkl)$.

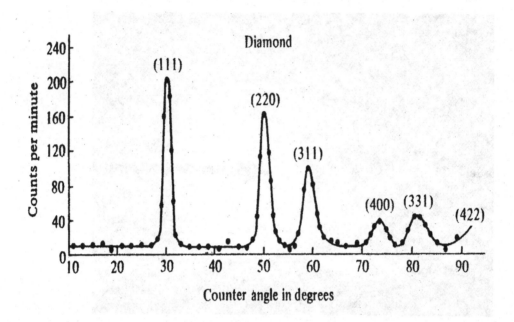

**Figure 3.22:** A neutron diffraction pattern due to powdered diamond.

The three radiations: X-rays, neutrons and electrons discussed in this chapter provide three distinctive windows to understanding crystal structure. X-rays are not charged and have no magnetic moment, neutrons are not charged but have a magnetic moment, electrons are negatively charged. Thus, these projectiles provide complementary information about crystals.

## 3.6   Part II: Interference

Electromagnetic waves and other waves such as water waves, sound waves, under certain conditions, are known to exhibit the phenomenon of *interference* when two or more such waves travel and overlap in a region of space. The condition for interference to occur is that the overlapping beams must be coherent.

Such interference is known to occur by two major ways: (1) Interference by Division of Wavefront and (2) Interference by Division of Amplitude. These two methods can be achieved in several ways as will be studied in this part of the chapter.

## 3.7   Interference by Division of Wavefront

### 3.7.1   Young's Double Slit Experiment

Interference in Young's double slit experiment is illustrated in Figure 3.13, and was discussed earlier in section 3.4.4 with respect to diffraction. The reason why we are revisiting this system is to emphasize that each of the two slits produces *diffraction* effects, and subsequently, the two diffracted intensities will superimpose to produce the resulting *interference* pattern, otherwise the analysis is as was discussed earlier.

### 3.7.2   Lloyd's mirror

Interference in Lloyd's mirror is as illustrated in Figure 3.23, where light from the primary source 1, which is referred as $S_1$ is reflected from the mirror, and source 2 (referred as $S_2$ is its virtual image, such that $S_1$ and $S_2$ act as the pair of coherent sources, and the distance from $S_1$ to $S_2$ is $2a = 2h = d$.

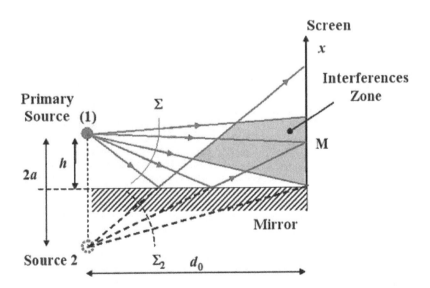

**Figure 3.23:** Interference by Lloyd's Mirror.

The observation screen is placed a distance $d_0 = D$ in contact with the far end of the mirror. Interference occurs between the direct rays from the source $S_1$ and the reflected rays from the mirror appearing to come from a virtual source $S_2$, resulting into a pattern of equally spaced dark and bright fringes on the screen. The reflected ray from $S_2$ undergoes a phase change of $\pi$, whereas the direct rays from $S_1$ remain unchanged. Thus, the conditions for bright and dark fringes are interchanged from those for the double slit experiment, and the conditions for destructive and constructive interference in terms of the optical path difference $\Delta r$ of two rays in Lloyd's mirror experiment are:

$$d\sin\theta \;=\; m\lambda, \qquad \text{where} \quad m = 0, \pm1, \pm2, \ldots, \qquad \text{(Dark fringe)} \qquad (3.77)$$

$$d\sin\theta \;=\; (m + \frac{1}{2})\lambda, \qquad \text{where} \quad m = 0, \pm1, \pm2, \ldots, \qquad \text{(Bright fringe)} \qquad (3.78)$$

The fringe at the edge of the mirror, corresponding to the central, $0^{th}$ order bright fringe in double slit interference pattern, is a dark interference fringe.

From above, noting that $2a = 2h = d$ and $d_0 = D$, the fringe separation $x$ which is also referred as 4yY, is obtained as

$$y = \frac{\lambda D}{d} \qquad (3.79)$$

and the separation $\Delta y_{bd}(\Delta y_{db})$ between adjacent bright and dark fringes is given as:

$$\Delta y_{bd} = \Delta y_{db} = \frac{1}{2}\frac{\lambda D}{d} \qquad (3.80)$$

### 3.7.3   Fresnel's double mirrors

Interference in Fresnel's double mirrors is as illustrated in Figure 3.24, where the setup consists of two plane mirrors inclined to each other at a very small angle ($\sim 1^{\circ}$). The incident wavefront from a

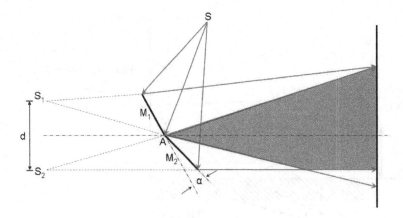

**Figure 3.24:** Interference by Fresnel's Double Mirrors.

source $S$ on reflection from the mirrors is divided producing virtual images $S_1$ and $S_2$, which act as the virtual coherent pair of sources, separated by distance $d$. Interference fringes are observed in area of the screen in which the reflected rays from both the mirrors overlap. The angular separation $2\alpha$ between $S_1$ and $S_2$ is twice the angle $\alpha$ between the mirrors, and their distance from the mirrors is the same as that of source $S$, which can be calculated. Both portions of the wavefront on reflection undergo a phase change of $\pi$, and thus remain in phase even after reflection.

The fringe separation (between bright and bright or dark and dark) is obtained as

$$y = \frac{\lambda D}{d} \qquad (3.81)$$

### 3.7.4  Fresnel's biprism

Interference by the Fresnel's biprism is illustrated in Figure 3.25, an experimental setup attributed to Fresnel around 1814. The biprism is a very thin glass prism of a very large prism angle ($\sim 179^o$) with the two prism surfaces almost along the same straight line. Screen $M$ with a wide slit is used to block the rays from the edges of the biprism from reaching the observation screen, which produce diffraction pattern from the straight edges $P$ and $Q$ of the biprism on either side of the interference pattern. A wavefront of monochromatic light from a source $S$ on incidence on the prism is divided into two. Portion of the wavefront refracting through the upper part of the prism is deviated downwards, and the portion that passes through the lower portion of the prism deviates upwards. The refracted rays from two parts of the biprism appear to diverge from virtual images $S_1$ and $S_2$, separated by distance $d$, which act as the two virtual coherent sources producing the interference pattern. An interference pattern of equi-spaced interference fringes is observed in region $bc$ on the screen in which the refracted rays from two portion of the biprism superimpose.

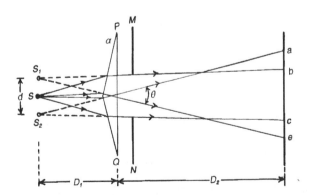

**Figure 3.25:** Experimental setup for the Fresnel's biprism interference.

If the separation between the two virtual sources $S_1$ and $S_2$ is denoted $d$, and $D = D_1 + D_2$ as the distance between the screen and the plane of the sources $S_1$ and $S_2$, the analysis for the fringe separation gives

$$\Delta y_{bb} = \Delta y_{dd} = \frac{\lambda(D_1 + D_2)}{d} \tag{3.82}$$

and the separation $\Delta y_{bd}(\Delta y_{db})$ between adjacent bright and dark fringes is given as:

$$\Delta y_{bd} = \Delta y_{db} = \frac{1}{2}\frac{\lambda(D_1 + D_2)}{d} \tag{3.83}$$

## 3.8   Interference by Division of Amplitude

### 3.8.1   Thin film interference

Inteference by a thin film illustrated in Figure 3.26,

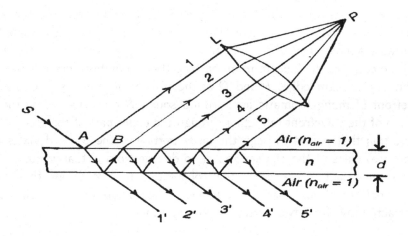

**Figure 3.26:** Interference from a plane-parallel film.

An analysis of the summation of the reflected fields and transmitted fields was studied earlier in chapter 2, and are summarised below for ease of reference.

**The reflected amplitude**, $R_A$ and **The reflected intensity**, $I_R$ are given by

$$R_A = \left[\frac{r_1 + r_2 e^{-2i\delta_1}}{1 + r_1 r_2 e^{-2i\delta_1}}\right]$$

$$I_R = \left[\frac{(r_1 + r_2)^2 - 4r_1 r_2 \sin^2 \delta_1}{(1 + r_1 r_2)^2 - 4r_1 r_2 \sin^2 \delta_1}\right]$$

while the **The transmitted amplitude**, $T_A$ and **The transmitted intensity**, $I_T$ are given by

$$T_A = \left[\frac{t_1 t_2 e^{-i\delta_1}}{1 + r_1 r_2 e^{-2i\delta_1}}\right]$$

$$I_T = \frac{n_3}{n_1}\frac{t_1^2 t_2^2}{(1 + r_1 r_2)^2 - 4r_1 r_2 \sin^2 \delta_1}$$

### 3.8.2 Wedge shaped film interference

Interference by a wedge shaped film is illustrated in figure 3.27, where there are two plane glass plates, such as microscope slides, placed in contact with each other at one end and making an angle $\theta$, using a spacer. A medium of refractive index $n$ (could also be air) is introduced between the glass plates. A monochromatic ray of light of wavelength $\lambda$ incident on the upper glass plate is reflected from the surfaces of the glass plates that constitute the upper and lower surfaces of the wedge as shown by the odd numbered rays 1, 3, 5,... and even numbered rays 2, 4, 6,... respectively.

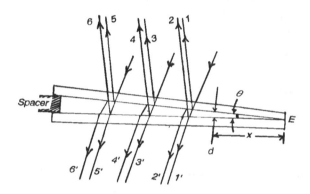

**Figure 3.27:** Interference from a wedge shaped film.

If $d$ is the thickness of the film, at a lateral distance $x$ from the edge of the wedge, then $d = x\tan\theta \sim x\,\theta$, and the optical path difference between the pair of rays reflected from the two surfaces of the film (for example rays *1* and *2*) is $2\,n\,d$, provided the observation is made almost normally to the film. The ray which is reflected from the lower surface of the film undergoes a phase change of $\pi$. Therefore the two rays interfere constructively to produce a bright fringe if:

$$2nd = 2nx\theta = \left(m + \frac{1}{2}\right)\lambda \tag{3.84}$$

and they will interfere destructively resulting in a dark fringe if:

$$2nd = 2nx\theta = m\lambda \tag{3.85}$$

where $m = 0, 1, 2, 3, \dots$. Rays from the two surfaces of the film are not parallel, but appear to diverge from a point within the film, and the fringes appear to be formed in the film itself. Such fringes are known as *localized fringes*.

At the edge of the film, the thickness of the film is zero, and the condition of destructive interference is satisfied. At this point one sees a dark fringe of zeroth order followed by a pattern of bright and dark fringes. The dark fringes are not perfectly dark because the intensity of the interfering waves are unequal. The locus of the same thickness of the film is a straight line parallel to the edge of the film. Therefore, the fringes are straight lines parallel to the edge of the film. The lateral separation $\Delta x$ between two successive bright or dark fringes is given by:

$$2n\Delta d = \lambda = 2n\Delta x\,\theta \qquad\qquad (3.86)$$

or

$$\Delta x = \frac{\lambda}{2n\,\theta} \qquad\qquad (3.87)$$

Thus the fringes are equally spaced.

Lastly, if white light is used, all colours satisfy the condition of a dark fringe at the edge of the film, and the edge is again a dark fringe. After that the condition of bright fringe for the shortest wavelength, *i.e.*, violet is satisfied, and the first bright fringe is violet followed by other colour fringes. Red with the longest wavelength is the last bright fringe in the series. After only a few distinct colour fringes, the interference pattern due to different wavelengths begins to overlap, and within a short distance from the edge the pattern changes first to a hue of colours, and then to nearly uniform white light at large thickness of the wedge. Why?

**Interference in transmitted waves**

Interference can also be seen in the rays transmitted from the other side of the film, such as rays *1'* and *2'* in Figure3.27. Ray *1'* is the straight transmitted ray without any reflection whereas ray *2'* undergoes two reflection from the glass surfaces and its phase changes by $2\pi$. Therefore, rays *1'* and *2'* are in phase, and the conditions of their constructive and destructive interference are reverse of those given by equations (3.84) and (3.85) respectively. Thus, in the transmitted waves the edge of the film is a bright fringe followed by alternate dark and bright fringes, and the interference pattern is complimentary to the pattern seen from the upper surface of the film. The dark fringes, once again, from the upper surface of the thin film, are not perfectly dark, rather partly bright.

### 3.8.3   Newton's rings interference

The experimental setup for observing interference due to by Newton's rings is illustrated in Figure 3.28. Light reflected from the upper and lower surfaces of the circular-wedge interferes to form an interference pattern that comprises concentric circular dark and bright localized fringes. The central spot, the point of contact between the lens and the glass plate is a dark fringe.

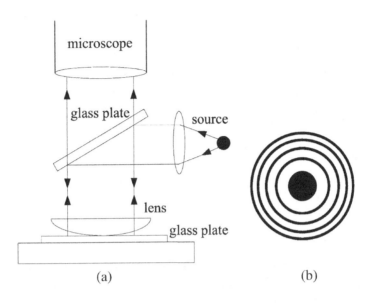

**Figure 3.28:** Experimental setup for observation of Newton's Rings interference.

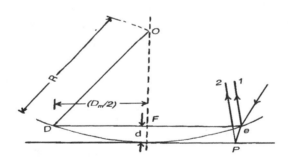

**Figure 3.29:** Path difference of two rays producing Newton's Rings.

From Figure 3.29, the thickness $d$ of the film is related to the diameter $D_m$ of the $m^{th}$ dark or bright interference, where the radius $x$ is $x = D + m/2$. If $R$ is the radius of curvature of the surface of the lens in contact with the glass plate, then

$$
\begin{aligned}
OD^2 &= FD^2 + OF^2 \\
R^2 &= x^2 + (R - d)^2 \\
x^2 &= R^2 - (R - d)^2 \\
&= R^2 - (R^2 + d^2 - 2Rd) \\
&= 2Rd - d^2
\end{aligned}
$$

$$\text{if} \quad d \ll R, d^2 \ll 2Rd$$

$$x^2 \;=\; \frac{D_m^2}{4} = 2Rd$$

$$D_m^2 \;=\; 8Rd \tag{3.88}$$

From equations (3.84) and (3.85) the conditions of bright and dark Newton's rings in terms of their diameters are:

$$(D_m)^2 \;=\; 8Rd = \frac{4R}{n}(2nd) = \frac{4R}{n}\left(m + \frac{1}{2}\right)\lambda, \qquad (m^{th} \quad bright \quad rings) \tag{3.89}$$

$$(D_m)^2 \;=\; 8Rd = \frac{4R}{n}(2nd) = \frac{4R}{n}m\lambda, \qquad (m^{th} \quad dark \quad rings) \tag{3.90}$$

where $n$ is the refractive index of the film, and $m = 0, 1, 2, 3, ....$

### 3.8.4   Michelson's Interferometer

The schematic illustration of the Michelson interferometer is illustrated in Figure 3.30, where A collimated beam of monochromatic light is incident at an angle at an optically parallel glass plate $P_1$ half silvered at the back, known as the *beam splitter*. The incident beam is split into two beams *1* and *2* of equal amplitude by partial reflection and transmission from the silvered face. There in no change of phase of beam *1* due to reflection. Why? $M_1$ and $M_2$ are two front silvered mirrors of high optical quality mounted with their reflecting surfaces perpendicular to each other, facing plate $P_1$. Mirror $M_1$ is mounted on a precision screw carriage, and can be moved forward or backwards keeping perfectly parallel to itself. The movement can be measured up to a micrometer. Mirror $M_2$ can be tilted with screws mounted at the back so that it can be set perpendicular to $M_1$, or at an angle as may be required for a particular experiment (application).

The two coherent waves *1* and *2* traveling perpendicular to each other from the beam splitter are incident on the mirrors normally, and reflected back to plate $P_1$. The waves are then directed by $P_1$ in to a telescope *T*, and superimposed to produce the interference pattern. Beam *1* directed towards mirror $M_1$ passes through the glass plate $P_1$ twice before it interferes with beam *1*, whereas beam *2* does not pass through the plate even once. To compensate for the optical path difference between the two beams introduced by plate $P_1$, a second optically identical glass plate $P_2$, known as *compensator*, is introduced parallel to $P_1$ in the path of beam *2*. The compensator plate is not an essential component of the interferometer as far as viewing of fringes is concerned, but it is indispensable for analytical work to nullify the effect of $P_1$ on the optical path difference of the two beams. With $P_2$ in place, the optical path difference between the two beams is simply equal to the physical path difference between the two beams measured from the point on $P_1$ at which the incident beam is divided.

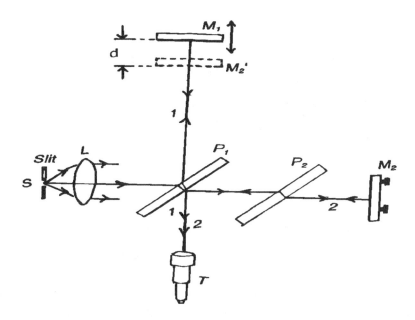

**Figure 3.30:** Michelson interferometer.

### Interference from Michelson Interferometer

In order to see an interference pattern from Michelson interferometer the following conditions must be met:

- The light source must be extended, and not a point source or a slit so that the entire area of the mirrors are filled with light. This is achieved by one of many possible ways which include using a large sodium or mercury lamp, or by introducing a lens between a point source and the interferometer such that the source is at the focal point of the lens, or by simply introducing a ground glass plate in front of the source.

- The light must be monochromatic or nearly so, for example the sodium light. This is particularly important if the distances of the mirrors from the beam splitter are not nearly equal.

### 1. Circular fringes

In the most common mode of the use of Michelson interferometer, the two mirrors are made perpendicular to each other with the help of screws at the back of mirror $M_2$, and their distances from $P_1$ are made nearly equal, to within millimeters by moving mirror $M_1$ in the required direction. Once these conditions have been fulfilled, a circular interference fringe pattern pops up in the telescope. In order to analyze the interference pattern, and to understand formation of the fringes, consider Figure 3.31.

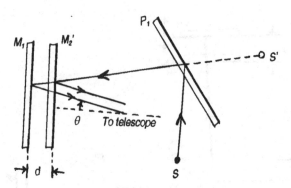

**Figure 3.31:** Optics of circular fringes in Michelson interferometer.

$M_2'$ is the virtual image of mirror $M_2$ reflected in $P_1$. Since $M_1$ and $M_2$ are perpendicular to each other, and if $d$ is the difference between their distances from $P_1$, then $M_1$ and $M_2'$ are parallel to each other separated by a distance $d$. The space between $M_1$ and $M_2'$ acts like a thin plane parallel film of air of thickness $d$, and the interference between the rays reflected from the two mirrors is the same as if the two rays were reflected from the two surfaces of a plane parallel thin film of thickness $d$ and refractive index $n = n_{air} = 1$. If $\theta$ is the angle at which the rays are reflected from the mirrors, then the path difference of the two rays is $2d\,cos\theta$ as in the case of a plane parallel film. Both rays undergo a phase change of $\pi$ on reflection from the mirror surfaces. Thus the change in phase of both the rays due to reflection cancel out, and the conditions of interference are:

$$2d\,cos\theta \;=\; m\lambda, \qquad (Bright \quad fringes) \qquad\qquad (3.91)$$

$$2d\,cos\theta \;=\; \left(m + \frac{1}{2}\right)\lambda, \qquad (Dark fringes) \qquad\qquad (3.92)$$

where $m = 0, 1, 2, 3, ...$ Locus of the same $\theta$ is a circle, so that the alternate dark and bright fringes are concentric circles. The reflected rays are parallel that appear to converge to infinity and the fringes are formed at infinity. A special case arises when the distances of both the mirrors are optically equal giving $d = 0$, which is undefinable for a physical thin parallel film of refractive material. In this case the condition of constructive interference is satisfied, and the field of view appears to be filled with uniform intensity. If white light is used, we shall also get uniform white intensity in the field of view if $d = 0$ or if $d >> \lambda_{visible}$. Why? But if $d$ is small, comparable to the average visible wavelength, a few colour circular fringes close to the center shall be observed, followed by haze of colours and then uniform white light resulting from the over lap of many bright fringes of different colours at each point.

## 2. Localized fringes

If the two mirrors are not perfectly parallel, and are at small angle to each other, the 'virtual' thin film between $M_1$ and $M_2'$ is wedge shaped. The interference pattern seen is the same as for a wedge shaped film of refractive index equal to 1, and the fringes are localized. If the angle between the mirrors is small, and for small thickness of the film fringes appear straight parallel to the edge of the wedge. If the angle between the mirrors is large, or at large thickness of the air film, the fringes are curved, convex towards the edge of the wedge, because the variation in the path difference with angle $\theta$.

### 3.8.5 Fabry-Perot Interferometer

The Fabry-Perot Interferometer is illustrated in Figure 3.32. It consists of two parallel glass plates $P_1$ and $P_2$, partly silvered on one side such that the reflectivity is approximately $80-90\%$. One of the plate is fixed while the other can be moved parallel to itself, changing the thickness of the air film. If the plates of the Fabry-Perot interferometer are mounted at a fixed separation, it is called a Fabry-Perot etalon, and is used for various spectroscopic applications.

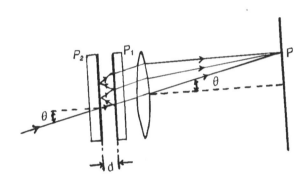

**Figure 3.32:** Interference by Fabry-Perot interferometer.

To understand the principle of the Fabry-Perot interferometer, we apply the theory studied in chapter 2 with regard to the analysis of the summation of the reflected fields and transmitted field, and the results are summarized in section 3.1 above.

Applying these results to the Fabry-Perot Interferometer, where if $n_3 = n_1$, we obtain

$$r_1 = -r_2 \Rightarrow r_1^2 = r_2^2 \tag{3.93}$$

$$t_1 = -t_2 \Rightarrow t_1^2 = t_2^2 \tag{3.94}$$

$$I_T = \frac{a^2 t_1^4}{1 + r_1^4 - 2r_1^2 \cos 2\delta_1}$$

$$= \frac{a^2 t_1^4}{[1 - r_1^2]^2} \frac{1}{\left\{1 + \frac{4r_1^2}{(1-r_1^2)^2} \sin^2 \frac{\delta}{2}\right\}} \tag{3.95}$$

$$\tag{3.96}$$

If there is negligible absorption, $1 - r_1^2 = t_1^2$, we obtain $I = I_T$ as

$$I = \frac{I_0}{\left\{1 + \frac{4r_1^2}{(1-r_1^2)^2} \sin^2 \frac{\delta}{2}\right\}} \tag{3.97}$$

$$I = \frac{I_0}{\left\{1 + F \sin^2 \frac{\delta}{2}\right\}} \tag{3.98}$$

where   $F = \dfrac{4r_1^2}{(1 - r_1^2)^2}$   is known as the FABRY COEFFICIENT OF FINESSE (3.99)

## 3.9 Exercises

**3.1.** Fraunhofer diffraction is observed using a spectrometer illuminated with a sodium lamp (wavelength, $\lambda = 589$ nm) and with a slit of variable aperture width $d$ inserted between the collimator and the telescope. The collimator lens has a focal length of 10 cm. If the collimator slit is replaced by double narrow slits separated by 1 mm, calculate the minimum width for which the double slits can be seen just separate when observed through the telescope.

**3.2.** (a) By making reasonable approximations, show that the intensity of the $m^{th}$ maxima of the diffraction pattern through a single slit is given by

$$I_m = I(0) \left[ \frac{1}{(m + \frac{1}{2}\pi)} \right]^2$$

and hence calculate the ratio of the intensity of the second maxima to the central maximum.

(b) With light of 5000 Å, the angular width of the central maxima in single slit diffraction pattern is $40°$. Calculate the width of the slit. What are the angular positions of the second and the third minima?

**3.3.** A plane diffraction grating has a grating constant of $1.79 \times 10^{-4}$ cm. Calculate the wavelength of monochromatic light for which the first and second order diffraction maxima are observed at $18°40'$ and $39°48'$ respectively.

**3.4** Three wave motions given below, simultaneously travel through a medium.

$$
\begin{aligned}
y_1 &= 3\sin(\omega t) \\
y_2 &= 5\sin(\omega t + \frac{\pi}{3}) \\
\text{and } y_3 &= 7\sin(\omega t + \frac{2\pi}{3})
\end{aligned}
$$

(i) Find the amplitude and phase of the resultant wave motion.
(ii) Write down the equation of the resultant wave motion, and
(iii) On one single diagram plot the phasor diagrams of all the wave motions.

**3.5** In a Young's double slit experiment interference pattern is observed using sodium light. The slits separation is 0.25 mm, and the observation screen is placed at a distance of 1m from the slits. Calculate the separation of the tenth bright fringe on either side of the central fringe due to both the wavelengths in sodium light. Repeat calculations for light of wavelengths given in table 6.1.

**3.6** In a Fresnel biprism experiment, the screen is placed 1 m away from the slit, and the separation between successive bright fringes is measured to be 0.1002 cm. A convex lens is used to measure the separation between the two virtual coherent sources. For two possible locations of the lens between the screen and the biprism, images of the coherent sources are formed on the screen which are 2.86

mm and 4.13 mm apart. Calculate *(i)* the separation between the pair of the coherent sources, and *(ii)* wavelength of light.

**3.7** In a Lloyd mirror experiment the source of sodium light is placed 1 mm above the mirror surface, and the mirror is 20 cm long. Find the separation of two successive bright fringes if the screen is placed *(i)* at the far edge of the mirror, *(ii)* 70 cm from the far edge of the mirror. What wavelength shall produce *(iii)* the same fringe separation at 70 cm as that of sodium light fringes at the edge of the mirror, *(iv)* the same fringe separation at the edge of the mirror as that of sodium light fringes at 70 cm location of the screen? Do these wavelengths lie in the visible region of spectra?

**3.8** A plane parallel film of thickness 0.0035 cm has a refractive index of 1.35. It is illuminated with sodium light at $\theta = 0^\circ, 30^\circ, 45^\circ, 60^\circ$, and $75^\circ$ Calculate the order of interference in each case, and state if the interference is constructive or destructive. For the same film, calculate the angles of incident for the first four maxima, and the first four minima in the interference with red light $(\lambda_R = 700nm)$, assuming the refractive index to be the same as given above.

**3.9** In Newton's rings experiment with light of wavelength 590 nm, the diameters of two consecutive rings in reflected light are measured to be 2.00 cm and 2.02 cm. Calculate the radius of curvature of the lens. Now a drop of water $(n = 1.33)$ is introduced between the lens and the glass plate. Calculate the diameters of the same rings. What orders of rings with water are observed that have the same diameters *i.e.* 2.00 cm and 2.02 cm respectively?

**3.10** (a) Using light of wavelength 600 nm, interference pattern of circular fringes is produced with a tube of 5 cm length in the path of one of the beams in Michelson interferometer. Now the tube is slowly evacuated while counting the interference fringes that shift. If 49 fringes have shifted with the full evacuation of the tube, calculate the refractive index of air. Now the tube is slowly filled with another gas, and 250 fringes are observed to shift. What is the refractive index of the gas? By how much should the mirror be moved to restore the interference pattern to exactly to the same state that was observed with the air in the tube?

(b) A Febry-Perot interferometer is used to determine the difference of two close wavelengths, the shorter one of which is 546.074 nm. If the consonance of fringes due to two wavelengths occur at the separations of the plates equal to 0.649 mm, 1.829 mm, and 3.009 mm, calculate the longer wavelength.

# Chapter 4

# Introduction to Optical Fiber Communications

## 4.1 Introduction

Optical fiber communication is used for several applications in telecommunications and broadcasting. It carries voice, data and vide signals nationally and internationally. In this section we describe optical fiber cables in Tanzania, around Africa and globally. The terrestrial optical fiber in Tanzania is known as the NICTBB (National ICT Broadband Backbone), illustrated in Figure 4.1.

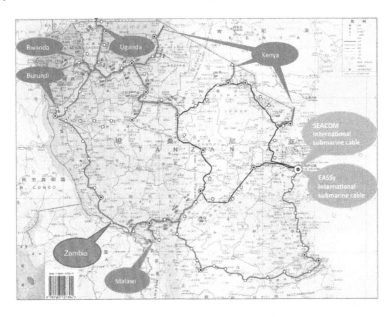

Figure 4.1: The NICTBB in Tanzania (Source: Ministry responsible for communications, Tanzania Government).

There are several submarine optical fiber cables around the African continent, as illustrated in Figure 4.2. Along the East African coast there are SEACOM, EASSY, TEAMS. There are several optical

83

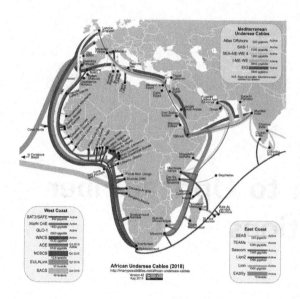

Figure 4.2: Submarine Optical Fiber Cables around Africa (Source: www.wikipedia.org)

fiber cables linking all the continents globally, as illustrated in Figure 4.3.

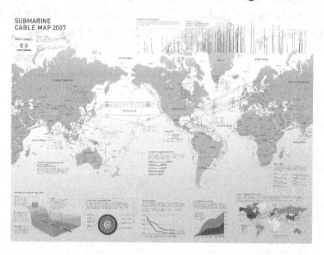

Figure 4.3: Submarine Global Optical Fibers (Source: www.wikipedia.org)

## 4.2 Optical Fiber Communication System

A block diagram of an optical fiber communication system is shown in Figure 4.4. The optical

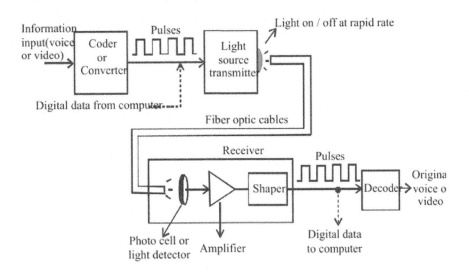

Figure 4.4: Optical Fiber Communication System Block Diagram

source in an optical fiber communication system is normally a light emitting diode (LED) or a laser, illustrated in Figure 4.5.

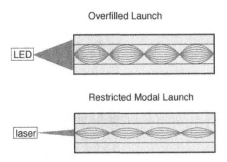

Figure 4.5: Optical sources for an Optical Fiber Communication System.

## 4.3 Structure of an Optical Fiber

An optical fiber is a cylindrical optical waveguide which consists of the core, cladding and coating as illustrated in Figure 4.6 to 4.8. The radius of the core is $a$ while the radius of the cladding is $b$. Standard fibers are normally manufactured so that the diameters $2a/2b = 8/125, 50/125, 62.5/125, 85/125$ and $100/140$, in units of $\mu m$.

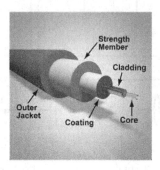

Figure 4.6: Structure of an Optical Fiber

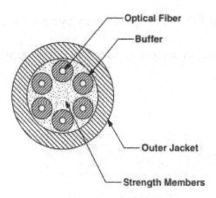

Figure 4.7: Optical Fiber Cables

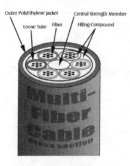

Figure 4.8: Optical Fiber Cables

## 4.4   Acceptance Angle, the Numerical Aperture and the $V$-parameter

When light enters and otical fiber, only rays within an acceptance angle will be totally internally reflected within the core. The *Acceptance angle* is the maximum angle which incident rays to the core will be able to undergo total internal reflection so that they are successfully transmitted through the fiber. The refractive indices of the core and cladding are $n_1$ and $n_2$ respectively, such that $n_1 > n_2$ so that total internal reflection occurs at the core claddng interface and propagation occurs within the core, as illustrated in Figure 4.9.

The parameter that is used as a measure of accepted rays is known as the *Numerical Aperture*, which is derived below.

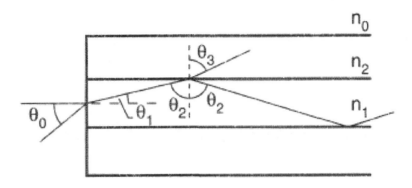

Figure 4.9: Propagation of light in an optical fiber

At the air/interface,

$$n_0 \sin \theta_0 = n_1 \sin \theta_1$$

At the core/cladding interface,

$$n_1 \sin \theta_2 = n_2 \sin 90$$

for the critical angle, or

$$\frac{n_2}{n_1} = \sin \theta_2$$

But

$$n_0 \sin \theta_0 = n_1 \sin(90 - \theta_2)$$
$$n_0 \sin \theta_0 = n_1 \cos \theta_2$$

Using $\sin^2 \theta_2 + \cos^2 \theta_2 = 1$, and straight forward algebra, we obtain, the *Numerical aperture*, $NA = \sin \theta_0$, is given by

$$NA = \frac{1}{n_0} \sqrt{n_1^2 - n_2^2} \tag{4.1}$$

$$= \sqrt{n_1^2 - n_2^2} \text{ for } n_0 = 1, \text{ air}$$

The $V$-parameter is defined in terms of $NA$ as

$$\begin{aligned} V &= NA \cdot k_0 \cdot a \\ &= \frac{2\pi a}{\lambda_0} NA \\ &= \frac{2\pi a}{\lambda_0} \sqrt{n_1^2 - n_2^2} \end{aligned} \qquad (4.2)$$

There is cutoff wavelength, $\lambda_c$ which gives a cutoff $V$-parameter, $V_{cutoff}$ given by

$$V_{cutoff} = \frac{2\pi a}{\lambda_c} \sqrt{n_1^2 - n_2^2} = 2.405 \qquad (4.3)$$

which gives the number of modes, $M$, where $V \gg 2.405$, as

$$M \approx \frac{V^2}{2}$$

## 4.5  Types of Optical Fibers and Refractive Index Profile

There are two main types of fibers:

- *Multimode Fibers* and

- *Single Mode Fibers.*

The Multimode fibers are further subdivided into *Multimode Step Index (MMSI) Fibers* and *Multimode Graded Index (MMGI) Fibers*, as illustrated in Figure 4.10.

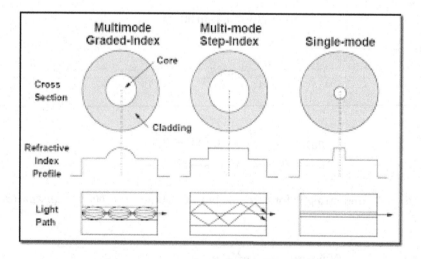

Figure 4.10: Types of Optical Fibers

## Multi-Mode Step Index (MMSI) Fibers

The refractive indices of the core and cladding are $n_1$ and $n_2$ respectively, such that $n_1 > n_2$ so that total internal reflection occurs at the core claddng interface and propagation occurs within the core. The radius of the core is $a$ while the radius of the cladding is $b$. The refractive indices of the core and claddng differ only slightly so that the fractional refractive index change is very small Most optical fibers currently in use are made of fused silica glass, $SiO_2$, of high purity, with the refractive index, $n_1$, ranges from 1.44 to 1.46, depending on the wavelength.

## Multi-Mode Graded Index (MMGI) Fibers

The refractive index, $n_(r)$, of the core varies as a function of the radial distance $r$, such that it is highest at the centre of the core, $n(0) = n_1$, and varies to its lowest value, $n(a) = n_2$, where the core meets the cladding. The cladding has a constant refractive index, $n_2$. The refractive index has a profile, illustrated in Figure 4.10.

$$n^2(r) = n_1^2 \left[ 1 - 2(\frac{r}{a})^p \Delta \right], r \leq a \tag{4.4}$$

where

$$\Delta = \frac{n_1^2 - n_2^2}{2n_1^2} \approx \frac{n_1 - n_2}{n_1} \tag{4.5}$$

and $\Delta$ normally lies between 0.001 and 0.02.

## Single Mode Step Index (SMSI)

In a Single Mode Step Index Fiber, light takes a single path through the core of the fiber. Single mode fibers have relatively narrow diameter, typically of the order of $8\mu m$ so that only one mode is transmitted. Single mode fibers normally carry higher bandwidth than multimode fibers, but requires a light source with a narrow spectral width.

## 4.6 Electromagnetic waves in an Optical Fiber

### Maxwell's Equations in a cylindrical optical fiber

Consider a material with a dielectric function $\epsilon(\omega)$ or refractive index, $n(\omega)$, which are related by the relation $\epsilon(\omega) = n^2(\omega)$, or $n(\omega) = \sqrt{\epsilon(\omega)}$. The electric field $\mathbf{E}$ and magnetic field $\mathbf{B}$ satisfy Maxwell's equations which are given below in the differential form.

$$\nabla \cdot \mathbf{E} = 0 \tag{4.6}$$

$$\nabla \cdot \mathbf{B} = 0 \tag{4.7}$$

$$\nabla \wedge \mathbf{E} = -\frac{\partial \mathbf{B}}{\partial t} \tag{4.8}$$

$$\nabla \wedge \mathbf{B} = \mu_0 \frac{\partial \mathbf{D}}{\partial t} = \mu_0 \frac{\partial \epsilon_0 \epsilon(\omega) \mathbf{E}}{\partial t} \tag{4.9}$$

where all symbols are as defined before. It can be seen from Maxwell's equations that

$$
\begin{aligned}
\nabla \wedge (\nabla \wedge \mathbf{E}) &= \nabla \wedge \left(-\frac{\partial \mathbf{B}}{\partial t}\right) \\
&= -\frac{\partial}{\partial t}(\nabla \wedge \mathbf{B}) \\
&= -\frac{\partial}{\partial t}\left(\mu_0 \frac{\partial}{\partial t}\epsilon_0 \epsilon(\omega)\mathbf{E}\right) \\
\nabla(\nabla \cdot \mathbf{E}) - \nabla^2 \mathbf{E} &= -\mu_0 \epsilon_0 \epsilon(\omega)\frac{\partial^2 \mathbf{E}}{\partial t^2} \tag{4.10}
\end{aligned}
$$

Consider a solution of the electric field of the form

$$
\begin{aligned}
E &= E_0 e^{i(\mathbf{k}\cdot\mathbf{r}-\omega t)} \tag{4.11} \\
\frac{\partial \mathbf{E}}{\partial t} &= -i\omega \mathbf{E} \\
\frac{\partial^2}{\partial t^2}\mathbf{E} &= (-i\omega)(-i\omega \mathbf{E}) = -\omega^2 \mathbf{E}
\end{aligned}
$$

$$
\begin{aligned}
-\nabla^2 \mathbf{E} &= -\mu_0 \epsilon_0 \epsilon(\omega)(-\omega^2)\mathbf{E} \\
&= -\frac{\epsilon(\omega)}{c^2}(-\omega^2)\mathbf{E} \\
&= \epsilon(\omega)\frac{\omega^2}{c^2}\mathbf{E}
\end{aligned}
$$

$$
\nabla^2 \mathbf{E} + \epsilon(\omega)\frac{\omega^2}{c^2}\mathbf{E} = 0 \tag{4.12}
$$

which is known as *Helmholtz Equation*, which can also be written in terms of the refractive index as

$$
\nabla^2 \mathbf{E} + n^2(\omega)\frac{\omega^2}{c^2}\mathbf{E} = 0 \tag{4.13}
$$

### Introduction of cylindrical coordinate system

The symmetry of the problem of propagation of electromagnetic waves in a cylindrical optical fiber cable requires that we solve the Helmholtz equation using cylindrical coordinates. The cylindrical coordinate system is illustrated in Figure 4.11.

$$
\begin{aligned}
x &= r\cos\theta \tag{4.14} \\
y &= r\sin\theta \tag{4.15} \\
z &= z \tag{4.16}
\end{aligned}
$$

The Laplacian Operator, $\nabla^2$, in cylindrical coordinates is given by

$$
\begin{aligned}
\nabla^2 &= \frac{1}{r}\frac{\partial}{\partial r}\left(r\frac{\partial}{\partial r}\right) + \frac{1}{r^2}\frac{\partial^2}{\partial \theta^2} + \frac{\partial^2}{\partial z^2} \tag{4.17} \\
&= \frac{\partial^2}{\partial r^2} + \frac{1}{r}\frac{\partial}{\partial r} + \frac{1}{r^2}\frac{\partial^2}{\partial \theta^2} + \frac{\partial^2}{\partial z^2} \tag{4.18}
\end{aligned}
$$

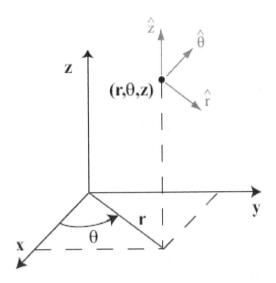

Figure 4.11: Cylindrical Coordinate System

Hence, the Helmholtz equation in cylindrical coordinates becomes

$$\left\{\frac{\partial^2}{\partial r^2}+\frac{1}{r}\frac{\partial}{\partial r}+\frac{1}{r^2}\frac{\partial^2}{\partial\theta^2}+\frac{\partial^2}{\partial z^2}\right\}E(r,\theta,z)+n^2(\omega)\frac{\omega^2}{c^2}E(r,\theta,z)\ =\ (4.19)$$

$$\frac{\partial^2}{\partial r^2}E(r,\theta,z)+\frac{1}{r}\frac{\partial}{\partial r}E(r,\theta,z)+\frac{1}{r^2}\frac{\partial^2}{\partial\theta^2}E(r,\theta,z)+\frac{\partial^2}{\partial z^2}E(r,\theta,z)+n^2(\omega)\frac{\omega^2}{c^2}E(r,\theta,z)\ =\ (4.20)$$

## Separation of Variables

By separation of variables, the electric field $E(r,\theta,z)$ can be separated into three functions, each involving a single independent variable, that is, the wave function can be cast as a product of three functions,

$$E(r,\theta,z)=R(r)P(\theta)F(z)=RPF \qquad (4.21)$$

where $R(r)=R$ is a function of $r$ only, $P(\theta)=P$ is a function of $\theta$ only, and $F(z)=F$ is a function of $z$ only. The function $R(r)$ will show the variation of the electric field along a radius vector from the nucleus, when $\theta$ and $z$ are constant. The function $P(\theta)$ will show the variation of the electric field with the $\theta$, and the function $F(z)$ will show the variation of with $z$. Substituting $E$ by $RPF$, we obtain

$$\frac{\partial^2}{\partial r^2}RPF+\frac{1}{r}\frac{\partial}{\partial r}RPF+\frac{1}{r^2}\frac{\partial^2}{\partial\theta^2}RPF+\frac{\partial^2}{\partial z^2}RPF+n^2(\omega)\frac{\omega^2}{c^2}RPF\ =\ (4.22)$$

$$\frac{PF}{RPF}\frac{d^2}{dr^2}R+\frac{PF}{RPF}\frac{1}{r}\frac{d}{dr}R+\frac{RF}{RPF}\frac{1}{r^2}\frac{d^2}{d\theta^2}P+\frac{RP}{RPF}\frac{d^2}{dz^2}F+\frac{1}{RPF}n^2(\omega)\frac{\omega^2}{c^2}RPF\ =\ (4.23)$$

In the above equation we have used ordinary derivatives instead of partial derivatives because each function depends on only one variable. Now, cancelling like terms in the above equation, we obtain

$$\frac{1}{R}\frac{d^2}{dr^2}R+\frac{1}{R}\frac{1}{r}\frac{d}{dr}R+\frac{1}{P}\frac{1}{r^2}\frac{d^2}{d\theta^2}P+\frac{1}{F}\frac{d^2}{dz^2}F+n^2(\omega)\frac{\omega^2}{c^2}\ =\ 0$$

$$-\left\{\frac{1}{R}\frac{d^2}{dr^2}R + \frac{1}{R}\frac{1}{r}\frac{d}{dr}R + \frac{1}{P}\frac{1}{r^2}\frac{d^2}{d\theta^2}P + n^2(\omega)\frac{\omega^2}{c^2}\right\} = \frac{1}{F}\frac{d^2}{dz^2}F \qquad (4.24)$$

The left hand side of the above equation is a function of $r$ and $\theta$ only, while the right hand side is a function of $z$ only. Thus each side can be varied independently of the other. Therefore, these equations are equal if and only if they are both equal to same constant, known as a separation constant and denoted by $\gamma^2$, and $\gamma = \alpha + i\beta$, a complex number. Thus we have,

$$\frac{1}{F}\frac{d^2}{dz^2}F = \gamma^2$$

or

$$\frac{d^2}{dz^2}F - \gamma^2 F = 0 \qquad (4.25)$$

which can be identified as the $z$-Equation, whose solution is discussed below. This value can be inserted above, rearrange, to obtain

$$-\left\{\frac{1}{R}\frac{d^2}{dr^2}R + \frac{1}{R}\frac{1}{r}\frac{d}{dr}R + \gamma^2 + n^2(\omega)\frac{\omega^2}{c^2}\right\} = \frac{1}{P}\frac{1}{r^2}\frac{d^2}{d\theta^2}P \qquad (4.26)$$

Multiplying by $r^2$ throughout, we obtain

$$-\left\{\frac{r^2}{R}\frac{d^2}{dr^2}R + \frac{r}{R}\frac{d}{dr}R + \gamma^2 r^2 + r^2 n^2(\omega)\frac{\omega^2}{c^2}\right\} = \frac{1}{P}\frac{d^2}{d\theta^2}P \qquad (4.27)$$

The left hand side of the above equation is a function of $r$ only, while the right hand side is a function of $\theta$ only. Thus each side can be varied independently of the other. Therefore, these equations are equal if and only if they are both equal to same constant, known as a separation constant, denoted by $-m^2$. Thus we have

$$\frac{1}{P}\frac{d^2}{d\theta^2}P = -m^2$$

or

$$\frac{d^2}{d\theta^2}P + m^2 P = 0 \qquad (4.28)$$

which can be identified as the $\theta$-Equation, whose solution is discussed below, and we also have

$$-\left\{\frac{r^2}{R}\frac{d^2}{dr^2}R + \frac{r}{R}\frac{d}{dr}R + \gamma^2 r^2 + r^2 n^2(\omega)\frac{\omega^2}{c^2} - m^2\right\} = 0 \qquad (4.29)$$

Multiply by $R$ and dividing by $r^2$, we obtain the radial equation

$$\frac{d^2}{dr^2}R + \frac{1}{r}\frac{d}{dr}R + \left(\gamma^2 + n^2(\omega)\frac{\omega^2}{c^2} - \frac{m^2}{r^2}\right)R = 0$$

$$\frac{d^2}{dr^2}R + \frac{1}{r}\frac{d}{dr}R + \left(h^2 - \frac{m^2}{r^2}\right)R = 0 \qquad (4.30)$$

where $h^2$ has been introduced defined by

$$\gamma^2 + n^2(\omega)\frac{\omega^2}{c^2} = h^2 \tag{4.31}$$

and the above second order differential equation is the *Radial Equation* and can be identified as the *Bessel Equation*, whose solution is discussed below.

### The $z$ Equation

$$\frac{d^2}{dz^2}F - \gamma^2 F = 0 \tag{4.32}$$

The solution of the above equation is of the form

$$F(z) = C_1 e^{-\gamma z} + C_2 e^{\gamma z} \tag{4.33}$$

where the constants $C_1$ and $C_2$ can be determined by using boundary conditions. Since we are only interested in a wave propagating in the positive $z$-direction, we can set $C_2 = 0$.

### The $\theta$ Equation

$$\frac{d^2}{d\theta^2}P + m^2 P = 0 \tag{4.34}$$

The solution of the above equation is of the form

$$P(\phi) = C_3 \cos(m\theta) + C_4 \sin(m\theta) \tag{4.35}$$

### The Radial Equation

$$\frac{d^2}{dr^2}R + \frac{1}{r}\frac{d}{dr}R + \left[h^2 - \frac{m^2}{r^2}\right]R = 0 \tag{4.36}$$

This equation shall be applied to two cases of physical interest: Case (1) in the core ($r < a$) and Case (2) in the cladding, ($r > a$), where $a$ is the radius of the core.

### Case (1) in the core with $r < a$,

In this case, $h$ must be real, with $h = h_1$, and we seek the solution of

$$\frac{d^2}{dr^2}R + \frac{1}{r}\frac{d}{dr}R + \left[h_1^2 - \frac{m^2}{r^2}\right]R = 0$$

The solution of the Radial equation, which is a Bessel equation, is of the form

$$R(r) = C_5 J_m(h_1 r) + C_6 Y_m(h_1 r) \tag{4.37}$$

where $J_m$ is a *Bessel function of the first kind and order m*, and $Y_m$ a *Bessel function of the second kind of order m, also known as Neumann Functions*. The Bessel functions are illustrated in Figure 4.12 and 4.13. It is noted that for the Bessel Function of the second kind, as $r \to 0, Y_\infty \to -\infty$, which has no physical meaning, and hence we set $C_6 = 0$. Hence we have

$$R(r) = C_5 J_m(h_1 r) \tag{4.38}$$

The electric field $E(r, \phi, z) = RPF$ is given by

$$E_z = J_m(h_1 r) \left[ C_3 \cos(m\theta) + C_4 \sin(m\theta) \right] e^{-\gamma z} \tag{4.39}$$

Similarly, the magnetic field, $H_z$ is given by

$$H_z = J_m(h_1 r) \left[ F_1 \cos(m\theta) + G_1 \sin(m\theta) \right] e^{-\gamma z} \tag{4.40}$$

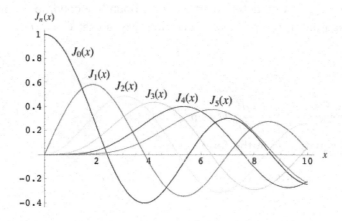

Figure 4.12: Bessel functions of the first kind

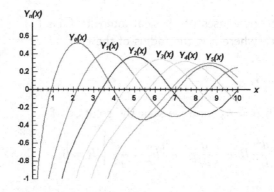

Figure 4.13: Bessel functions of the second kind

**Case (2) in the cladding with $r > a$,**

In this case, $h$ must be imaginary, with $h = ih_2$, and we seek the solution of

$$\frac{d^2}{dr^2}R + \frac{1}{r}\frac{d}{dr}R - \left[h_2^2 + \frac{m^2}{r^2}\right]R = 0 \tag{4.41}$$

The solution of this Radial equation in the cladding, is of the form

$$R(r) = C_7 I_m(h_2 r) + C_8 K_m(h_2 r) \tag{4.42}$$

where $I_m$ is a *Modified Bessel function of the first kind and order m*, and $K_m$ a *Modified Bessel function of the second kind of order m*. The Modified Bessel functions are illustrated in Figure 4.14 and 4.15. It is noted that for the Modified Bessel Function of the second kind, as $r \to \infty$, $I_m \to +\infty$, which has no physical meaning, and hence we set $C_7 = 0$. Hence we have, in the cladding,

$$R(r) = C_8 K_m(h_2 r) \tag{4.43}$$

The electric field $E(r, \phi, z) = RPF$ is given by

$$E_z = K_m(h_2 r)\left[A_2 \cos(m\theta) + B_2 \sin(m\theta)\right] e^{-\gamma z} \tag{4.44}$$

Similarly, the magnetic field, $H_z$ is given by

$$H_z = K_m(h_2 r)\left[F_2 \cos(m\theta) + G_2 \sin(m\theta)\right] e^{-\gamma z} \tag{4.45}$$

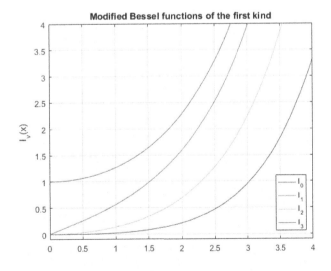

Figure 4.14: Modified Bessel Functions of the First Kind

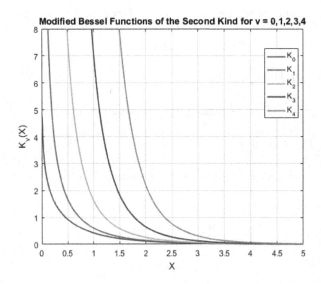

Figure 4.15: Modified Bessel Functions of the second kind.

## 4.7 Attenuation in Optical Fibers

When an electromagnetic wave propagated through a material it becomes *attenuated* in the direction of propagation for several reasons, for example, absorption, scattering, dispersion, fiber bending and others. A graph of attenuation vs wavelength for a typical silica-based fiber is illustrated in Figure 4.16. Note that there are marked attenuation peaks at $1.24\mu$m, $1.38\mu$m due to hydroxil ions, $OH^-$. These arise due to hydroxyl impurities in the fiber glass material and hydrogen atoms which diffuse into the structure.

There are three optical windows: First window at 850nm, Second window at 1300nm, and the third window at 1550nm.
There is a background attenuation process that decreases with wavelength due to Rayleigh scattering as result of refractive index variation with wavelength, of the form

$$\text{Rayleigh scattering} \quad \propto \quad \frac{1}{\lambda^4}$$
$$= \quad \frac{A_R}{\lambda^4} \tag{4.46}$$

where $A_R \approx 0.90$dB km$^{-1}$.

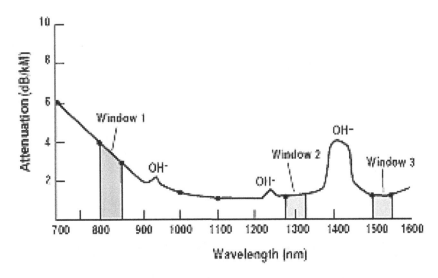

Figure 4.16: Attenuation in Optical Fibers

## 4.8   Exercises

**4.1.**(a) Describe the structure of an optical fiber labelling all the parts clearly.
(b) Describe how optical fiber cables in the past few years have impacted on the communication sector globally.
(c) Describe the difference between Multimode Step Index (MMSI) fiber, Multimode Graded Index (MMGI) Fiber and the Single Mode Step Index (SMSI), and how light propagates in each of these fibers.

**4.2** Distingush between submarine optical fiber cables and terrestrial optical fber cables and discuss their applications.
(b) Describe the following submarine optical fiber cables in East Africa:  SEACOM, EASSY and TEAMS.

**4.3**. (a) Define (i) the Acceptance angle (ii) Numerical aperture
(b) Light from a medium of refractive index $n_0$ enters an optical fiber whose core has refractive index $n_1$ and is surrounded by a cladding with refractive index $n_2$ such that $n_1 > n_2$.
(i) Show that the Numerical aperture is given by

$$\frac{1}{n_0}\sqrt{n_1^2 - n_2^2}$$

(ii) Calculate the acceptance angle and Numerical aperture if $n_0 = 1.0, n_1 = 1.48$ and $n_2 = 1.44$.
(iii) Calculate the critical ange at the core-cladding interface if $n_1$ and $n_2$ are as given in (ii) above.

**4.4**. (a) Define the V-parameter of an optical fiber whose core has radius $a$ and refractive index $n_1$, and a cladding of radius $b$, and light of wavelength $\lambda$ is propagating in it.
(b) Light of wavelength $1.55\mu$m is propagatng in an optical fiber consisting of a core with diameter $8.2\mu$m and refractive index is 1.4531 while the cladding refractive index is 1.4483.
(i) Calculate the critical angle at the core-cladding interface.
(ii) Calculate the V-parameter.

**4.5**. An optical fiber of 86.5%SiO$-$2-13.5%GeO$_2$ core of diameter 6 $\mu$m and refractive index 1.47 and a cladding of refractive index 1.46, is operated at 1300 nm with a laser source of width 2 nm.
(a) Calculate the V-number.
(b) What is the maximum allowed diameter of the core that will allow operation as a single mode?
(c) What is the wavelength below which the fiber will operate as a multimode?
(d) Calculate the Numerical aperture.
(e) Calculate the maximum acceptance angle.

**4.6** A multimode fiber with a core diameter 60 $\mu$m, core refractive index of 1.47, and a cladding refractive index of 1.45, both at an operating wavelength of 870 nm.
(a) Calculate the Numerical aperture.
(b) Calculate the V-number and estimate the number of guided modes.
(c) What is the maximum allowed diameter of the core that will allow operation as a single mode?

(d) Calculate the modal dispersion.?

**4.7**. (a) Explain what is meant attenuation of light as it propagates in an optical fiber and discuss some of the causes of attenuation in the fiber.
(b) The attenuation coefficient $\alpha_{dB}$ of a signal in dB per unit length is given by

$$\alpha_{dB} = \frac{1}{L} 10 \log \left( \frac{P_{in}}{P_{out}} \right)$$

The optical power launched into a single-mode optical fiber from a laser diode is approximately 1mW. The photodetector at the output requires a minimum power of 1 nW to provide a clear signal above noise. The fiber operates at 1.31 $\mu$m and has an effective attenuation coefficient of 0.40 dB km$^{-1}$. Calculate the length of the fiber that can be used without inserting a repeater to regenerate the signal.

**4.8**. (a) Write down the Helmholtz equation for an electromagnetic wave propagating in an optical fiber for an electric field $E(r, \theta, z)$.
(b) Assuming the method of separation of variables, where the electric field can be written as

$$E(r, \theta, z) = R(r)P(\theta)F(z) = RPF$$

Show that the functions $F, P$ and $R$ satisfy
(i) The $z$-Equation, of the form

$$\frac{d^2 F}{dz^2} - \gamma^2 F = 0$$

(ii) The $\theta$-equation, of the form

$$\frac{d^2}{d\theta^2} P + m^2 P = 0$$

(iii) The Radial Equation, which is a *Bessel Equation*, of the form

$$\frac{d^2}{dr^2} R + \frac{1}{r} \frac{d}{dr} R + \left( h^2 - \frac{m^2}{r^2} \right) R = 0$$

where $h^2$ is given by

$$h^2 = \gamma^2 + n^2(\omega) \frac{\omega^2}{c^2}$$

and all symbols are in usual notation.
(c) State without proof the solutions, $F, P$ and $R$ to the above three equations, explaining all the terms clearly.

**4.9**. One of the causes of attenuation in optical fibers is due to *Rayleigh scattering*, which leads to an attenuation given by expressions in two forms.

$$\sigma_R^A = \frac{8\pi^3}{3\lambda^4} n^8 p^2 \beta_T k_B T_f$$

and

$$\sigma_R^B = \frac{8\pi^3}{3\lambda^4}(n^2 - 1)\beta_T k_B T_f$$

where
$\lambda$ is the free space wavelength

$n$ is the refractive at $\lambda$

$T_f$ is a temperature known as fictive temperature

$\beta_T$ is the isothermal compressibility at $T_F$

$k_B$ is the Boltzmann constant,

Calculate $\sigma_R^A$ and $\sigma_R^B$ due to Rayleigh scattering at $\lambda = 155\mu$m, given the following parameters: $T_f = 1180°C$, $\beta_T \approx 7 \times 10^{-11}$m$^2$N$^{-1}$, $n = 1.5$, $p = 0.28$.

**4.10**. Describe the lightwave guidance mechanism in solid and hollow core photonic crystal fibers (PCFs).

# Chapter 5

# Introduction to Satellite Communications

## 5.1 Introduction

Satellite communications has developed significantly over the past few years since the first artificial satellite, Sputnik 1, was launched by the Soviet Union on 4th October 1957, followed by the Explorer 1 launched by the United States on 31st January 1958. Artificial satellites are distinguished from natural satellites such as the moon since they are put into orbit by human endeavor. Each satellite that is put into space is a result of a mission that has a purpose: make scientific measurements, weather forecasting, astronomical investigations, global fixed and mobile telephony, radio and TV broadcasting, internet, navigation, military, rural connectivity and others. The satellite mission determines the satellite payload and orbit, where the payload may be communications equipment, scientific equipment (telescopes such as the Hubble Space Telescope launched by NASA in 1990), electrical power source, thermal control, orbital Space control and other items of practical interest. Space control and other items of practical interest.

Some primary satellite launch sites include:

USA: Cape Canaveral in orida, Vandenberg in California.

Russia: Svobodny, Vostochny.

China:Jiuquan, Taiyuan, Wenchang, Xichang.

Japan: Akita, Uchinoura.

India: Thumba.

and others.

## 5.2 Satellite orbital mechanics

### 5.2.1 Central force orbital mechanics

A **central force** is a force whose line of action passes through a defined center and whose magnitude depends on the distance from that center. With respect to satellite motion, it is relevant to consider one example of central forces, which is the Gravitational force, $F$, given by the Newton's Universal

Law of Gravitation which states that the gravitational force is proportional to the product of masses $m_1$ and $m_2$ and inversely proportional to the square of the separation distance $r$ between the masses.

$$
\begin{aligned}
F &= \frac{Gm_1m_2}{r^2} \\
&= f(r)\hat{\mathbf{r}} \\
\text{where} \quad f(r) &= \frac{k}{r^2} \\
\text{where} \quad k &= Gm_1m_2 \\
\text{where} \quad G &= 6.67 \times 10^{-11}\text{Nm}^2\text{kg}^{-2}
\end{aligned}
\tag{5.1}
$$

is the universal gravitational constant.

Consider the **position** of a satellite given by polar coordinates $(\mathbf{r}, \theta)$, where $\mathbf{r} = r\hat{\mathbf{r}}$

The **velocity** of the satellite is given by $\mathbf{v}$, where

$$
\begin{aligned}
\mathbf{v} &= \frac{d\mathbf{r}}{dt} \\
&= \frac{d(r\hat{\mathbf{r}})}{dt} \\
&= \dot{r}\hat{\mathbf{r}} + r\frac{d\hat{\mathbf{r}}}{dt} \\
\mathbf{v} &= \dot{r}\hat{\mathbf{r}} + r\dot{\theta}\hat{\theta}
\end{aligned}
\tag{5.2}
$$

The **acceleration** of the satellite is given by $\mathbf{a}$, where

$$
\begin{aligned}
\mathbf{a} &= \frac{d\mathbf{v}}{dt} \\
&= \frac{d\dot{r}\hat{\mathbf{r}} + r\dot{\theta}\hat{\theta}}{dt} \\
\mathbf{a} &= (\ddot{r} - r\dot{\theta}^2)\hat{\mathbf{r}} + (r\ddot{\theta} + 2\dot{r}\dot{\theta})\hat{\theta} \\
\mathbf{a} &= a_r\hat{\mathbf{r}} + a_\theta\hat{\theta}
\end{aligned}
\tag{5.3}
\tag{5.4}
$$

where

$$
\begin{aligned}
a_r &= (\ddot{r} - r\dot{\theta}^2) \quad \text{is the Radial component of the acceleration} \tag{5.5}\\
a_\theta &= r\ddot{\theta} + 2\dot{r}\dot{\theta} \tag{5.6}\\
&= \frac{1}{r}\frac{d(r^2\dot{\theta})}{dt} \quad \text{is the Normal (or Transverse) component of the acceleration} \tag{5.7}
\end{aligned}
$$

The **Angular momentum** of the satellite is given by $\mathbf{L}$, which has magnitude $L$, given by

$$
\begin{aligned}
L &= |\mathbf{r} \wedge m\mathbf{v}| \\
&= |\mathbf{r} \wedge m(\dot{r}\hat{\mathbf{r}} + r\dot{\theta}\hat{\theta})| \\
L &= mr^2\dot{\theta} \quad \text{is the magnitude of the Angular momentum}
\end{aligned}
\tag{5.8}
$$

The **Equation of motion** for the satellite is given by $m\mathbf{a} = \mathbf{F}$, where

$$
\begin{aligned}
m\mathbf{a} &= \mathbf{F} \\
m\left[(\ddot{r} - r\dot{\theta}^2)\hat{\mathbf{r}} + (r\ddot{\theta} + 2\dot{r}\dot{\theta})\hat{\theta}\right] &= f(r)\hat{\mathbf{r}}
\end{aligned}
$$

On equating components,

$$
\begin{aligned}
m\left((\ddot{r} - r\dot{\theta}^2)\right) &= f(r)\hat{\mathbf{r}} \quad \text{is the Radial equation} & (5.9) \\
m\left((r\ddot{\theta} + 2\dot{r}\dot{\theta})\right) &= 0 \quad\quad \text{is the Transverse equation} & (5.10)
\end{aligned}
$$

which on equating components, gives the Radial equation and Transverse equation shown above, and are discussed below.

**Transverse equation**

$$
\begin{aligned}
m\left[r\ddot{\theta} + 2\dot{r}\dot{\theta}\right] &= 0 \\
\frac{m}{r}\frac{d(r^2\dot{\theta})}{dt} &= 0 \\
\frac{d(r^2\dot{\theta})}{dt} &= 0 \\
r^2\dot{\theta} &= \text{constant} = h = \frac{L}{m}
\end{aligned}
$$

where $h = \frac{L}{m}$ is the angular momentum per unit mass.

**Radial equation**

$$
\begin{aligned}
m\left[\ddot{r} - r\dot{\theta}^2\right] &= f(r) \\
\text{Let} \quad r = \frac{1}{u} &\Rightarrow r = u^{-1} \\
\dot{r} = -u^{-2}\dot{u} &= -\frac{1}{u^2}\frac{du}{dt} = -\frac{1}{u^2}\frac{du}{d\theta}\frac{d\theta}{dt} \\
&= -r^2\dot{\theta}\frac{du}{d\theta} \\
&= -h\frac{du}{d\theta}
\end{aligned}
$$

Differentiating again,

$$
\begin{aligned}
\ddot{r} &= -h\frac{d}{dt}\left(\frac{du}{d\theta}\right) \\
&= -h\frac{d\theta}{dt}\frac{d}{d\theta}\left(\frac{du}{d\theta}\right) \\
&= -h\dot{\theta}\frac{d^2u}{d\theta^2} \\
&= -h^2u^2\frac{d^2u}{d\theta^2}
\end{aligned}
$$

Note that from the transverse equation, we obtained $r^2\dot{\theta} = h, \Rightarrow r\dot{\theta}^2 = \frac{1}{u}h^2u^4$

The Radial equation can be cast in the form

$$m\left[-h^2u^2\frac{d^2u}{d\theta^2} - \frac{1}{u}h^2u^4\right] = f(u^{-1})$$

$$\frac{d^2u}{d\theta^2} + u = -\frac{f(u^{-1})}{mh^2u^2} \quad \text{where} \quad f(u^{-1}) = -ku^2 \quad (5.11)$$

$$\frac{d^2u}{d\theta^2} + u = \frac{k}{mh^2} \quad (5.12)$$

which can be identified as a Second Order Inhomogeneous Differential Equation of the orbit, a form of the Binet equation. The solution of this equation is the sum of the complementary function plus a particular integral.

$$u = A\cos(\theta - \theta_0) + \frac{k}{mh^2}$$

The boundary conditions are such that, initially, $\theta_0 = 0$,

$$u = A\cos\theta + \frac{k}{mh^2}$$

$$r = \frac{1}{u} = \frac{1}{A\cos\theta + \frac{k}{mh^2}}$$

$$= \frac{mh^2}{k\left(1 + \frac{Amh^2}{k}\cos\theta\right)}$$

$$= \frac{mh^2}{k(1 + e\cos\theta)} \quad (5.13)$$

where $e = \dfrac{Amh^2}{k}$ is the eccentricity of the orbit, and for an elliptical orbit, $e < 1$. (5.14)

$$r = \frac{mh^2}{k(1+e)}\frac{(1+e)}{(1+e\cos\theta)}$$

$$= \frac{r_0(1+e)}{(1+e\cos\theta)} \quad (5.15)$$

where $r_0 = \dfrac{mh^2}{k(1+e)}$ for $\theta = 0$ (5.16)

and $r_1 = r_0\dfrac{(1+e)}{(1+e)}$ for $\theta = \pi$ (5.17)

The elliptical orbit for an artificial satellite is illustrated in Figure 5.1, showing:

$a$: Semi-major axis
$b$: Semi-minor axis
$c$: Distance between centre of the ellipse and the focal point$=ae$
Apogee: at this position the satellite is farthest from the earth and the velocity vector is perpendicular

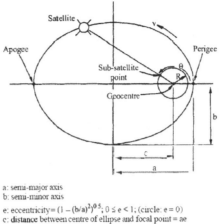

a: semi-major axis
b: semi-minor axis
e: eccentricity= $(1 - (b/a)^2)^{0.5}$; $0 \le e < 1$; (circle: e = 0)
c: distance between centre of ellipse and focal point = ae
R: mean radius of Earth
r, $\theta$ polar coordinates of satellite; $\theta$ (the true anomaly) is measured
    from perigee

Figure 5.1: The Elliptical orbit for an Artificial Satellite

to the radial vector

Perigee: at this position the satellite is closets from the earth and the velocity vector is perpendicular to the radial vector.

### 5.2.2   Kepler's Laws

Although Kepler's laws were developed for the solar system, they are also valid for satellite systems. There are three Kepler's laws.

**First Kepler's law: Law of Orbits**
The planets move in elliptical orbits with the sun at one of the focus, and satisfy the equation

$$r = \frac{r_0(1 + e)}{(1 + e \cos \theta)}$$

where $e < 1$ is the eccentricity of the orbit, as was shown in the previous section, equation 5.14.

**Second Kepler's law: Law of Areas**

For any planet, the radius vector sweeps out equal areas in equal times (see Figure 5.2). This can easily be proved.

$$\text{Area,} \quad \Delta A = \frac{1}{2} |\mathbf{r} \wedge \Delta \mathbf{r}|$$
$$\frac{dA}{dt} = \frac{1}{2} |\mathbf{r} \wedge \frac{d\mathbf{r}}{dt}|$$

$$= \frac{1}{2m}|\mathbf{r} \wedge m\mathbf{v}|$$

$$= \frac{L}{2m} \tag{5.18}$$

which implies that since $L$ is a constant, $\frac{dA}{dt}$ is a constant, that is, the planet, and in this case, the satellite radius vector sweeps out equal areas in equal times.

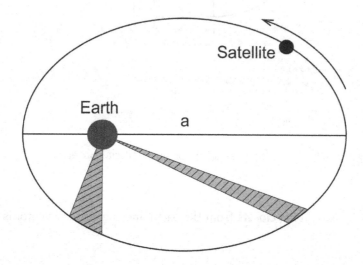

Figure 5.2: The satellite radius vector sweeps out equal areas in equal time.

### Third Kepler's law: Law of Periods

The square of the period is proportional to the cube of the semimajor axis of the orbit. this can easily be proved as shown below.

From Kepler's Law of areas,

$$\frac{dA}{dt} = \frac{L}{2m}$$

$$\int_1^2 dA = \int_1^2 \frac{L}{2m} dt$$

$$A_{12} = t_{12}\frac{L}{2m}$$

$$= t_{12}\frac{h}{2}$$

$$\text{Hence} \quad t_{12} = \frac{2}{h}A_{12}$$

Integrating, we obtain

$$\text{Period,} \quad T = \frac{2}{h}\pi ab \quad \text{where} \quad \pi ab \text{ is the area of the ellipse}$$

$$\text{But} \quad \frac{b}{a} = \sqrt{1-e^2} \quad \text{for an ellipse}$$

$$\text{Hence Period,} \quad T = \frac{2}{h}\pi a^2\sqrt{1-e^2}$$

$$\text{But} \quad a = \frac{r_0}{(1-e)} = \frac{mh^2}{k(1-e^2)} \quad \text{for an ellipse}$$

$$\text{Hence,} \quad T^2 = \frac{4\pi^2 a^4}{h^2}(1-e^2)$$

$$= \frac{4\pi^2 a^4}{k^2}\frac{mh^2}{ka}$$

$$T^2 = \frac{4\pi^2 m}{k}a^3 = Ca^3 \tag{5.19}$$

which is Kepler's third law, where

$$C = \frac{4\pi^2 m}{k} = \frac{4\pi^2}{GM} \tag{5.20}$$

which is the same for all planets (or satellites in this case).

### 5.2.3 The Rocket equation

The rocket which carries the satellite into orbit emits matter and thus one is essentially dealing with a variable mass problem. Consider a rocket of mass $M$ moving with velocity $v$, and that a small mass $\delta m$ is emitted, moving with velocity $v_{dm} = v - u$, and the rocket remaining mass $M - \delta m$ increases velocity to $v + \delta v$, as illustrated in Figure 5.3.

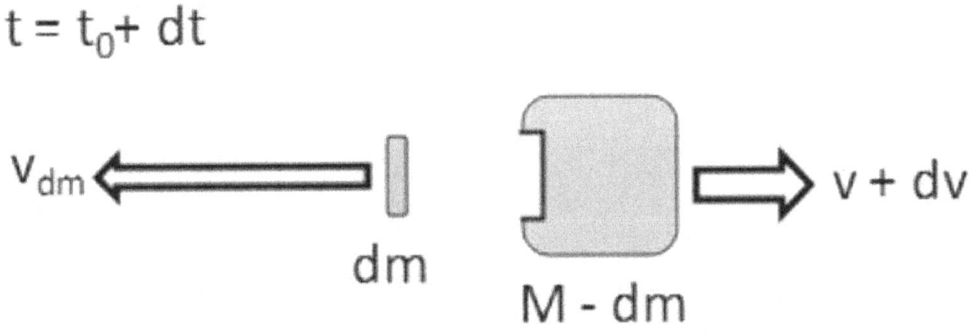

Figure 5.3: Rocket carrying a satellite loses mass

By conservation of momentum,

$$Mv = (M-dm)(v+dv) + dm(v-u)$$

$$Mv + Mdv - vdu - dmdv - dmdv + vdm - udu = Mv$$

Neglecting second order small terms

$$
\begin{aligned}
M\,dv &= u\,dm \\
\text{But} \quad dM &= -dm \\
\frac{dv}{u} &= -\frac{dM}{M} \\
\text{Integrating} \\
\frac{v}{u} &= -\ln\frac{M}{M_0} \\
M &= M_0 e^{-v/u}
\end{aligned}
\tag{5.21}
$$

which is the Rocket Equation.

## 5.3   Satellite Communication Resources

### 5.3.1   Spectrum

Spectrum is an important resource in satellite communications, without which the satellite communications industry would not operate. Spectrum communications uses the radio frequency in the C and Ku band as summarised in Table 5.1 for Region 1 defined by the ITU to comprise Europe, Africa and the Middle East west of the Persian Gulf.

**Table 5.1:** Satellite Communication Frequency Bands.

| Frequency Band | Frequency Range (GHz) | Downlink Band (GHz) | Uplink Band (GHz) |
|---|---|---|---|
| C | 4 - 8 | 3.7 - 4.2 | 5.925 - 6.425 |
| Ku | 12 - 18 | 10.9 - 12.75 | 14 - 14.5 |

### 5.3.2 Orbits: GEO, LEO, MEO and HEO

The ITU plays an important role in the management of global orbital resources. There are four major types of orbits: GEO, LEO, MEO and HEO, as illustrated in Figure 5.4 and are discussed below.

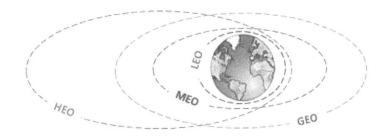

Figure 5.4: Types of Satellites Orbits: GEO, LEO, MEO and HEO

**Geostationary (or Geosynchronous) Earth Orbit (GEO):**

A satellite is in a geostationary or geosynchronous orbit if it appears to be at rest relative to the earth. For this to happen, it must have a period of one day, that is $T = 1$ day $= 24$ hours $= 1440$ minutes $= 8.84 \times 10^4 s$.

Hence by Kepler's third law, the height $H$ above the earth of radius $R_E$ and mass $M_E$ can be calculated

$$
\begin{aligned}
T^2 &= C(H + R_E)^3 \\
&= \frac{4\pi^2}{GM_E}(H + R_E)^3 \\
T &= \frac{2\pi}{\sqrt{GM_E}}(H + R_E)^{3/2} \\
\text{Hence,} \quad H &= \left(\frac{T}{2\pi}\right)^{2/3} (GM_E)^{\frac{1}{2} \times \frac{2}{3}} - R_E \\
H &= 3.58 \times 10^4 \text{km}
\end{aligned}
\tag{5.22}
$$

which shows the GEO satelltes are at an altitude of approximately 35,800 km.

**Low Earth Orbit (LEO):**

The LEO satelltes are at an altitude of approximately 100 to 1,500 km.

**Medium Earth Orbit (MEO):**

The MEO satelltes are at an altitude of approximately 5000 to 10,000 km.

**Highly Elliptical Orbit (HEO):**

The HEO satelltes is an elliptic orbit with a low-altitude often under 1000 km perigee and over 35,786 km apogee. The term "highly elliptical" refers to ecentricity of the orbit rather than the apogee altitude.

## 5.4   Satellite Communication Infrastructure

The Satellite communication infrastructure consists of the Space segment and Ground segment as illustrated in Figure 5.5.

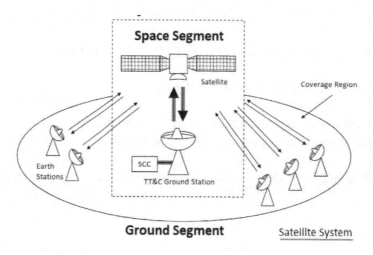

Figure 5.5: Satellite infrastructure showing Space segment and Ground segment

### 5.4.1   Space segment

The space segment consists of the Payload subsystem and Bus subsystems.
The payload subsystems consists of

- Antennas

- Repeater (Receivers, Multiplexers, Amplifiers, Processing ans Switching Units)

The Bus subsystems consists of

- Tracking, Telemetry, Command and Ranging (TTC&R)

- Solar panels

- Batteries

- Reaction control system

- Altitude and spacecraft control processing

- Thermal control and structure

### 5.4.2  Ground segment

The Ground segment consists of

- Earth station which may be one or several, containing transmitting and receiving equipment.

- Antennas

- Users' equipment

- Links to terrestrial networks

- Reaction control system

- Altitude and spacecraft control processing

- Thermal control and structure

## 5.5  Satellite Applications

Satellites have many appications as summarised below.

- Telecommunications (mobile and fixed): The role of satellite communications has increased in recent years as evidenced by the growth of both mobile and fixed telephony.

- Broadcasting (TV and Radio)

- Internet

- Weather forecasts: There are satellites that are put into space to monitor the climate of the earth as well as predict weather conditions in different regions.

- Scientific research

- GPS

- Distance education

- Aviation

- Marine

- Military operations

- Disaster preparedness

## 5.6  Exercises

**5.1** (a) The frequency range of the $Ku$-band is 12 GHz to 18 GHz. Express this as a wavelength range.
(b) State the typical downlink and uplink frequency bands for the $Ku$-band.

**5.2** (a) The frequency range of the $C$-band is 4 GHz to 8 GHz. Express this as a wavelength range.
(b) State the typical downlink and uplink frequency bands for the $C$-band.

**5.3**.(a) (i) Explain what is meant by a central force.
(ii) Is the force between the earth and a satellite a central force? Explain.

(b) Show that the motion of a satellite around the earth satisfies a second order differential equation of the form

$$\frac{d^2u}{d\theta^2} + u = \frac{k}{mh^2}$$

where $u = 1/r$ and other symbols are in the usual notation.

(c) Give a solution to the second order differential equation in (b) and show that it leads to an equation for the satellite elliptical orbit of the form

$$r = \frac{mh^2}{k(1 + e\cos\theta)}$$

where $e < 1$ is the eccentricity for the ellipse.

**5.4**. (a) Write down an expression for the total energy $E$ of a satellite of mass $m$ and coordinates $(r, \theta)$ moving in a central force field satisfying an inverse square law.

(b) Introducing a parameter $u = \frac{1}{r}$, show that the particle described above satisfies

$$\theta = \int \frac{du}{\sqrt{\frac{2E}{mh^2} + \frac{2ku}{mh^2} - u^2}}$$

where $h$ is the angular momentum per unit mass and $k$ is a constant in the force law.

(c) By solving the integral given above, show that

$$r = \frac{mh^2}{k(1 + e\cos\theta)}$$

where $e < 1$ is known as the eccentricity.

Hint: $\int \frac{dx}{\sqrt{ax^2+bx+c}} = \frac{1}{\sqrt{-a}} \sin^{-1}\left\{ \frac{-2ax-b}{\sqrt{b^2-4ac}} \right\}$ and choose the constant of integration as $-\frac{\pi}{2}$.

**5.5**. (a) Describe the following types of satellite orbits: LEO, MEO, GEO and HEO.

(b) Estimate the magnitude of the height above the earth for which the satellite would appear stationary above the earth.
(c) What is the minimum number of stationary satellites is required so that every point along the equator would be in view of the satellites?

**5.6**. A point on a satellite orbit where its velocity is perpendicular to its radial vector is called an *apse*.
(a) Show that at an apse
(i) $\dot{r} = 0$
(ii) $r^2 + \frac{k}{E}r - \frac{L^2}{2mE} = 0$

where      $r$ is the position at an apse

               $k$ is a constant in the force law

               $E$ is the total energy of a satellite

               $L$ is the angular momentum of the satellite

               $m$ is the mass of the satellite

(b) A satellite is moving in a circular orbit of radius $2R_E$ around the earth of radius $R_E$. At a certain instant of time, the direction of the motion of the satellite is changed through an angle $\alpha$ without change of speed, ready for landing. Find tha angle $\alpha$ in order that the satellite just touches the earth.

**5.7**. A satellite of mass 10 kg is launched into orbit 50 km above the surface of the earth.
(a) Calculate the potential energy of the earth-satellite system at the surface of the earth.
(b) With what speed should the satellite be launched?

**5.8.** (a) If the residual mass of a rocket, without payload or fuel, is a given fraction $\lambda$ of the initial mass of a rocket plus fuel, show that the total take-off mass required to accelerate a payload $m$ to velocity $v$ is given by

$$M_0 = m\frac{1-\lambda}{e^{-v/u} - \lambda}$$

(b) (i) Derive the corresponding expression for a two stage rocket, in which each stage produces the same velocity impulse.
(ii) What is the minimum number of stages required if $v = 4u$ and $\lambda = 0.15$?

**5.9**. A satellite of mass $m = 50$ kg is initially placed in a circular orbit of radius $r_i = 2R_E$ around the Earth. It loses 20% of its energy and descends to a new circular orbit, of radius $r_f$.

(a) Calculate the initial energy.
(b) Calculate the new orbit, $r_f$.
(c) Did the speed increase or decrease because of the losses of satellite's energy? Determine the percentage change in the speed'

**5.10.** A shuttle of mass $10^4$ kg is sent into orbit to repair an ailing telecommunications satellite in a geosynchronous orbit. It lands in an orbit of radius $r$ which is 5% smaller than the geosynchronous orbit $r_s$. The pilot of the shuttle uses boosters to get into the orbit of radius $r_s$. How much energy must the boosters provide?

# Chapter 6

# Introduction to Mobile Cellular Communications

## 6.1 Introduction

A mobile cellular communication system or mobile network is a communication network where the last link is wireless, using the radio frequency part of the electromagnetic spectrum. The network is distributed over areas called cells, each served by at least one fixed-location base station. This base station provides the cell with the network coverage which can be used for transmission of different types of services such as voice, SMS, data and others. In recent years, as a result of developments in information and communication technologies (ICTs), mobile communication systems have become ubiquitous as evidenced by digital lifestyles through mobile phones, digital financial services, digital music, digital television, online newspapers, digital medical equipment, e-applications (e-commerce, e-learning, e-health, e-government) and the internet.

The growth of the communication sector has been phenomenal over the past few years. There are many reasons for this. *First*, technological developments such as *convergence* of technologies, for example the convergence of telecommunications, broadcasting and computers, as illustrated in Figure 6.1. *Secondly*, regulation of the communication sector, a subject we shall discuss in section 6.8.4, which arose primarily because of liberalization from monopoly companies to competition between service providers. *Thirdly*, applications which have had a big impact on digital lifestyles (see Figure 6.2). Another example of convergence is the mobile networks and the internet, mobile networks and banks to provide mobile financial services. This has had a large impact on the economy of the country.

## 6.2 Mobile Communications and Electromagnetic Waves

### 6.2.1 Maxwell's equations

Mobile communications uses electromagnetic waves whose frequencies are in the radio frequency (RF) range. Electromagnetic waves satisfy Maxwell's equations which were studied in chapter 1.

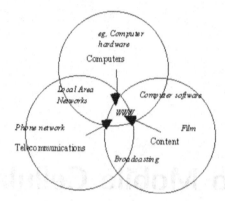

Figure 6.1: Convergence of telecommunications, broadcasting and computers.

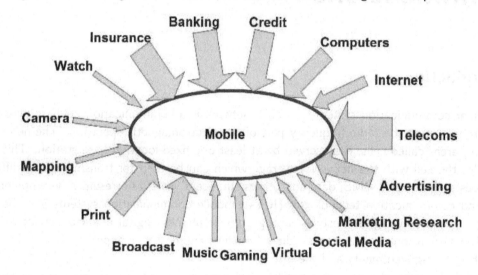

Figure 6.2: Mobile communication applications. (Source: Tomi T. Ahonen, Communities-dominate.blogs.com)

Recall Maxwell's equations in free space ($\rho = 0, j = 0$), summarised below.

$$\nabla \cdot \mathbf{E} = 0$$
$$\nabla \cdot \mathbf{B} = 0$$
$$\nabla \wedge \mathbf{E} = -\frac{\partial \mathbf{B}}{\partial t}$$
$$\nabla \wedge \mathbf{B} = \mu_0 \epsilon_0 \frac{\partial \mathbf{E}}{\partial t}$$

Taking the curl on both sides of the third equation,

$$\nabla \wedge (\nabla \wedge \mathbf{E}) = \nabla \wedge \left(-\frac{\partial \mathbf{B}}{\partial t}\right)$$
$$\text{Using} \quad \mathbf{A} \wedge (\mathbf{B} \wedge \mathbf{C}) = \mathbf{B}(\mathbf{A} \cdot \mathbf{C}) - \mathbf{C}(\mathbf{A} \cdot \mathbf{B})$$

$$\nabla(\nabla \cdot \mathbf{E}) - \nabla^2 \mathbf{E} = -\frac{\partial}{\partial t}(\nabla \wedge \mathbf{B})$$

$$= -\frac{\partial}{\partial t}\left(\mu_0 \epsilon_0 \frac{\partial}{\partial t}\mathbf{E}\right)$$

$$\nabla(\nabla \cdot \mathbf{E}) - \nabla^2 \mathbf{E} = -\mu_0 \epsilon_0 \frac{\partial^2 \mathbf{E}}{\partial t^2}$$

$$-\nabla^2 \mathbf{E} = -\mu_0 \epsilon_0 \frac{\partial^2 \mathbf{E}}{\partial t^2}$$

$$\nabla^2 \mathbf{E} = \frac{1}{c^2}\frac{\partial^2 \mathbf{E}}{\partial t^2} \tag{6.1}$$

which is known as the Wave Equation, and

$$c^2 = \frac{1}{\mu_0 \epsilon_0}$$

$$c = \frac{1}{\sqrt{\mu_0 \epsilon_0}} = 3 \times 10^8 \qquad \text{ms}^{-1} \tag{6.2}$$

which is the velocity of light in free space. Considering $c = \lambda\nu$, there are several ranges of frequencies or wavelengths within the radio frequencies.

### 6.2.2 Properties of Electromagnetic Waves

The electromagnetic waves which are used in mobile networks satisfy, like other waves, the following properties: Reflection, Refraction, Polarisation, Dispersion, Diffraction, Interference, Absorption and Scattering.

In chapter 2, we studied *reflection and refraction* of electromagnetic waves at a boundary, and Fresnels equations were derived for p-waves and s-waves, where the Reflection Coefficients $R_p$ and $R_s$ as well as Transmission Coefficients $T_p$ and $T_s$ were given. These coefficients can also be written as below, by introducing dielectric constants $\epsilon_1, \epsilon_2$ for mediums 1 and 2 respectively, where $\sqrt{\epsilon_1} = \frac{ck_1}{w}$, $\sqrt{\epsilon_2} = \frac{ck_2}{w}$ in accordance with Maxwell's equations.

$$R_p = |r_p|^2 = |\left[\frac{\sqrt{\epsilon_1}\cos\theta_0 - \sqrt{\epsilon_2}\cos\theta_2}{\sqrt{\epsilon_1}\cos\theta_0 + \sqrt{\epsilon_2}\cos\theta_2}\right]|^2 \tag{6.3}$$

$$R_s = |r_s|^2 = |\left[\frac{2\sqrt{\epsilon_1}\cos\theta_0}{\sqrt{\epsilon_1}\cos\theta_0 + \sqrt{\epsilon_2}\cos\theta_2}\right]|^2 \tag{6.4}$$

$$T_p = \frac{\sqrt{\epsilon_2}}{\sqrt{\epsilon_1}}|t_p|^2 = \frac{\sqrt{\epsilon_2}}{\sqrt{\epsilon_1}}|\left[\frac{\sqrt{\epsilon_2}\cos\theta_0 - \sqrt{\epsilon_1}\cos\theta_2}{\sqrt{\epsilon_2}\cos\theta_0 + \sqrt{\epsilon_1}\cos\theta_2}\right]|^2 \tag{6.5}$$

$$T_s = \frac{\sqrt{\epsilon_2}}{\sqrt{\epsilon_1}}|t_s|^2 = \frac{\sqrt{\epsilon_2}}{\sqrt{\epsilon_1}}|\left[\frac{2\sqrt{\epsilon_2}\cos\theta_0}{\sqrt{\epsilon_2}\cos\theta_0 + \sqrt{\epsilon_1}\cos\theta_2}\right]|^2 \tag{6.6}$$

The above equations can be expressed just in terms of the angle of incidence $\theta_0$ by simplifying the equations using Snell's law $k_1 \sin\theta_0 = k_2 \sin\theta_2$ (See exercise 6.2).

In chapter 3, a model for *diffraction* of electromagnetic waves was studied. It was shown that if a transmitter T and receiver R are separated by a direct path $d_1 + d_2$, and there is an obstruction of height $h$, then the path difference is given by (See Figure 6.3)

$$\delta = \frac{h^2(d_1 + d_2)}{2d_1d_2}$$

Hence, Phase difference $= \frac{2\pi}{\lambda}\delta = \frac{2\pi h^2(d_1 + d_2)}{2\lambda d_1 d_2}$

$$(6.7)$$

The Fresnel-Kirchoff diffraction parameter, $\nu_{FK}$, is given by

$$\nu_{FK} = h\sqrt{\frac{2(d_1 + d_2)}{\lambda d_1 d_2}}$$

$$(6.8)$$

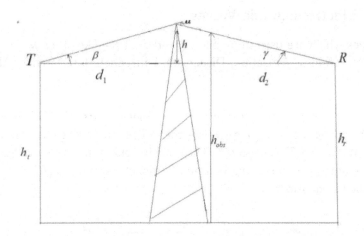

Figure 6.3: Diffraction of electromagnetic waves in mobile infrastructure.

Another important property is *Absorption* of electromagnetic waves by material medium, which occurs when the material through which electromagnetic waves are propagating has a complex refractive index, $n(\omega)$ and hence a complex dielectric function, $\epsilon(\omega)$. Consider an electric field $E$ given by

$$E(z, t) = E_0 e^{i(kz - \omega t)}$$

$$(6.9)$$

The wavevector $k$ satisfies

$$\frac{c^2 k^2}{\omega^2} = \epsilon(\omega) = \epsilon'(\omega) + i\epsilon''(\omega)$$

$$(6.10)$$

where $\epsilon'(\omega)$ and $\epsilon''(\omega)$ are the real and imaginary parts of the dielectric function. Also, we have

$$\frac{ck}{\omega} = n(\omega) = \eta(\omega) + i\kappa(\omega) \tag{6.11}$$

where $\eta(\omega)$ is the real part of the refractive index and $\kappa(\omega)$ is the imaginary part, usually called the extinction coefficient. The refractive index is the square root of the dielectric function.

$$n(\omega) = \sqrt{\epsilon(\omega)} \tag{6.12}$$

or

$$n^2(\omega) = \epsilon(\omega)$$

From equations (6.10) and (6.11), we obtain

$$[\eta(\omega) + i\kappa(\omega)]^2 = [\epsilon'(\omega) + i\epsilon''(\omega)]$$

which can be cast in the form

$$\eta^2(\omega) - \kappa^2(\omega) + i2\eta(\omega)\kappa(\omega) = \epsilon'(\omega) + i\epsilon''(\omega)$$

Separating the real and imaginary parts, we have

$$\epsilon'(\omega) = \eta^2(\omega) - \kappa^2(\omega) \tag{6.13}$$

$$\epsilon''(\omega) = 2\eta(\omega)\kappa(\omega) \tag{6.14}$$

With a little algebra (see exercise 1.10 in chapter 1, equations (6.13) and (6.14) can be cast as quadratic equations for $\eta^2(\omega)$ and $\kappa^2(\omega)$ in the form

$$4\eta^4(\omega) - 4\epsilon'(\omega)\eta^2(\omega) - \epsilon''(\omega)^2 = 0$$

and

$$4\kappa^4(\omega) + 4\epsilon'(\omega)\kappa^2(\omega) - \epsilon''(\omega)^2 = 0$$

The above quadratic equations have the following solutions

$$\eta(\omega) = \{\frac{1}{2}[\epsilon'(\omega) + \sqrt{(\epsilon'(\omega)^2 + \epsilon''(\omega)^2)}]\}^{1/2} \tag{6.15}$$

$$\kappa(\omega) = \{\frac{1}{2}[-\epsilon'(\omega) + \sqrt{(\epsilon'(\omega)^2 + \epsilon''(\omega)^2)}]\}^{1/2} \tag{6.16}$$

Thus, the propagating wave, from equations (6.29), (6.31), (6.35) and (6.36) is of the form

$$E(z,t) = E_0 e^{-\frac{\omega}{c}\kappa z} e^{i(\frac{\omega}{c}\eta z - \omega t)} \tag{6.17}$$

which shows that the amplitude will be *decaying exponentially* due to absorption. The quantities $\eta(\omega), \kappa(\omega)$, or equivalently $\epsilon'(\omega), \epsilon''(\omega)$ are important in studying optical properties of materials.

## 6.3   Mobile Communication Standards

There are several standards that have evolved in mobile cellular communication, for example: the Global System for Mobile (GSM), the Code Division multiple Access (CDMA) as discussed below.

### 6.3.1   GSM: Global System for Mobile

The GSM is the most widely used communication standard globally with approximately 80% to 85% market share, which allows roaming by subscribers. The evolution of the GSM standard is illustrated in Figure 6.4, from 1G as TACS in the 1980s, through 2G as GPRS/EDGE from late 1990s, followed by 3G as WCDMA/HSPA/HSPA+ from early 2000. These developments led to 4G as LTE/LTE Advanced from around 2010, which will lead to 5G. The acronyms are spelt out below.

TACS: Total Access Communication System.
GPRS: General Packet Radio Service.
EDGE: Enhanced Data for GSM Evolution.
WCDMA: Wide-band Code Division multiple Access
HSPA: High Speed Packet Access.
LTE: Long Term evolution.

### 6.3.2   CDMA: Code Division multiple Access

The evolution of the CDMA standard is illustrated in Figure 6.5, from IS-95 (CDMA One) released in 1995, followed by 3G CDMA 2000 1x, CDMA 1xEV-DO, brannched to eHRPD and UMTS, led to 4G LTE around 2010, and later 5G, The acronyms are spelt out below.

IS-95: Interim Standard released in 1995.
GPRS: General Packet Radio Service.
EV-DO: Evolution Data Optimised.
eHRPD: Evolved High Rate Packet Data.
UMTS: Universal Mobile Telecommunications System.
LTE: Long Term evolution.

### 6.3.3   Other standards

**WiMAX** (Worldwide Interoperability for Microwave Access) is a standard dedicated to the advancement of broadband wireless access.

**Wifi** is a technology that allows electronic devices to connect through a wireless local area network (WLAN), mainly using the 2.4 GHz or 5 GHz ISM radio bands.

**Bluetooth** is a technology that allows electronic devices to connect through a personal area networks (PAN), mainly using the 2.4 GHz ISM radio bands.

**ZigBee** is a technology that allows electronic devices to connect through a personal area network,

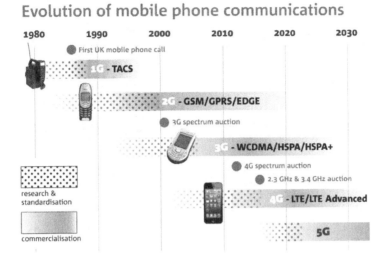

Figure 6.4: Evolution of the GSM standard. (Source: https://commsbusiness.co.uk)

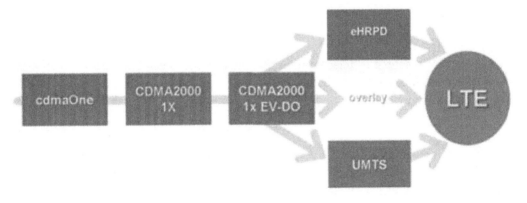

Figure 6.5: Evolution of the CDMA standard (Source: https://andajun.wordpress.com)

mainly using the 2.4 GHz ISM radio bands.

## 6.4 Mobile Communication Resources

### 6.4.1 Spectrum

Spectrum is an important resource in telecommunications, without which the communications industry would not operate. Spectrum in the radio frequency range, 3 kHz to 300 GHz is what is used in the communications sector. Different frequency ranges are used for different applications, as illustrated in Figure 6.6.

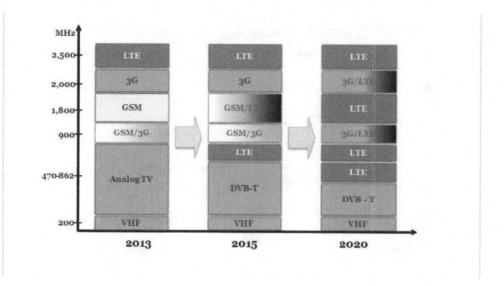

Figure 6.6: Evolution of spectrum applications

### 6.4.2   Numbers

Numbers are an important resource in telecommunications, without which the communications industry would not operate. We shall discuss below, the E.164 numbering plan established by the ITU and USSD (also known as short codes).

#### (i) E.164

The ITU has established a comprehensive numbering plan known as E.164, summarised in Figure 6.7. E.164 numbering plan provides a maximum of 15 digits, of which the Country Code (CC) is up to 3 digits, followed by the National Destination Code (NDC) and the Subscriber Number (SN).

#### (ii) Unstructured Supplementary Service Data (USSD)

Unstructured Supplementary Service Data (USSD), also known as "Short codes" or "Quick codes" is a protocol used by GSM networks for communication between a subscriber and the service provider's computers to provide a given service.

## 6.5   Multiple Access Techniques

Mobile operators are allocated spectrum which is a limited resource. In view of this, it is important to use the allocated spectrum efficiently, and multiple access techniques have been devised to achieve this. The main multiple access techniques are: Frequency Division Multiple Access (FDMA), Time Division Multiple Access (TDMA), Code Division Multiple Access (CDMA), as illustrated in Figure 6.8, and discussed below.

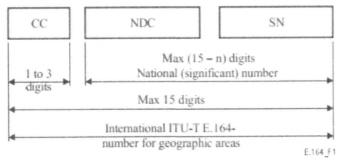

CC    Country Code for geographic area
NDC   National Destination Code
SN    Subscriber Number
n     Number of digits in the country code

NOTE – National and international prefixes are not part of the international
ITU-T E.164-number for geographic areas.

Figure 6.7: Numbering plan with a maximum of 15 digits. (Source: ITU)

## 6.5.1 Frequency Division Multiple Access (FDMA)

FDMA is a technology whereby the allocated bandwidth is subdivided into individual channels in terms of frequencies, each of which can carry voice, data or video service for a user, as illustrated in Figure 6.8. Below, the number of FDMA channels for GSM900, GSM1800 and GSM1900/PCS1900 is calculated.

For **GSM900**, with uplink from 890 MHz to 915MHz. If the spacing between two carriers is 200 kHz, then

$$
\begin{aligned}
\text{Total bandwidth (Uplink),} \quad &= \quad 915 - 890 = 25 \quad \text{MHz} \\
\text{Spacing between two carriers,} \quad &= \quad 200 \quad \text{kHz} \\
\text{Number of carriers,} \quad &= \quad \frac{\text{Bandwidth}}{\text{Spacing}} = \frac{25 \times 10^6}{200 \times 10^3} = 125 \\
&\Rightarrow \quad 125 - 1 = 124
\end{aligned}
$$

Since each carrier can have 8 channels, total number of channels, $= 124 \times 8 = 992$ Channels

For **GSM1800**, with uplink from 1710 MHz to 1785 MHz. If the spacing between two carriers is 200 kHz, then

$$
\begin{aligned}
\text{Total bandwidth (Uplink),} \quad &= \quad 1785 - 1710 = 75 \quad \text{MHz} \\
\text{Spacing between two carriers,} \quad &= \quad 200 \quad \text{kHz} \\
\text{Number of carriers,} \quad &= \quad \frac{\text{Bandwidth}}{\text{Spacing}} = \frac{75 \times 10^6}{200 \times 10^3} = 375 \\
&\Rightarrow \quad 375 - 1 = 374
\end{aligned}
$$

Since each carrier can have 8 channels, total number of channels, $= 374 \times 8 = 2992$ Channels

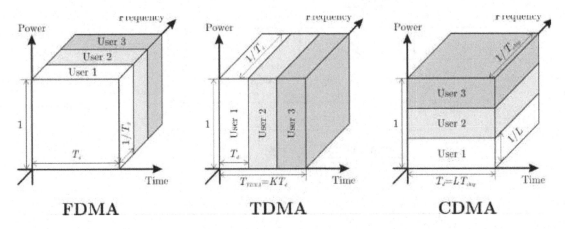

Figure 6.8: Multiple Access Techniques: FDMA, TDMA and CDMA.

For **GSM1900/PCS1900**, with uplink from 1850 MHz to 1910 MHz. If the spacing between two carriers is 200 kHz, then

$$\text{Total bandwidth (Uplink),} \quad = \quad 1910 - 1850 = 60 \quad \text{MHz}$$
$$\text{Spacing between two carriers,} \quad = \quad 200 \quad \text{kHz}$$
$$\text{Number of carriers,} \quad = \quad \frac{\text{Bandwidth}}{\text{Spacing}} = \frac{60 \times 10^6}{200 \times 10^3} = 300$$
$$\Rightarrow \quad 300 - 1 = 299$$

Since each carrier can have 8 channels, total number of channels, $\quad = \quad 299 \times 8 = 2392 \quad \text{Channels}$

### 6.5.2   Time Division Multiple Access (TDMA)

TDMA is a technology whereby the allocated bandwidth is subdivided into individual channels in terms of time slots, each of which can carry voice, data or video service for a user, as illustrated in Figure 6.8. As an example in GSM, the radio spectrum is divided into 200 KHz bands and then uses time division techniques to get eight time slots. The users, each using the allocated time slot, transmit in rapid succession one after the other. GSM defines how long a time slot as follows.

$$\text{A GSM Multiframe is} \quad = \quad 120 \text{ ms long} = 26 \text{ GSM Frames}$$
$$\text{A GSM Frame is} \quad = \quad \frac{120}{26} \text{ ms long} = 4.615 \text{ ms long} \quad = 8 \text{ Time slots}$$
$$\text{A GSM Time slot is} \quad = \quad \frac{4.615}{8} \text{ ms long} = 576.92 \text{ } \mu s \text{ long}$$
$$\text{A GSM Frame has} \quad = \quad 1248 \text{ bits}$$
$$\text{A GSM Time slot has} \quad = \quad \frac{1248}{8} = 156 \text{ bits}$$

### 6.5.3 Code Division Multiple Access (CDMA)

CDMA is a technology whereby the allocated bandwidth is subdivided into individual channels in terms of unique codes, each of which can carry voice, data or video service for a user, as illustrated in Figure 6.8.

CDMA uses two important types of codes to channelize users.

- Walsh codes: these codes channelize users on the forward link (BTS to mobile)

- Pseudorandom Noise (PN) codes: these codes channelize users on the reverse link (mobile to BTS)

## 6.6 Cellular concept

Cellular telephony derives its name from the partition of a geographic area into small parts called "cells". Cells which are next to each other use different set of frequencies to avoid interference and provide guaranteed quality of service within each cell. Noncontiguous cells can reuse the same frequency.

### 6.6.1 Reuse distance

The reuse distance $D$ between clusters is illustrated 6.9.

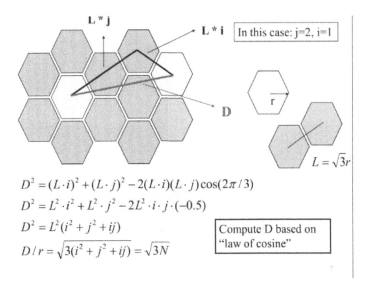

$$D^2 = (L \cdot i)^2 + (L \cdot j)^2 - 2(L \cdot i)(L \cdot j)\cos(2\pi/3)$$
$$D^2 = L^2 \cdot i^2 + L^2 \cdot j^2 - 2L^2 \cdot i \cdot j \cdot (-0.5)$$
$$D^2 = L^2(i^2 + j^2 + ij)$$
$$D/r = \sqrt{3(i^2 + j^2 + ij)} = \sqrt{3N}$$

Figure 6.9: Reuse distance $D$ between mobile clusters

The distance $D$ between clusters can easily be calculated using the Cosine Rule, as illustrated in Figure 6.9.

$$
\begin{aligned}
D^2 &= (L \cdot i)^2 + (L \cdot j)^2 - 2(L \cdot i)(L \cdot j)\cos(\frac{2\pi}{3}) \\
&= L^2 i^2 + L^2 j^2 - 2L^2 \cdot i \cdot j \cdot (-\frac{1}{2}) \\
&= L^2 (i^2 + ij + j^2) \\
D^2 &= L^2 N \qquad\qquad\qquad (6.18) \\
\text{where} \quad N &= i^2 + ij + j^2 \qquad\qquad\qquad (6.19)
\end{aligned}
$$

Note that

For $i = 1, j = 1, N = 3$
For $i = 2, j = 0, N = 4$
For $i = 1, j = 2, N = 7$
For $i = 2, j = 2, N = 12$
For $i = 2, j = 3, N = 19$

Systems with Cluster size, $N = 1, 3, 4, 7, 12, 13$ and $13$ are illustrated in Figure 6.10.

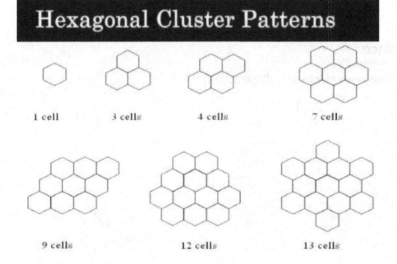

Figure 6.10: Systems with cluster sizes, N = 1, 3, 4, 7, 9, 12 and 13.

## 6.6.2   Frequency reuse

Frequency reuse is extremely important in mobile cellular communications. Frequenciesare allocated in such a way that where neighbouring cells must use different frequencies, but cells at a certain distances (reuse distance) are allowed to re-use frequencies, in a regular pattern of areas, called 'cells', each covered by one base station. In mobile-telephone networks these cells are usually hexagonal, as illustrated in Figure 6.9, 6.10 and 6.11.

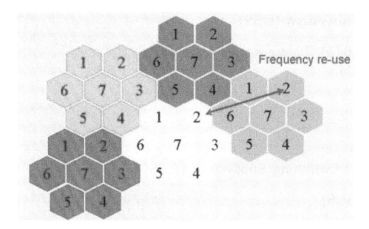

Figure 6.11: Frequency reuse in a cluster with N =7.

## 6.7 Mobile Communication Infrastructure

In a typical mobile cellular communication network, there are three interconnected subsystems illustrated in Figure 6.12, and these are: the Base Station Subsystem (BSS), the Networking and Switching Subsystem (NSS) and the Operational Support Subsystem (OSS), discussed below.

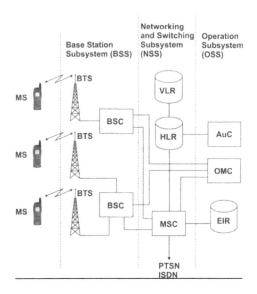

Figure 6.12: Mobile Cellular Communication Network

### 6.7.1   Base Station Subsystem (BSS)

Base Transceiver Station (BTS): Transceiver and Antennas
Base Station Controller (BSC)
Receivers
Transmitters
Power supply

### 6.7.2   Networking and Switching Subsystem (NSS)

Visitor Location Register (VLR)
Home Location Register (HLR)
Mobile Switching Center (MSC)

### 6.7.3   Operational Support Subsystem (OSS)

Authentication Centre (AuC)
Operations and Maintenance Center (OMC)
Equipment Identification Reg. (EIR)

### 6.7.4   Mobile Station: Cellphone/Tablet/PC

A mobile phone must have amongst several parts including at least a display, keyboard, speaker, microphone, a circuit board, battery, antenna, SIM and an operating system. A typical cell phone is ilustrated in Figure 6.13.

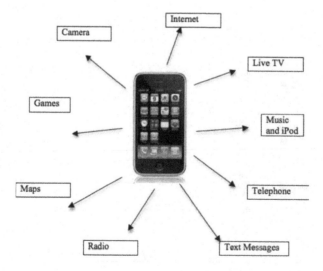

Figure 6.13: A mobile phone has multiple uses as a result of convergence.

## 6.8 Mobile Communications Industry stakeholders

In this section we discuss the growth of the communications. It will be realised that the growth can be attributed to interplay of so many stakeholders: Government, Operators also known as service providers and technology vendors, consumers (the public) and the Regulators, as illustrated in Figure 6.14.

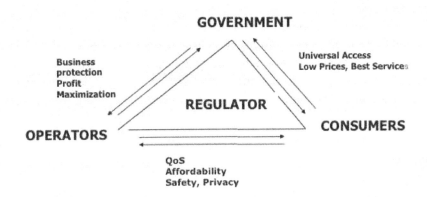

Figure 6.14: Stakeholders in the communication sector

### 6.8.1 Government

The success of the communications sector is dependent on an effective Government that has a Ministry responsible for communications that has developed a properly articulated policy, a communications sector legislation passed by parliament and regulations to be enforced by the regulator and operators. An example of a communications legislation is the Electronic and Postal Communications Act (EPOCA) of 2010 in Tanzania. The contents of this legislation are summarized below.

- Part I: Preliminary

- Part II: Electronic

- Part III: Postal

- Part IV: Competition

- Part V: Enforcement and Compliance

- Part VI: Offences and Penalties

- Part VII: Miscellaneous Provisions

- Part VIII: Transitional Provisions

- Part IX: Consequential Amendments

### 6.8.2  Operators

Operators of telecommunication services, also known as telecommunication service providers, are of several types.  They are mostly private companies who may be multinationals, operating in several countries, for example Vodafone, MTN, Airtel or Milicom (operating as Tigo), or may be government owned.  There are also Over the top (OTT) Operators such as Google, Facebook, twitter, Skype, Whatsapp, Viber, YouTube and others.

Mobile communication operators invest in building the necessary infrastructure to enabe them provide the necessary services and applications for their subscribers.  These companies are normally licensed to carry out their services, and in Tanzania, for example, they are licensed by the regulator TCRA, and have to comply with the communication regulations, including license regulations whereby operators have the following licenses which were introduced by the Converged Licensing Framework of 2005:

- Network Facility License

- Network Services License

- Applications Services License

while for radio and TV broadcasting there is the

- Content Service License

Operators aspire to provide excellent Grade of Service (GoS) or Quality of service (QoS). They are usually guided by teletraffic principles, such as Erlang B and Erlang C formula for telecommunications traffic.

 An Erlang is a dimensionless unit of telephone traffic. If there are a given number of calls per given time, say $\lambda$, and the average time of a phone call is $h$, then the number of Erlangs, $E$, is given by

$$E = \lambda h \tag{6.20}$$

where $h$ and $\lambda$ must be in the same units: calls per second and average seconds per call, calls per minute and average minutes per call, calls per hour and average hours per call, calls per day and average in days per call, and so on.

Let us calculate the number of Erlangs as an example.

If there are 30 calls per hour and an average of each call is 5 minutes, then

$$
\begin{aligned}
\text{Number of Erlangs} \;&=\; \text{Number of calls per hour} \times \text{Average time per call in hours} \\
&=\; 30 \times \frac{5}{60} \\
&=\; 2.5 \text{ Erlangs}
\end{aligned}
$$

Another example. If there are 720 calls per day and an average of each call is 5 minutes, then

$$
\begin{aligned}
\text{Number of Erlangs} \;&=\; \text{Number of calls per hour} \times \text{Average time per call in hours} \\
&=\; 720 \times \frac{5}{(24 \times 60)} \\
&=\; 2.5 \text{ Erlangs}
\end{aligned}
$$

According to Erlang B Formula,

$$
\text{Grade of Service (GoS)} \;=\; \frac{\left(\frac{A^N}{N!}\right)}{\left(\sum_{k=0}^{N} \frac{A^k}{k!}\right)} \tag{6.21}
$$

$$
\begin{aligned}
\text{where } A \quad &\text{is} \quad \text{the expected traffic in Erlangs} \\
N \quad &\text{is} \quad \text{the number of circuits or phone lines}
\end{aligned}
$$

If the system places calls in queue rather than dropping them, then Erlang C Formula is used, given by

$$
P_W \;=\; \frac{\left(\frac{A^N}{N!}\frac{N}{N-A}\right)}{\left(\sum_{i=0}^{N-1} \frac{A^i}{i!}\right) + \frac{A^N}{N!}\frac{N}{N-A}} \tag{6.22}
$$

$$
\begin{aligned}
\text{where } P_W \quad &\text{is} \quad \text{the probability that a customer has to wait for service} \\
A \quad &\text{is} \quad \text{the total traffic in Erlangs} \\
N \quad &\text{is} \quad \text{the number of servers}
\end{aligned}
$$

### 6.8.3 Consumers

The term "Consumers" refers to the public who use the services of operators or service providers, and these consumers are also referred as subscribers, since they subscribe to use the services. There have been unprecedented mobile subscribers growth globally. In Figure 6.15, world mobile subscribers for the period 1980 show an increasing trend, with forecast to 2020.

It can be noted in Figure 6.16 that the mobile world subscriptions overtook fixed subscriptions around 2001, thus making mobile more dominant than fixed communication. Figure 6.17 shows global ICT developments for the period 2001 to 2016, including fixed, mobile, internet, broadband (fixed) and broadband (mobile)..

The global growth in ICTs is also mirrored in individual countries, for example, in Figure 6.18 is shown Tanzania's mobile and fixed subscriptions during 2010 to 2015, and Tanzania's internet penetration is shown in Figure 6.19. TCRA's website (www.tcra.go.tz) shows updated statistics.

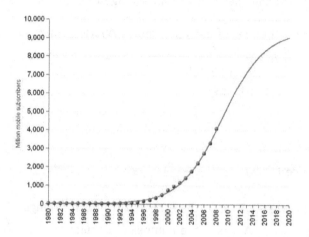

Figure 6.15: World mobile subscribers from 1980 including forecast to 2020 (Source: ITU).

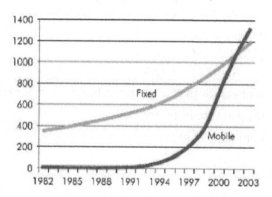

Figure 6.16: World Telephone Subscribers: Fixed and Mobile (Source: ITU).

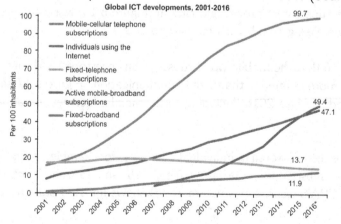

Figure 6.17: Global ICT Developments, 2001 - 2016 (Source: ITU)

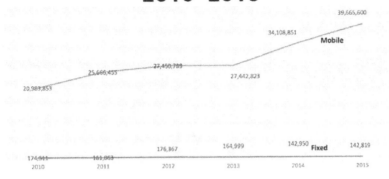

Figure 6.18: Tanzania Mobile and Fixed Subscriptions 2010 - 2015 (Source: TCRA)

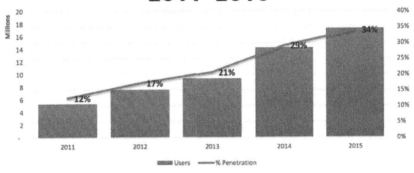

Figure 6.19: Tanzania Internet Penetration (Source: TCRA)

## 6.8.4    Regulator

Regulation has played an important role in the growth of mobile communications.  There is an international regulator, the ITU as well National regulatory bodies, such as the the FCC (Federal Communications Commission) in USA, the OFCOM (Office of Communications) in UK and TCRA (Tanzania Communications Regulatory Authority) in Tanzania.

### International Regulation: The ITU

The global telecommunications/ICT regulator is the International Telecommunications Union (ITU), a specialized UN agency whose structure is shown in Figure 6.20.

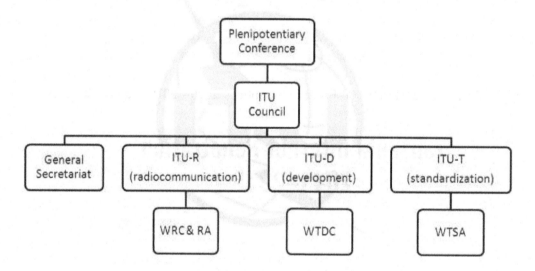

Figure 6.20: The ITU structure (Source: ITU).

### National Communication Regulatory Authority

A National Communications Regulator is an instittition responsible for regulating the communications sector.  Examples are the FCC in USA, the OFCOM in UK and TCRA in Tanzania.

### Duties and Functions of a Regulator

Typical duties and functions of a regulatory body, for example, TCRA, in Tanzania include the following(as listed in the TCRA Act of 2003):

### Duties

- Promoting effective competition and encourage efficiency
- Protect interests of consumers
- Protect financial viability of efficient suppliers

- Promote the availability of regulated services

- Enhance public knowledge

- Protect the environment

**Functions**

- Issue, renew and cancel licences

- Standards

- Regulate rates and charges

- Manage frequency spectrum

- Manage numbering resource

**Regulations**

The working tool for regulators is to enforce regulations which are issued under the communications law promulgated by parliament. For example, the TCRA has the following regulations:

- Consumer protection regulation

- Tariff regulation

- Interconnection regulation

- Radio communication, Frequency spectrum regulations

- Access, Co-location and Infrastructure sharing

- Quality of Service regulations

- Standards regulations

- Numbering and addressing regulations

- Mobile Number Portability (MNP) regulations

- Central Equipment Identification Register (CEIR) regulations

- Computer Emergency Response Team (CERT) regulations

- Telecommunications Traffic Monitoring System (TTMS) regulations

- Account Separation

- Digital Broadcasting regulations

- Content regulations

- Postal regulations

- Licensing regulations

- Competition regulations

## Mobile Communication Emerging Issues

As technology develops very fast, there are several emerging issues which the Regulator has to be alert to.

- Internet of things (IOT)

- Big Data

- Smart cities

- Spectrum challenge

- Cybersecurity

- Impact of OTT services

## 6.9  Exercises

**6.1**. (a) Explain what is meant by *convergence* in the mobile communication industry, and give at least three examples.

(b) Discuss the *applications* of mobile communications with respect to voice, SMS, email, internet, mobile money transfer and social media including opportunities and challenges.

(c) Discuss *cybersecurity* with respect to mobile communications networks.

**6.2** (a) A mobile operator has been allocated spectrum which allows it to propagate electromagnetic waves which undergo reflection and refraction from the ground. Show that the Reflection Coefficients, $R_p$ and $R_s$ for p-waves and s-waves respectively are given by

$$R_p = |r_p|^2 = |\left[\frac{\sqrt{\epsilon_1}\cos\theta_0 - \sqrt{\epsilon_2 - \epsilon_1\sin^2\theta_0}}{\sqrt{\epsilon_1}\cos\theta_0 + \sqrt{\epsilon_2 - \epsilon_1\sin^2\theta_0}}\right]|^2$$

$$R_s = |r_s|^2 = |\left[\frac{2\sqrt{\epsilon_1}\cos\theta_0}{\sqrt{\epsilon_1}\cos\theta_0 + \sqrt{\epsilon_2 - \epsilon_1\sin^2\theta_0}}\right]|^2$$

(b) Assuming medium 1 is air (take $\epsilon_1 \approx 1.0$) and medium 2 is ground (take $\epsilon_2 \approx 15.0$), calculate $R_p$ and $R_s$ for an angle of incidence of $\theta_0 = 15°$.

(c) Assuming medium 1 is air (take $\epsilon_1 \approx 1.0$) and medium 2 is water (take $\epsilon_2 \approx 81.0$), calculate $R_p$ and $R_s$ for an angle of incidence of $\theta_0 = 15°$.

**6.3**. (a) Show that the Fresnel-Kirchoff diffraction parameter, $\nu_{FK}$, is given by

$$\nu_{FK} = h\sqrt{\frac{2(d_1 + d_2)}{\lambda d_1 d_2}}$$

where all symbols are inthe usual notation.

(b) A cellular base station antenna is mounted on a tower above a building. The line of sight path from the antenna tothe nearest user passes near the edge of the building. The distance from the antenna to the edge is 3 m and from the edge to the user is 100 m. The system operates at 900 MHz.
(i) By how much must the line-of-sight path clear the edge of the building to ensure that diffraction effects are negligible?
(ii) Calculate the the Fresnel-Kirchoff diffraction parameter, $\nu$, and explain its physical significance.

**6.4** The mobile industry has developed several technologies. Complete the following table in terms of the frequency range and calculate the corresponding wavelength range.

| Frequency Band | Frequency Range | Downlink Band | Uplink Band |
|---|---|---|---|
| GSM | | | |
| WCDMA (or UMTS) | | | |
| CDMA | | | |

**6.5**. In Table 6.1, telecommunications operators and their allocated frequencies, are given. Calculate the respective wavelengths and complete Table 6.1.

**Table 6.1:** Telecommunications operators and their allocated frequencies.

| Operator | Allocated Frequency | Wavelength, $\lambda$ (Units ?) |
|---|---|---|
| LTE Operator | 800 MHz | ? |
| 2G GSM Operator/HSPA | 900 MHz | ? |
| GSM Operator/HSPA | 1800 MHz | ? |
| HSPA Operator | 2100 MHz | ? |
| WCS Operator | 2.3 GHz | ? |
| 4G Operator/LTE | 2.6 GHz | ? |

**6.6**. (a) Describe the following multiple access techniques: FDMA, TDMA and CDMA.

(b) A GSM 900 MHz network operates using uplink bandwidth 890 MHz to 915 MHz. Assuming FDMA is used with carriers 200 KHz, 8 channels each. Calculate the number of channels.

(c) A GSM network operates with TDMA such that a frame is 4. 615 ms long and has 8 time slots with 1248 bits. Calculate how long a time slot is and how many bits does a time slot have? What is therate (in kps)?

**6.7**. (a) Explain the concept of "cellular" in mobile cellular communications.

(b) Consider a hexagon as shown below, with distance from the center to the vertex as $R$.

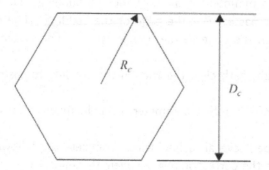

Figure 6.21: A hexagon cell with distance from the center to the vertex as $R_C$.

(i) Show that the area of the hexagon is given by

$$\frac{3}{2}\sqrt{3}R_C^2 = 2.6R_C^2$$

(ii) Show that the distance $D_C = R_C\sqrt{3}$
(c) Show that the distance $D$ between clusters is given by

$$D = L^2(i^2 + ij + j^2)$$

and explain all the symbols used.

**6.8**. (a) Describe the E.164 Numbering plan as established by the ITU.
(b) Which portions of the E.164 are allocated by the ITU and which are allocated the National Regulator?

**6.9**. (a) Explain what is meant by USSD.
(b) In the numbering plan established by TCRA in Tanzania, the short code for financial services is $*150 * XY\#$, where $X$ and $Y$ are integers $0, 1, 2, \cdots, 9$
(i) What is the maximum number of service providers that can be allocated this code?
(ii) What would you suggest be done when the maximum number in (i) above is exceeded?

**6.10**. (a) Explain what is meant by an "Erlang" in telecommunications traffic.

(b) Suppose a call center has 10 phone lines, receives 480 calls per day, and the average duration of a call is 15 minutes.
(i) Calculate the number of Erlangs.
(ii) Hence calculate the GoS.

(c) In a certain call centre, 300 calls are received per hour, and the average call duration is 5 minutes. Suppose it is desired to achieve a GoS of 0.025.
(i) Calculate the number of Erlangs.
(ii)Calculate the number of phone lines if the desired GoS is to be achieved.

# Chapter 7

# Wave-Particle Duality

Light is commonly observed in everyday life. However, the question is what is light? Is light made up of waves? Is light made up of particles? Is light a wave or particle or both? The answer to these questions is through experimentation. In sections 7.1 below, the wave nature of light is discussed, while in section 7.2 the particle nature is discussed. This means that light can behave as either a wave, or as a particle, and hence light shows wave-particle duality as we shall see in section 7.3.

## 7.1 Wave nature of light and the Electromagnetic spectrum

The wave nature of light is best illustrated by waves comprising the electromagnetic spectrum. The electromagnetic spectrum is a family of a wide range of waves which differ in terms of their wavelength or frequency. These ranges, in order of increasing frequency (or decreasing wavelength) include Radio waves, Microwaves, Infrared, Visible, Ultraviolet, X-rays and Gamma-rays. The classification of the various ranges does not have sharp boundaries. The electromagnetic spectrum is illustrated in Figure 7.1.

The relation between the speed of light in vacuum, $c$, wavelength, $\lambda$ and frequency, $\nu$ is

$$c = \lambda \nu \tag{7.1}$$

The various types of waves comprising the electromagnetic spectrum are described below:

- *Radio waves:* These waves have wavelengths ranging from a few kilometers to 0.3 m, and the corresponding frequency range is from a few Hz to $10^9$ Hz. These waves are generated by electronic devices and are utilised in Radio and Television broadcasting systems.

- *Microwaves:* These waves have wavelengths ranging from 0.3 m down to $10^{-3}$ m and the corresponding frequency range is from $10^9$ Hz to $3 \times 10^{11}$ Hz. These waves are used in radar and other communication devices.

- *Infrared:* These waves have wavelengths ranging from $10^{-3}$ m down to $7.8 \times 10^{-7}$ m (or $7800 \mathring{A}$), and the corresponding frequency range is from $3 \times 10^{11}$ Hz to $4 \times 10^{14}$ Hz. These waves are generated by molecules and hot bodies and have many applications in astronomy, industry and medicine.

- *Visible:* These waves are the ones to which the retina in our eyes is sensitive, and the wavelength range is from $7.8 \times 10^{-7}$ m (or 7800Å) down to $3.8 \times 10^{-7}$ m (or 3800Å), and the corresponding frequency range is from $4 \times 10^{14}$ Hz to $8 \times 10^{14}$ Hz. Visible light is produced by atoms and molecules as a result of electronic transitions. The different sensations that light produces on the eye is what is referred to as *colours* in everyday language. In Table 7.1, the wavelength and frequency ranges are shown for various colours.

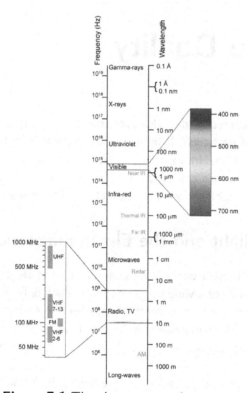

**Figure 7.1** The electromagnetic spectrum.

**Table 7.1:** Ranges of wavelengths and frequencies in the visible portion of the electromagnetic spectrum.

| Colour | Wavelength, $\lambda$ (m) | Frequency, $\nu$ (Hz) |
|--------|---------------------------|------------------------|
| Violet | 3.99 to $4.44 \times 10^{-7}$ | 7.69 to $6.59 \times 10^{14}$ |
| Blue   | 4.55 to $4.92 \times 10^{-7}$ | 6.59 to $6.10 \times 10^{14}$ |
| Green  | 4.92 to $5.77 \times 10^{-7}$ | 6.10 to $5.20 \times 10^{14}$ |
| Yellow | 5.77 to $5.97 \times 10^{-7}$ | 5.20 to $5.03 \times 10^{14}$ |
| Orange | 5.97 to $6.22 \times 10^{-7}$ | 5.03 to $4.82 \times 10^{14}$ |
| Red    | 6.22 to $7.80 \times 10^{-7}$ | 4.82 to $3.84 \times 10^{14}$ |

- *Ultraviolet:* These waves have wavelengths ranging from $3.8 \times 10^{-7}$ m down to $6 \times 10^{-10}$ m, and the corresponding frequency range is from $8 \times 10^{14}$ Hz to $3 \times 10^{17}$ Hz. Overexposure to these rays from the sun is known to cause skin cancer.

- *X-rays:* These waves have wavelengths ranging from $10^{-9}$ m down to $6 \times 10^{-12}$ m, and the corresponding frequency range is from $3 \times 10^{17}$ Hz to $5 \times 10^{19}$ Hz. These rays are used in hospitals for medical diagnostics.

- *Gamma-rays:* These waves have wavelengths ranging from $10^{-10}$ m down to $10^{-14}$ m, and the corresponding frequency range is from $3 \times 10^{18}$ Hz to more than $3 \times 10^{22}$ Hz. Gamma-rays are nuclear in origin and are produced by many radioactive substances. These rays are used for the treatment of certain types of cancers.

There are several experiments which demonstrate the wave nature of Electromagnetic waves such as:

- Reflection

- Refraction

- Dispersion

- Interference

- Diffraction and

- Polarisation

In Table 7.2, telecommunications operators and their allocated frequencies, are given. The various acronyms refer to different types of mobile technologies, such as **GSM**: Global System for Mobile communications, **LTE**: Long Term Evolution, **HSPA**: High Speed Packet Access, **WCS**: Wireless Communication Service.

**Table 7.2:** Telecommunications operators and their allocated frequencies.

| Operator | Allocated Frequency |
|---|---|
| LTE Operator | 800 MHz |
| 2G GSM Operator/HSPA | 900 MHz |
| GSM Operator/HSPA | 1800 MHz |
| HSPA Operator | 2100 MHz |
| WCS Operator | 2.3 GHz |
| 4G Operator/LTE | 2.6 GHz |

(b) In Table 7.3, broadcasting operators and their allocated frequencies, are given. The various acronyms refer to different types of broadcasting technologies, such as **FM**: Frequency Modulation, **MW**: Medium Wave, **DTT**: Digital Terrestrial Television, **C band downlink** for Satellite Television.

144 CHAPTER 7. WAVE-PARTICLE DUALITY

**Table 7.3:** Broadcasting operators and their allocated frequencies.

| Operator | Allocated Frequency |
| --- | --- |
| FM Radio | 88.6 MHz |
| FM Radio | 90.7 MHz |
| MW Radio | 621 KHz |
| MW Radio | 972 KHz |
| DTT TV Station | 470 MHz |
| C band downlink | 4.2 GHz |

The frequency allocations in Tanzania are a mandate for the Tanzania Communications Regulatory Authority (TCRA), in accordance with global standards mandated by the International Telecommunications Union (ITU), as was discussed in chapter 6.

## Wave nature of light and Electromagnetic theory

According to electromagnetic theory, light is composed of the electric field $\mathbf{E}$ and the magnetic field $\mathbf{B}$. The electric field $\mathbf{E}$ and magnetic field $\mathbf{B}$ are perpendicular (*orthogonal*) to each other and are perpendicular (*transverse wave*) to the direction of propagation, as illustrated in Figure 7.2.

In free space, the electric field $\mathbf{E}$ and magnetic field $\mathbf{B}$ satisfy the wave equation given by

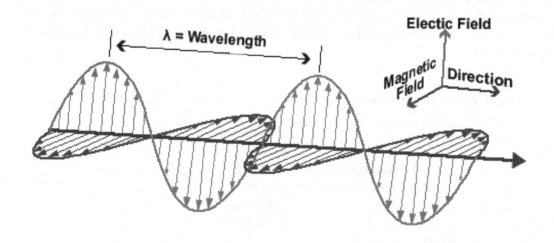

**Figure 7.2: Electromagnetic waves**

$$\nabla^2 \mathbf{E} = \frac{1}{c^2} \frac{\partial^2 \mathbf{E}}{\partial t^2} \qquad (7.2)$$

and

$$\nabla^2 \mathbf{B} = \frac{1}{c^2} \frac{\partial^2 \mathbf{B}}{\partial t^2} \qquad (7.3)$$

where $c$ is the velocity of light, given by

$$c = \frac{1}{\sqrt{\mu_0 \epsilon_0}} = 3.0 \times 10^8 ms^{-1}$$

The one dimensional version of equation 7.2, say along $x$-axis is

$$\frac{\partial^2 \mathbf{E}}{\partial x^2} = \frac{1}{c^2} \frac{\partial^2 \mathbf{E}}{\partial t^2}$$

which has a solution of a wave propagating along the x - axis given by any of the following forms:

$$E = E_0 e^{i(kx - \omega t)} \tag{7.4}$$

or

$$E = E_0 \sin(kx - \omega t)$$

or

$$E = E_0 \cos(kx - \omega t)$$

which is a plane wave of amplitude $E_0$ with wave vector $k(= 2\pi/\lambda)$ and angular frequency $\omega(= 2\pi\nu)$. Similar equations for the magnetic field vector $\mathbf{B}$ can be obtained, which is left as an exercise for the student.

## 7.2 Particle nature of light and Quantum theory

The particle nature of light can be demonstrated by several experimental observations, for example: the radiation spectrum, the photoelectric effect, X-ray production and the Compton effect. Each of these observation is described briefly below.

### 7.2.1 The radiation spectrum

Experimental observation of the radiation spectrum show that:

(i) Hot solids, for example, heated metals, heated coal, electric bulb filament etc emit radiation. As the temperature rises, the dominant frequencies increase (or wavelength decreases). A typical example is a heated object becomes red hot and then bluish white on increase in temperature.

(ii) The total power emitted by a hot body is proportional to $T^4$, where $T$ is absolute temperature.

(iii) The shape of the radiation spectrum is such that it increases at low frequencies and decreases at high frequencies, with a peak in between, as illustrated in Figure 7.3.

The above observations of the radiation spectrum can not be accounted for by classical physics. Max Planck in 1900 was able to explain the radiation spectrum by developing the following model:

(i) The atoms of the walls of a hot body or a cavity emit radiation in bundles of light known as quanta, of energy, $E$, given by

$$E = nh\nu \text{ where } n = 1, 2, 3, \cdots \tag{7.5}$$

where $h = 6.626 \times 10^{-34}$ J.s. is Planck's constant and $\nu$ is frequency of the electromagnetic radiation. This is what is now referred to as the *Planck's quantum hypothesis*.

(ii) An ingeneous interpolation between the high frequency radiation spectrum and the low frequency radiation spectrum to obtain a radiation spectrum over the whole frequency range in terms of an energy density $U(\nu, T)$ per unit frequency interval or the radiant power per unit frequency interval was done by Planck, and the functions are given in several forms, such as

$$\frac{dU(\nu, T)}{d\nu} = \frac{8\pi h\nu^3}{c^3} \frac{1}{e^{h\nu/k_B T} - 1} \tag{7.6}$$

as the *energy density* of a photon gas, or as

$$\frac{dW(\nu, T)}{d\nu} = \frac{2\pi h\nu^3}{c^2} \frac{1}{e^{h\nu/k_B T} - 1} \tag{7.7}$$

as the *radiant power* of a black body, and it can be noted that equations (7.6) and (7.7) differ by a factor of $c/4$.

Equations (7.6) and (7.7) are different forms of *Planck's radiation law*, where $k_B = 1.381 \times 10^{-23}$ J.K$^{-1}$ is the Boltzmann constant and all other symbols are as defined earlier. The frequency dependence of the energy density and the radiant power are illustrated in Figure 7.3 for three different temperatures.

We have somewhat given a historical development of the ideas developed to understand the black body radiation. Following the development of quantum mechanics in the early part of the 20th Century, equations (7.6) or (7.7) can now be proved explicitly by invoking ideas of quantum theory whereby light is made of particles known as *photons* and that these particles behave as *bosons* satisfying Bose-Einstein statistics.

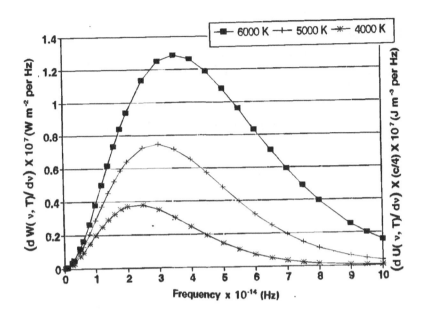

**Figure 7.3:** The frequency dependence of the energy density and radiant power at three different temperatures.

All the observations of the radiation spectrum which could not be explained by classical physics can be explained by the Planck's radiation law. The peak in the radiation spectrum is found to satisfy

$$\lambda_m T = 2.9 \times 10^{-3} \text{mK} \tag{7.8}$$

which is known as *Wien's displacement law*, and explains observation (i) above. This means that if there are peaks at wavelengths $\lambda_1, \lambda_2$ and $\lambda_3$ at temperatures $T_1, T_2$ and $T_3$ respectively, then

$$\lambda_1 T_1 = \lambda_2 T_2 = \lambda_3 T_3 = \text{Constant} = 2.9 \times 10^{-3} \text{mK}$$

and that if $T_1 < T_2 < T_3$, then $\lambda_1 > \lambda_2 > \lambda_3$ (or $\nu_1 < \nu_2 < \nu_3$ in terms of frequencies).

The total energy density of a photon gas is the area under the radiation spectrum which is obtained by performing an integration, and the result is given by

$$U(T) = \sigma_g T^4 \tag{7.9}$$

and the total radiant power is obtained as

$$P = A\epsilon\sigma T^4 \tag{7.10}$$

where it can be noted that

$$\sigma = \frac{c}{4}\sigma_g$$

Equations (7.9) and (7.10) are alternative forms of *Stefan-Boltzmann law* , where $A$ is the surface area of the emitting body, $\epsilon$ is the emissivity $(0 < \epsilon \leq 1)$, $\sigma = 5.67 \times 10^{-8}$ Wm$^{-2}$K$^{-4}$ is the Stephan-Boltzmann constant, and $T$ is the absolute temperature. This explains observation (ii) above. The shape of the radiation spectrum is explained by equation (7.6) or (7.7) which invokes the particle nature of light.

It is worth noting the following limits of physical interest. In the low frequency range, the radiation spectrum reduces to:

$$
\begin{aligned}
U(\nu, T) &= \frac{8\pi\nu^2}{c^3} \frac{h\nu}{\left[1 + \frac{h\nu}{k_B T} + \ldots - 1\right]} \\
&\approx \frac{8\pi\nu^2}{c^3} \frac{h\nu}{\left[1 + \frac{h\nu}{k_B T} - 1\right]} \\
&\approx \frac{8\pi\nu^2}{c^3} k_B T
\end{aligned}
\tag{7.11}
$$

and equation (7.11) is known as *Rayleigh-Jeans law* , first proposed in 1900. Rayleigh-Jeans law agrees with experimental observations fairly well at low frequencies but does not agree with experiment at high frequencies.

In the high frequency range, the radiation spectrum reduces to:

$$
\begin{aligned}
U(\nu, T) &\approx \frac{8\pi h\nu^3}{c^3} e^{-h\nu/k_B T} \\
&\approx \nu^3 \alpha e^{-\beta\nu/T}
\end{aligned}
\tag{7.12}
$$

where $\alpha$ and $\beta$ are constants, and equation (7.12) is known as *Wien's law*, first proposed in 1894. Wien's law agrees with experimental observations fairly well at high frequencies but does not agree with experiment at low frequencies.

It is interesting that Rayleigh-Jeans law and Wien's law were known long before quantum theory, but as discussed above these two laws were valid as approximations in the high and low frequency limits respectively. With the advent of the quantum theory, Planck's radiation law, given as equation (7.6) or (7.7), incorporated both these laws.

## 7.2.2   The photoelectric effect

In the *photoelectric effect*, when light is incident onto a metal, electrons are emitted under certain conditions. A typical experimental set up to observe the photoelectric effect is illustrated in Figure 7.4.

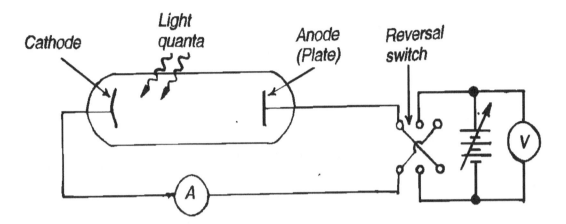

**Figure 7.4:** Experimental set up for observing the photoelectric effect. When light of frequency $\nu$ strikes the cathode plate, electrons are emitted and attracted to the anode plate, thus constituting an electric current.

Experimental observations show that:

(i) If the anode plate is made *positive* with respect to the cathode plate, electrons are accelerated towards the anode plate, and a photoelectric current is generated. This photoelectric current saturates to a constant value, $I$, depending on the intensity of the light. This implies that the intensity of incident light only determines the number of emitted photoelectrons, not their energy. Typical results are illustrated in Figure 7.5.

(ii) If the anode plate is made *negative* with respect to the cathode plate, electrons are repelled from the anode plate, and hence the current decreases rapidly until it becomes zero at some stopping voltage, $V_s$, independent of the intensity. Thus the maximum energy of a photoelectron is given by

$$eV_s = \frac{1}{2}mv_{max}^2 \qquad (7.13)$$

(iii)For each metal, there exists a threshold frequency, $\nu_0$, below which no electrons can be emitted no matter what the intensity of the incident radiation is. Typical results are illustrated in Figure 7.5. This implies that the energy of the photoelectrons emitted depends on the frequency of the incident radiation and the material constituting the cathode plate.

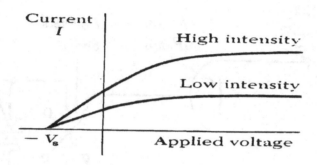

**Figure 7.5:** A graph of the photoelectric current, $I$ against the applied voltage, $V$, for two intensities of light. The current increases with intensity but reaches a saturation value for positive values of $V$. The current vanishes for voltages equal to or less than $-V_s$.

The above observations of the photoelectric effect can not be accounted for by classical physics. According to classical physics, an increase in the intensity of the incident light should cause more electrons to be emitted, and a decrease in intensity should cause less electrons to be emitted, in agreement with observation (i) above. However, the relation between frequency of the incident radiation and the energy of the emitted electrons can not be explained by classical physics. Also, the existence of the threshold frequency as in observation (iii) above can not be accounted for by classical physics.

The failure of classical physics led to Einstein's theory of the photoelectric effect developed in 1905, and this is summarised below.

(i) Electromagnetic radiation consists of *quanta* which are discrete particles of light known as *photons*, with energy $E$, given by

$$E = nh\nu \text{ where } n = 1, 2, 3, \cdots$$

as discussed earlier, in equation (7.5).

(ii) When electromagnetic radiation interacts with the metal, it behaves as consisting of photons which are *particle like*, with each photon delivering energy to an individual electron, not to the atom or the metal as a whole.

(iii) The threshold frequency occurs because a certain amount of minimum energy must be supplied to the electron in order to free it from the metal.

(iv) Conservation of energy is satisfied. An incident photon of energy $h\nu$ is expended to remove an electron from a metal by doing some work (work function $\phi$) and give the electron some kinetic energy $\frac{1}{2}mv^2$, and hence

$$h\nu \;=\; \phi + \frac{1}{2}mv^2 \qquad\qquad (7.14)$$

Using equation(7.13) gives

$$h\nu \;=\; \phi + eV_s$$

$$\text{or}$$

$$V_s \;=\; \frac{h}{e}\nu - \frac{\phi}{e} \qquad\qquad (7.15)$$

which can be plotted as $V_s$ against $\nu$, and will give a straight line of slope $= \frac{h}{e}$ and intercept $-\frac{\phi}{e}$. This is illustrated in Figure 7.6.

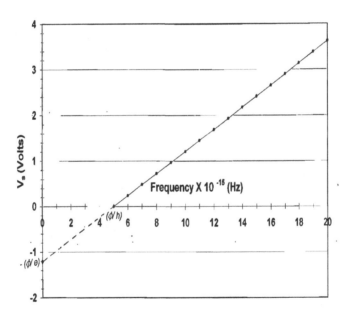

**Figure 7.6:** A graph of stopping voltage, $V_s$ against frequency, $\nu$. The straight line has a slope $= \frac{h}{e}$ and an intercept $-\frac{\phi}{e}$. Photons with frequency less that $\nu_0 = \frac{\phi}{h}$ do not have sufficient energy to eject electrons from plate.

### 7.2.3 X-ray production

In *X-ray production*, it is observed that when electrons bombard a metal target under certain conditions, X-rays (or *bremsstrahlung*, a word from German: *bremse* - brake and *strahlung* - radiation) are produced. A typical experimental set up for X-ray production is illustrated in Figure 7.7, where a metal coil C is heated and emits electrons of charge $e$ by thermionic emission. These electrons strike a metal target A, and X-rays are emitted. The target A is held at a high positive potential in the range of kV to impart high energy to the incident electrons.

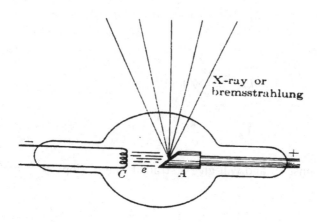

**Figure 7.7:** Experimental set up for X-ray (or bremsstrahlung) production.

Experimental observations show that:

(i) For each given electron energy, there is a minimum wavelength of the X-rays emitted.

(ii) The minimum wavelength of the X-rays emitted is independent of the type of metal used as a target material.

The above observations for X-ray production can not be explained by classical physics. According to classical physics there is no reason why there should be a minimum wavelength for the emitted X-rays.

The failure of classical physics to explain the minimum wavelength led to the *Duane-Hunt law*, developed in 1915, and this is summarised as follows. According to quantum theory, the minimum wavelength $\lambda_{min}$ corresponds to the conversion of the total kinetic energy of a single electron into a single photon of the emitted X-rays, related by

$$eV = h\nu_{max} = \frac{hc}{\lambda_{min}} \tag{7.16}$$

which can be expressed as

$$\lambda_{min} = \frac{hc}{eV} \tag{7.17}$$

which is known as Duane-Hunt law, where $h$ is Planck's constant, $c$ is the speed of light, $e$ is electronic charge and $V$ is the potential difference across the X-ray tube.

### 7.2.4 The Compton effect

The *Compton effect* is observed when X-rays or $\gamma$-rays are scattered by electrons in matter. A schematic diagram for the Compton effect is illustrated in Figure 7.8.

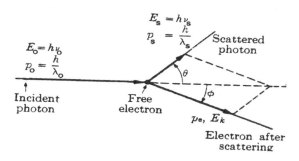

**Figure 7.8:** A schematic diagram illustrating the Compton effect.

Experimental observations show that:

(i) The scattered beam is observed to have two wavelengths, which are the incident wavelength $\lambda_0$ and the scattered wavelength $\lambda_s$, shifted from $\lambda_0$.

(ii) The shift, $\Delta\lambda = \lambda_s - \lambda_0$, is referred to as the *Compton shift*, and is independent of the target material, and only depends on the scattering angle $\theta$.

The above observations for the Compton effect can be accounted for when it is invoked that the scattering can be explained by assuming that x-ray particles (photons) collide with electrons as particles. The Compton shift was first explained by Compton in 1922, and using conservation of momentum and conservation of energy, it is given by

$$\Delta\lambda = \lambda_s - \lambda_0 = \frac{h}{m_0 c}(1 - \cos\theta) \tag{7.18}$$

where $h/(m_0 c)$ is known as the Compton wavelength.

## 7.3 Wave-Particle duality of light

In sections 1, the discussion showed that light is made up of waves. In section 2, the discussion concluded that light is made up of particles known as photons. These observations raise the question: what is light? Is light made up waves? Is light made up of particles? To reconcile these observations, light is actually *both*: at times it acts as if it is made up of waves while for some other physical phenomena it acts as if it is made up of particles, and hence the concept of *wave-particle duality*. What this means is that waves show particle-like properties and particles show wave-like properties. The concepts of De Broglie's hypothesis and Planck's energy quantisation are important in understanding the wave-particle duality.

The *De Broglies's hypothesis*, first postulated in 1924, states that momentum $p$ (a particle property) and wavelength $\lambda$ (a wave property) are related by

$$p = \frac{h}{\lambda} \tag{7.19}$$

Alternatively, this equation can be expressed as

$$p = \hbar k \tag{7.20}$$

where $\hbar = h/2\pi = 1.055 \times 10^{-34}$ J.s. and $k = 2\pi/\lambda$ is the wavenumber.

*Planck's energy quantisation* states that energy $E$ and frequency $\nu$ are related by

$$E = h\nu \tag{7.21}$$

Alternatively, this equation can be expressed as

$$E = \hbar\omega \tag{7.22}$$

where $\omega = 2\pi\nu$ is the angular frequency. There is yet another alternative form, after using equations (7.1) and (7.19) in equation (7.21): it is given by

$$E = pc \tag{7.23}$$

## 7.4   Complementarity, Correspondence and Uncertainty Principles

### 7.4.1   Complementarity Principle

The Complementarity principle says that the wave description and particle description are complementary. However, one never uses the wave and particle aspects at the same time to describe the same process.

### 7.4.2   Correspondence Principle

The Correspondence principle says that the results of quantum mechanics reproduces the results of classical mechanics in the limit of large quantum numbers.

### 7.4.3 Uncertainty Principle

The Uncertainty principle sets the fundamental limit to the precision with which certain pairs of physical properties can be made, for example position uncertainty $\Delta x$ and momentum uncertainty $\delta p_x$, or energy uncertainty $\Delta E$ and time uncertainty $\Delta t$, their products satify

$$\Delta E \Delta t \geq h$$

and

$$\Delta x \Delta p_x \geq h$$

## 7.5    Exercises

**7.1**. (a) State *Wien's displacement law.*

(b) When the Earth absorbs solar energy, it becomes heated, and subsequently re-radiates the energy into the atmosphere. At what wavelength is the radiation most intense if the Earth's surface temperature is 27 °C ?

(c) Calculate the temperature of a thermal source whose radiation has a maximum intensity when the wavelength is 1.8 $\mu$m.

(d) At what wavelength is the light emitted by the Sun most intense if the surface temperature of the Sun is approximately 5000 K?

**7.2**. (a) State *Stefan-Boltzmann law.*

(b) A metal ball 3 cm in radius is heated in a furnace up to 5000 C. If its emissivity is 0.5, at what rate does it radiate energy?

(c) The tungsten filament of an incandescent light bulb is a wire of diameter 0.08 mm and length 5.0 cm, and it is at a temperature of 3200 K. What is the power radiated by the filament assuming that it acts like a blackbody?

**7.3**. (a) Sun radiates like a blackbody at 5000 K. Calculate the total power radiated by the sun, and the power received per unit area on the earth's surface. What is the total power received on total surface of the earth?

(b) Calculate the rate at which sun is losing mass.

**7.4**. (a) State, explaining all symbols used, (i) *De Broglie's hypothesis* (ii) *Planck's quantum hypothesis*, and hence show that Energy $E$, momentum $p$ and speed of light $c$ are related by $E = pc$.

(b) Complete the following table

|          | Energy(eV) | $\lambda$ (m)        |
|----------|:----------:|:--------------------:|
|          | 1          | ?                    |
| proton   | ?          | $1 \times 10^{-10}$  |
|          | 1000       | ?                    |
|          | 1          | ?                    |
| electron | ?          | $1 \times 10^{-10}$  |
|          | 1000       | ?                    |
|          | 1          | ?                    |
| photon   | ?          | $1 \times 10^{-10}$  |
|          | 1000       | ?                    |

**7.5**. An experiment to observe the *photoelectric effect* uses a mercury arc lamp as the source of UV radiation shining on lithium. Several filters are used to isolate discrete wavelengths, $\lambda$, and the following stopping voltages, $V_s$, are recorded.

| Wavelength, $\lambda(\mathring{A})$ | $V_s$ (volts) |
|---|---|
| 2536 | 2.4 |
| 3132 | 1,5 |
| 3663 | 0.9 |
| 4358 | 0.35 |
| 5770 | (no photoelectrons) |

(a) Plot a graph of $V_s$ vs frequency, $\nu$

(b) Hence determine the work function, $\phi$, of lithium and the value of $h/e$.

**7.6**. (a) State the De Broglie's hypothesis.

(b) Show that if the energy of electrons is given in eV, then the wavelength in $\mathring{A}$ is given by

$$\lambda = \frac{12}{(eV)^{\frac{1}{2}}}\mathring{A}$$

**7.7**. (a) Can neutrons be diffracted? Discuss.

(b) Show that if the energy of neutrons is given in eV, then the wavelength in $\mathring{A}$ is given by

$$\lambda = \frac{0.28}{[E(eV)]^{\frac{1}{2}}}\mathring{A}$$

**7.8**. (a) Describe how x-rays are produced.

(b) Show that if the energy of X-rays is given in keV, then the wavelength in $\mathring{A}$ is given by:

$$\lambda(\mathring{A}) = \frac{12.4}{E(keV)}$$

**7.9**. (a) State the uncertainty principle in terms of position and momentum.

(b) The position of a free electron is determined by some optical means to within an uncertainty of $10^{-6}$ m.

(i) Calculate the uncertainty in terms of velocity.

(ii) After 10 s, how well can one know the position?

(iii) Are zero point oscillations consistent with the uncertainty principle? Discuss.

**7.10**. (a) State the uncertainty principle in terms of energy and time.

(b) An atom radiates a photon in approximately $10^{-9}$ s. What is the uncertainty in the energy of the photon?

# Chapter 8

# The Wave function and Solutions of the Schrödinger Equation in different systems

## 8.1 Introduction: The Wave function

In the previous chapter, we learnt that quantum particles show the characteristic of *wave-particle duality*. A quantum particle of momentum $p$ has wavelength $\lambda$ related by the De Broglie's hypothesis,

$$p = \frac{h}{\lambda} = \hbar k \tag{8.1}$$

where $\hbar = h/2\pi$ and $k$ is a wavenumber, $k = 2\pi/\lambda$.

Also, energy $E$ is related to frequency $\nu$ by the Planck's quantum hypothesis,

$$E = h\nu = \hbar\omega \tag{8.2}$$

where $\omega$ is angular frequency, $\omega = 2\pi\nu$.

A quantum particle is represented by a wave function, $\Psi(x, y, z, t)$, which has the following properties.

1. $\Psi$ contains all the measurable information about the particle.
2. $\Psi$ has a complex conjugate $\Psi^*$, and $\Psi^*\Psi$ gives the probability of finding a particle, such that

$$\int \Psi^*\Psi d\tau = 1 \tag{8.3}$$

that is, the probability of finding the particle somewhere must be one.
3. $\Psi$ must be single valued.
4. $\Psi$ must be well defined, that is, it shoud not have singularities.
5. The average value (also known as an expectation value) of an observable $A$ is given by

$$< A > = \frac{\int \Psi^* \hat{A} \Psi d\tau}{\int \Psi^* \Psi d\tau} \tag{8.4}$$

6. $\Psi$ is continuous across a boundary.

7. The first derivative of $\frac{\partial \Psi}{\partial x}$ is continuous across a boundary. The physical significance of this is conservation of momentum. Let us illustrate this by a wave function of the form

$$\Psi(x,t) = Ae^{i(kx-\omega t)} \tag{8.5}$$

Note that

$$
\begin{aligned}
\frac{\partial \Psi}{\partial x} &= ikAe^{i(kx-\omega t)} \\
&= ik\Psi \\
i\hbar \frac{\partial \Psi}{\partial x} &= (i\hbar)(ik)\Psi \\
&= -p\Psi \\
&\text{or} \\
p\Psi &= -i\hbar \frac{\partial \Psi}{\partial x}
\end{aligned}
$$

which gives an expression for momentum operator, $\hat{p}$ as

$$p \to -i\hbar \frac{\partial}{\partial x} \tag{8.6}$$

In 3-dimensions, we have

$$p \to -i\hbar \nabla$$

8. The expression for the energy operator, $E$ can be obtained as follows. Consider a wave function of the form

$$\Psi(x,t) = Ae^{i(kx-\omega t)}$$

Note that

$$
\begin{aligned}
\frac{\partial \Psi}{\partial t} &= -i\omega Ae^{i(kx-\omega t)} \\
&= -i\omega \Psi \\
i\hbar \frac{\partial \Psi}{\partial t} &= (i\hbar)(-i\omega)\Psi \\
&= \hbar\omega \Psi \\
&= E\Psi \\
&\text{or} \\
E\Psi &= i\hbar \frac{\partial \Psi}{\partial t}
\end{aligned}
$$

which gives an expression for energy operator, $\hat{E}$ as

$$E \to i\hbar \frac{\partial}{\partial t} \tag{8.7}$$

9. The wave function satifies an equation of motion. In classical mechanics, there is the principle of conservation of enegy, whereby the total energy comprising kinetic energy, $KE$, plus the potential energy, $PE$, is a constant.

$$
\begin{aligned}
E_{TOTAL} &= KE + PE \\
&= \frac{p^2}{2m} + V \\
&= Constant
\end{aligned}
\tag{8.8}
$$

In quantum mechanics, the total energy is replaced by the Hamiltonian, $H$, as the total energy operator, and the wave function, $\Psi$ satisfies the equation of motion

$$H\Psi = E\Psi \tag{8.9}$$

$$\left[ -\frac{\hbar^2}{2m}\frac{\partial^2}{\partial x^2} + V \right]\Psi = E\Psi \tag{8.10}$$

known as the *Schrödinger Time Independent Equation*, and

$$H\Psi = i\hbar\frac{\partial\Psi}{\partial t} \tag{8.11}$$

$$\left[ -\frac{\hbar^2}{2m}\frac{\partial^2}{\partial x^2} + V \right]\Psi = i\hbar\frac{\partial\Psi}{\partial t} \tag{8.12}$$

known as the *Schrödinger Time Dependent Equation*.
In three dimensions, the Schrödinger Time Independent Equation and Schrödinger Time Dependent Equation are of the following form, respectively

$$\left[ -\frac{\hbar^2}{2m}\nabla^2 + V \right]\Psi = E\Psi$$

$$\left[ -\frac{\hbar^2}{2m}\nabla^2 + V \right]\Psi = i\hbar\frac{\partial\Psi}{\partial t}$$

10. The wave function $\Psi$ and its complex conjugate, $\Psi^*$, satify the continuity equation

$$\nabla\cdot\mathbf{j} + \frac{\partial\rho}{\partial t} = 0$$

$$\text{where the probability current } \mathbf{j} = \frac{\hbar}{2mi}\left[ \Psi^*(\nabla\Psi) - \Psi(\nabla\Psi^*) \right]$$

$$\text{and the probability density function } \rho = \Psi^*\Psi = |\Psi|^2$$

See exercise 8.2 where the proof of the above is set as an exercise.

## 8.2 Solutions of the Schrödinger Equation in Different Systems

### 8.2.1 Potential Step

Consider quantum particles of mass $m$, energy $E$ moving such that they encounter a potential step, as illustrated in Figure 8.1, with the potential $V(x)$ varying as

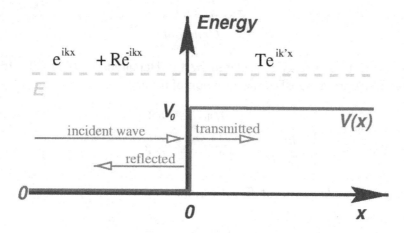

Figure 8.1: Potential Step

$$V(x) = \begin{cases} 0, & x < 0 \\ V_0, & x \geq 0 \end{cases} \tag{8.13}$$

The incident quantum particles will experience reflection and transmission at the boundary. The equation of motion of the quantum particles is the Schrödinger equation. There are two cases of physical interest.

*Case (a):* $E > V_0$

In region I $(x < 0)$, the equation of motion is given as

$$-\frac{\hbar^2}{2m}\frac{\partial^2 \Psi_1}{\partial x^2} = E\Psi_1$$

$$\frac{\partial^2 \Psi_1}{\partial x^2} + \frac{2mE}{\hbar^2}\Psi_1 = 0$$

which has a solution

$$\Psi_1(x) = Ae^{ik_1 x} + Be^{-ik_1 x} \tag{8.14}$$

where

$$k_1 = \frac{1}{\hbar}\sqrt{2mE}$$

In region II $(x > 0)$, the equation of motion is given as

$$-\frac{\hbar^2}{2m}\frac{\partial^2 \Psi_2}{\partial x^2} + V_0\Psi_2 = E\Psi_2$$

$$\frac{\partial^2 \Psi_2}{\partial x^2} + \frac{2m}{\hbar^2}(E - V_0)\Psi_2 = 0$$

which has a solution

$$\Psi_2(x) = Ce^{ik_2x} \tag{8.15}$$

where

$$k_2 = \frac{1}{\hbar}\sqrt{2m(E - V_0)}$$

Applying boundary conditions at $x = 0$, that the wave function is continuous and the first derivatives of the wave functions are continuous, we obtain

$$\Psi_1(x = 0) = \Psi_2(x = 0)$$

$$\left(\frac{\partial \Psi_1}{\partial x}\right)_{x=0} = \left(\frac{\partial \Psi_2}{\partial x}\right)_{x=0}$$

The above equations can be used to obtain the reflection coefficient $R$ and transmission coefficient $T$ in somewhat cumbersome but straight forward algebra. These are obtained as

$$R = |\frac{B}{A}|^2 = \left(\frac{k_1 - k_2}{k_1 + k_2}\right)^2$$

$$= \frac{[1 - (1 - V_0/E)^{1/2}]^2}{[1 + (1 - V_0/E)^{1/2}]^2} \tag{8.16}$$

$$T = \frac{k_2}{k_1}|\frac{C}{A}|^2 = \frac{4k_1k_2}{(k_1 + k_2)^2}$$

$$= \frac{4(1 - V_0/E)^{1/2}}{[1 + (1 - V_0/E)^{1/2}]^2} \tag{8.17}$$

It can be noted that

$$R + T = 1 \tag{8.18}$$

The profile of the wave function for a potential step for $E > V_0$ is illustrated below in Figure 8.2.
*Case (b):$E < V_0$*
The modification to the above wave functions is that

$$k_2 \rightarrow i\beta$$

and hence

$$R = |\frac{B}{A}|^2 = |\frac{k_1 - i\beta}{k_1 + i\beta}|^2$$

$$= |\frac{1 - i(V_0/E - 1)^{1/2}}{1 + i(V_0/E - 1)^{1/2}}|^2 = 1$$

$$T = 0$$

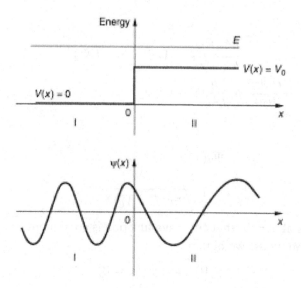

Figure 8.2: The wave function for a potential step for $E > V_0$

and again, $R + T = 1$.

The profile of the wave function for a potential step for $E < V_0$ is illustrated below in Figure 8.3.

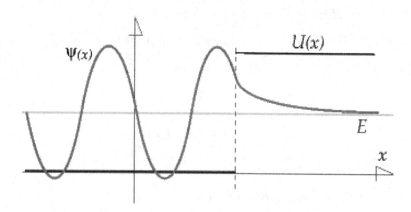

Figure 8.3: The wave function for a potential step for $E < V_0$.

## 8.2.2 Potential Barrier

Consider quantum particles of mass $m$, energy $E$ moving such that they encounter a potential barrier, as illustrated in Figure 8.4, with the potential $V(x)$ varying as

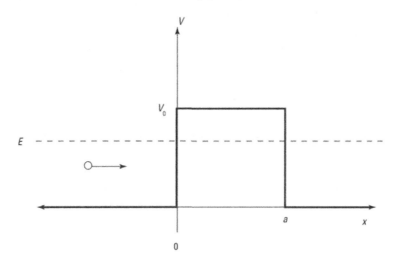

Figure 8.4: Potential Barrier

$$V(x) = \begin{cases} 0, & x < 0 \\ V_0, & 0 < x < a \\ 0, & x > a \end{cases} \qquad (8.19)$$

The incident quantum particles will experience reflection and transmission at the boundary $x = 0$ and $x = a$. There are two cases of physical interest.

*Case (a):*$E > V_0$

In region I $(x < 0)$, the equation of motion is given as

$$-\frac{\hbar^2}{2m}\frac{\partial^2 \Psi_1}{\partial x^2} = E\Psi_1$$

$$\frac{\partial^2 \Psi_1}{\partial x^2} + \frac{2mE}{\hbar^2}\Psi_1 = 0$$

which has a solution

$$\Psi_1(x) = A_0 e^{ik_1 x} + A_1 e^{-ik_1 x} \qquad (8.20)$$

where

$$k_1 = \frac{1}{\hbar}\sqrt{2mE}$$

In region II $(0 < x < a)$, the equation of motion is given as

$$-\frac{\hbar^2}{2m}\frac{\partial^2 \Psi_2}{\partial x^2} + V_0 \Psi_2 = E\Psi_2$$

$$\frac{\partial^2 \Psi_2}{\partial x^2} + \frac{2m}{\hbar^2}(E - V_0)\Psi_2 = 0$$

which has a solution

$$\Psi_2(x) = B_0 e^{ik_2 x} + B_1 e^{-ik_2 x} \tag{8.21}$$

where

$$k_2 = \frac{1}{\hbar}\sqrt{2m(E - V_0)}$$

In region III $(x > a)$, the equation of motion is given as

$$-\frac{\hbar^2}{2m}\frac{\partial^2 \Psi_3}{\partial x^2} = E\Psi_3$$

$$\frac{\partial^2 \Psi_3}{\partial x^2} + \frac{2m}{\hbar^2}E\Psi_3 = 0$$

which has a solution

$$\Psi_3(x) = C e^{ik_2 x} \tag{8.22}$$

where

$$k_3 = \frac{1}{\hbar}\sqrt{2mE}$$

Applying boundary conditions at $x = 0$ and $x = a$, that the wave functions are continuous and the first derivatives of the wave functions are continuous, we obtain

$$\Psi_1(x = 0) = \Psi_2(x = 0)$$
$$\left(\frac{\partial \Psi_1}{\partial x}\right)_{x=0} = \left(\frac{\partial \Psi_2}{\partial x}\right)_{x=0}$$
$$\Psi_2(x = a) = \Psi_3(x = a)$$
$$\left(\frac{\partial \Psi_2}{\partial x}\right)_{x=a} = \left(\frac{\partial \Psi_3}{\partial x}\right)_{x=a}$$

The above four equations provide relations between the five coefficients $A_0, A_1, B_0, B_1$ and $C$. After tedious but straight forward algebra, the above equations can be used to obtain the reflection coefficient $R$ and transmission coefficient $T$ as

$$R = |\frac{A_1}{A_0}|^2 = [1 + \frac{4(E/V_0)(E/V_0 - 1)}{\sin^2(k_2 a)}]^{-1} \tag{8.23}$$

$$T = |\frac{C}{A_0}|^2 = [1 + \frac{\sin^2(k_2 a)}{4(E/V_0)(E/V_0 - 1)}]^{-1} \tag{8.24}$$

with

$$k_2 a = \sqrt{\frac{2mV_0 a^2}{\hbar^2}(\frac{E}{V_0} - 1)}$$

and note that $R + T = 1$.

*Case (b)*:$E < V_0$
The wave functions $\Psi_I$ and $\Psi_{III}$ have the same form as in case (a), but

$$\Psi_{II}(x) = B_0 e^{-\beta x} + B_1 e^{\beta x} \tag{8.25}$$

with

$$\beta = \frac{1}{\hbar}\sqrt{2m(V_0 - E)}$$

and $R$ and $T$ become

$$R = [1 + \frac{4(E/V_0)(1 - V_0/E)}{\sinh^2 \beta a}]^{-1} \tag{8.26}$$

and

$$T = [1 + \frac{\sinh^2 \beta a}{4(E/V_0)(1 - V_0/E)}]^{-1} \tag{8.27}$$

with

$$\beta a = \sqrt{\frac{2mV_0 a^2}{\hbar^2}(1 - \frac{E}{V_0})}$$

It can be noted that $R + T = 1$ for all values of $E/V_0$, and there is an oscillating behaviour for $E/V_0 > 1$. Note the phenomenon of QUANTUM TUNNELLING for $E < V_0$ as illustrated in figure 8.5, whereby the particle tunnels through the barrier, a phenomenon which would not be possible in classical mechanics, where a particle starting with energy $mgH$ can not go beyond a barrier of height greater than $H$. The quantum mechanical prediction of tunnelling is observed in a number of

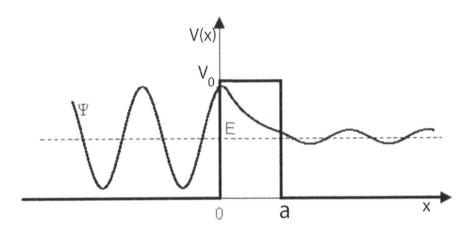

Figure 8.5: The wave function for a potential barrier for $E < V_0$.

situations, for example:

- electronic devices, such as the tunnel diode, quantum transistors, nanoelectronics

- the Scanning Electron Microscope (SEM)

- radioactivity in the nucleus

- quantum computing

### 8.2.3  Potential well

Consider quantum particles of mass $m$, energy $E$ moving such that they encounter a *potential well*, as illustrated in Figure 8.6, with the potential $V(x)$ varying as

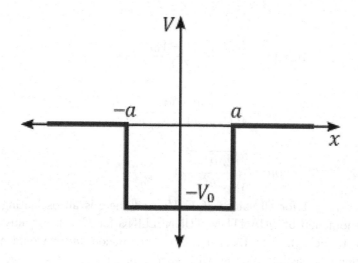

Figure 8.6: Potential Well

$$V(x) = \begin{cases} 0, & x < -a \\ -V_0, & -a < x < a \\ 0, & x > a \end{cases} \tag{8.28}$$

The incident quantum particles will experience reflection and transmission at the boundary $x = -a$ and $x = a$. There are two cases of physical interest.

*Case (a):* $E > V_0$

In region I $(x < -a)$, the equation of motion is given as

$$-\frac{\hbar^2}{2m}\frac{\partial^2 \Psi_1}{\partial x^2} = E\Psi_1$$

$$\frac{\partial^2 \Psi_1}{\partial x^2} + \frac{2mE}{\hbar^2}\Psi_1 = 0$$

which has a solution

$$\Psi_1(x) = A_0 e^{ik_1 x} + A_1 e^{-ik_1 x} \qquad (8.29)$$

where

$$k_1 = \frac{1}{\hbar}\sqrt{2mE}$$

In region II $(-a < x < a)$, the equation of motion is given as

$$-\frac{\hbar^2}{2m}\frac{\partial^2 \Psi_2}{\partial x^2} - V_0 \Psi_2 = E\Psi_2$$

$$\frac{\partial^2 \Psi_2}{\partial x^2} + \frac{2m}{\hbar^2}(E + V_0)\Psi_2 = 0$$

which has a solution

$$\Psi_2(x) = B_0 e^{ik_2 x} + B_1 e^{-ik_2 x} \qquad (8.30)$$

where

$$k_2 = \frac{1}{\hbar}\sqrt{2m(E + V_0)}$$

In region III $(x > a)$, the equation of motion is given as

$$-\frac{\hbar^2}{2m}\frac{\partial^2 \Psi_3}{\partial x^2} = E\Psi_3$$

$$\frac{\partial^2 \Psi_3}{\partial x^2} + \frac{2m}{\hbar^2}E\Psi_3 = 0$$

which has a solution

$$\Psi_3(x) = C e^{ik_1 x} \qquad (8.31)$$

where

$$k_1 = \frac{1}{\hbar}\sqrt{2mE}$$

Applying boundary conditions at $x = -a$ and $x = a$, that the wave functions are continuous and the first derivatives of the wave functions are continuous, we obtain

$$\Psi_1(x = -a) = \Psi_2(x = -a)$$

$$\left(\frac{\partial \Psi_1}{\partial x}\right)_{x=-a} = \left(\frac{\partial \Psi_2}{\partial x}\right)_{x=-a}$$

$$\Psi_2(x = a) = \Psi_3(x = a)$$

$$\left(\frac{\partial \Psi_2}{\partial x}\right)_{x=a} = \left(\frac{\partial \Psi_3}{\partial x}\right)_{x=a}$$

The above four equations provide relations between the five coefficients $A_0$, $A_1$, $B_0$, $B_1$ and $C$. After tedious but straight forward algebra, the equations can be used to obtain the reflection coefficient $R$ and transmission coefficient $T$ as

$$R = |\frac{A_1}{A_0}|^2 = [1 + \frac{4(E/V_0)(1 + E/V_0)}{\sin^2(2k_2a)}]^{-1} \tag{8.32}$$

and

$$T = |\frac{C}{A_0}|^2 = [1 + \frac{\sin^2(2k_2a)}{4(E/V_0)(1 + E/V_0)}]^{-1} \tag{8.33}$$

with

$$2k_2a = \sqrt{4\frac{2mV_0a^2}{\hbar^2}(1 + \frac{E}{V_0})} = k_2L$$

It can be noted that $R + T = 1$. Both $R$ and $T$ show oscillations, with certain values of $E/V_0$ giving $R = 0$ and $T = 1$. This means that for such energies the potential well is completely transparent to the incident particles, and the condition for this to happen is

$$k_2 2a = k_2L = n\pi$$

or, equivalently

$$L = n\frac{\lambda_2}{2}, \lambda_2 \text{ is the wavelength in region II}$$

which corresponds to the width $L = 2a$ of the potential well to be equal to an integral multiple of half wavelengths of the wave function inside the well. This behaviour is analogous to the selective transmission of light by thin films. The quantum mechanical transparency of a potential well is observed in the scattering of electrons by noble-gas atoms and is known as the "Ramsauer-Townsend effect" (Ramsauer 1921, Townsend 1922). The wave function for a potential well is illustrated in Figure 8.7.

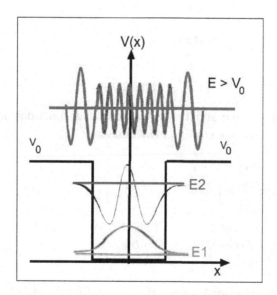

Figure 8.7: The wave functions for a potential well.

### 8.2.4 Infinite potential well

Consider quantum particles of mass $m$, energy $E$ moving such that they are confined in an infinite potential well, whose potential function $V(x)$ is such that

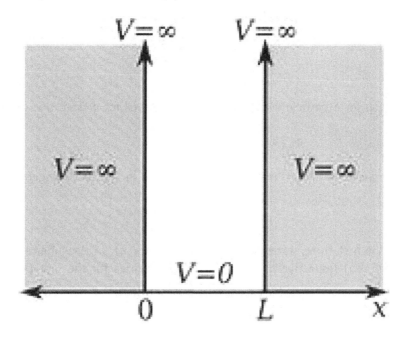

Figure 8.8: Infinite Potential Well

$$V(x) = \begin{cases} 0, & 0 < x < L \\ \infty, & x < 0 \text{ and } x > L \end{cases} \tag{8.34}$$

as illustrated in Figure 8.8. The wave function inside the infinite potential well is a solution of the Schrödinger equation, in the form

$$-\frac{\hbar^2}{2m}\frac{\partial^2 \Psi}{\partial x^2} = E\Psi$$

or, after rearranging

$$\frac{\partial^2 \Psi}{\partial x^2} + \frac{2mE\Psi}{\hbar^2} = 0$$

which is a second order differential equation, with a solution

$$\Psi(x) = A\cos(kx) + B\sin(kx) \tag{8.35}$$

Boundary conditions require that

$$\Psi(0) = 0 \Rightarrow A + 0 = 0 \Rightarrow A = 0$$

$$\Psi(L) = 0 \Rightarrow 0 + B\sin(kL) = 0 \Rightarrow kL = n\pi$$

Normalization requires that

$$\int_0^L \Psi^*(x)\Psi(x)dx = 1$$

which gives

$$B^2 \int_0^L \sin^2(\frac{n\pi x}{L}) = 1$$

and hence

$$\frac{1}{2}B^2L = 1 \Rightarrow B = \sqrt{\frac{2}{L}}$$

and thus the wave function for a particle in an infinite potential well for a state $n$ is given by

$$\Psi_n(x) = \sqrt{\frac{2}{L}}\sin(\frac{n\pi x}{L}) \tag{8.36}$$

The energy levels are obtained by using the above wave function in the Schrödinger equation, and are given by

$$E_n = \frac{n^2\hbar^2\pi^2}{2mL^2} \tag{8.37}$$

The wave functions for the first three states $\Psi_1(x)$, $\Psi_2(x)$ and $\Psi_3(x)$, of the infinite potential well are illustrated in Figure 8.9(a) respectively. The probability densities for the first three states are illustrated in Figure 8.9(b).

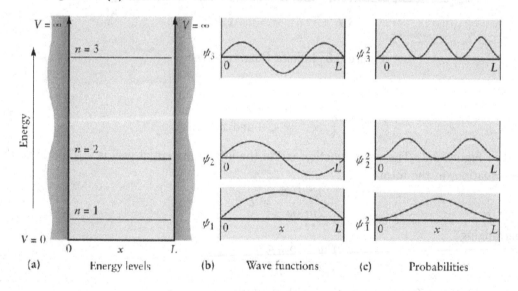

Figure 8.9: The wave functions and probability functions for an infinite potential well.

## 8.2.5  Particle in a box

Consider a quantum particle confined in a cubical box of side $L$, whose potential function is such that

$$V(x,y,z) = \begin{cases} 0, & \text{inside the box} \\ \infty, & \text{outside the box} \end{cases} \tag{8.38}$$

as illustrated in Figure 8.10. The Schrödinger equation becomes,

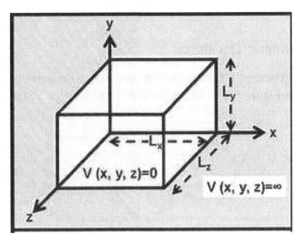

Figure 8.10: Particle in a box

$$-\frac{\hbar^2}{2m}\nabla^2\Psi(\mathbf{r}) = E\Psi(\mathbf{r})$$

$$-\frac{\hbar^2}{2m}\left(\frac{\partial^2}{\partial x^2} + \frac{\partial^2}{\partial y^2} + \frac{\partial^2}{\partial z^2}\right)\Psi(x,y,z) = E\Psi(x,y,z)$$

Consider a solution

$$\Psi(x) = Ae^{ik_x x} + Be^{-ik_x x} \tag{8.39}$$

Applying boundary conditions, $\Psi(x=0) = \Psi(x=L)$, one obtains $A + B = 0$, or $B = -A$, and hence

$$\Psi(x) = A(e^{ik_x x} - e^{-ik_x x}) = 2iA\sin k_x x = C\sin k_x x$$

which gives $k_x L = n_x \pi$ or $k_x = n_x \pi/L$, and hence $p_x = \hbar k_x = n_x \pi \hbar/L$. Similarly for $\Psi(y)$ and $\Psi(z)$. The wave function for a particle in the box is a solution of the Schrödinger equation , and is given by

$$\begin{aligned}\Psi(x,y,z) &= \sin k_x x \sin k_y y \sin k_z z \\ &= \sin(\frac{n_x \pi x}{L})\sin(\frac{n_y \pi y}{L})\sin(\frac{n_z \pi z}{L})\end{aligned} \tag{8.40}$$

Using this in the Schrödinger equation, one obtains

$$\frac{\hbar^2}{2m}\left(k_x^2 + k_y^2 + k_z^2\right)\Psi(x,y,z) = E\Psi(x,y,z)$$

which gives the energies as

$$\begin{aligned}E &= \frac{\hbar^2 k^2}{2m} \\ &= \frac{\hbar^2}{2m}\left(k_x^2 + k_y^2 + k_z^2\right) \\ E_n &= \frac{\pi^2 \hbar^2}{2mL^2}(n_x^2 + n_y^2 + n_z^2)\end{aligned} \tag{8.41}$$

where $n_x, n_y, n_z$ take integral values, and the total energy of the particle in a box for state $n$ is quantised.

## 8.2.6   The Quantum Harmonic Oscillator

Before we study the Quantum Harmonic Oscillator, let us summarise some results of the classical harmonic oscillator illustrated in Figure 8.11. The equation of motion for a classical simple harmonic

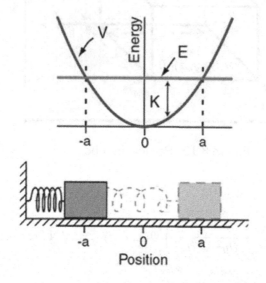

Figure 8.11: Classical Simple Harmonic Oscillator

oscillator is given by

$$
\begin{aligned}
ma &= \sum F \\
m\frac{d^2x}{dt^2} &= -kx \\
\frac{d^2x}{dt^2} + \frac{k}{m}x &= 0
\end{aligned}
$$

The Force $F$ is related to the potential $V$ by

$$F = -\frac{\partial V}{\partial x}$$

and hence

$$
\begin{aligned}
V &= -\int F dx \\
&= \int kx dx \\
V &= \frac{1}{2}kx^2
\end{aligned}
$$

which gives the total energy as the sum of the kinetic energy $KE = \frac{1}{2}mv^2 = \frac{p^2}{2m}$ and potential energy $PE = V$.

$$
\begin{aligned}
E_{TOTAL} &= KE + PE \\
&= \frac{p^2}{2m} + \frac{1}{2}kx^2
\end{aligned}
\tag{8.42}
$$

The Hamiltonian $H$ for *Quantum Harmonic Oscillator* is given by the kinetic energy operator $T$ plus the potential energy operator $V$

$$
\begin{aligned}
H &= T + V \tag{8.43} \\
&= -\frac{\hbar^2}{2m}\frac{\partial^2}{\partial x^2} + \frac{1}{2}kx^2 \tag{8.44}
\end{aligned}
$$

The Schrödinger Equation for the quantum harmonic oscillator becomes

$$
H\Psi = E\Psi
$$

$$
\left[-\frac{\hbar^2}{2m}\frac{\partial^2}{\partial x^2} + \frac{1}{2}kx^2\right]\Psi = E\Psi
$$

$$
\frac{\partial^2\Psi}{\partial x^2} + \frac{2m}{\hbar^2}\left(E - \frac{1}{2}kx^2\right)\Psi = 0
$$

$$
\frac{\partial^2\Psi}{\partial x^2} + 2\sqrt{\frac{m}{\hbar^2}}\left(\sqrt{\frac{m}{\hbar^2}}\sqrt{k}\right)\left[\frac{E}{\sqrt{k}} - \frac{1}{2}\sqrt{k}x^2\right]\Psi = 0
$$

$$
\frac{1}{\sqrt{\frac{mk}{\hbar^2}}}\frac{\partial^2\Psi}{\partial x^2} + \left[2\sqrt{\frac{m}{\hbar^2}}\frac{E}{\sqrt{k}} - 2\sqrt{\frac{m}{\hbar^2}}\frac{1}{2}\sqrt{k}x^2\right]\Psi = 0
$$

$$
\frac{1}{\sqrt{\frac{mk}{\hbar^2}}}\frac{\partial^2\Psi}{\partial x^2} + \left[2\sqrt{\frac{m}{\hbar^2 k}}E - \sqrt{\frac{mk}{\hbar^2}}x^2\right]\Psi = 0
\tag{8.45}
$$

Let two quantities, $\alpha$ and $\lambda$ be introduced, defined by

$$
\sqrt{\frac{mk}{\hbar^2}} = \alpha^2
$$

$$
2E\sqrt{\frac{m}{\hbar^2 k}} = \lambda
$$

Using the above definitions, the differential equation becomes

$$
\frac{1}{\alpha^2}\frac{\partial^2\Psi}{\partial x^2} + \left[\lambda - \alpha^2 x^2\right]\Psi = 0
$$

Introduce a variable $y = \alpha x$, where $\alpha$ is a constant, and we obtain

$$
\frac{\partial^2\Psi}{\partial(\alpha^2 x^2)} + \left[\lambda - \alpha^2 x^2\right]\Psi = 0
$$

$$
\frac{\partial^2\Psi}{\partial y^2} + \left[\lambda - y^2\right]\Psi = 0
\tag{8.46}
$$

The above differential equation is much simpler. Let us look for asymptotic solutions such tha $y \to \infty$ and thus $y \gg$ wavelength, $\lambda$, and thus

$$\frac{\partial^2 \Psi}{\partial y^2} - y^2 \Psi = 0 \tag{8.47}$$

which has a solution of the form $e^{\pm \frac{y^2}{2}}$, and we choose $e^{-\frac{y^2}{2}}$, and thus the function will be of the form

$$\Psi(y) = H(y)e^{-\frac{y^2}{2}} \tag{8.48}$$

and the question is what form does $H(y)$ take? This can be answered by putting the form of $\Psi$, $\frac{\partial \Psi}{\partial y}$ and $\frac{\partial^2 \Psi}{\partial y^2}$ and finding, after some algebra, that

$$\frac{\partial^2 H(y)}{\partial y^2} - 2y \frac{\partial H(y)}{\partial y} + (\lambda - 1)H(y) = 0 \tag{8.49}$$

which is a second order differential equation known as *Hermite Equation* whose solutions are *Hermite Polyomials* satifying

$$H_n(y) = (-1)^n e^{y^2} \frac{\partial^n}{\partial y^n} e^{-y^2} \tag{8.50}$$

with the requirement that

$$\begin{aligned}
\lambda &= 2E\sqrt{\frac{m}{\hbar^2 k}} = (2n+1) \text{ with } n = 0, 1, 2, \cdots \\
&= 2E\sqrt{\frac{m}{\hbar^2 m\omega^2}} = (2n+1) \text{ using } k = m\omega^2 \\
&= \frac{2E}{\hbar\omega} = (2n+1) \\
E &= \frac{\hbar\omega}{2}(2n+1) \\
E &= (n+\frac{1}{2})\hbar\omega
\end{aligned} \tag{8.51}$$

which gives the quantised energies of the quantum hamonic oscilator. Note that we can make the following observations. *First*, the lowest energy of the oscillator for $n = 0$ is $E_0 = \frac{1}{2}\hbar\omega$, is called the ground state or the zero point energy. This energy is a characterstic result of quantum mechanics. The existence of zero point energy is in agreement with experiment and is an important feature of quantum mechanics. *Secondly*, the energies $E_n$ depend only on one quantum number, $n$, thus all the energy levels of the oscillator are non-degenerate. *Thirdly*, successive energy levels are equally spaced, with the separation between two adjacent levels being $\hbar\omega$.

Note that the existence of the zero point energy is quite consistent with the uncertainty principle. The momentum of the particle in the ground state is $E_0 = \frac{p^2}{2m}$ or $p = \pm\sqrt{2mE_0}$ (the average momentum is of course zero), and the uncertainty in the knowledge of momentum is therefore $\Delta p \approx 2\sqrt{2mE_0}$. On the other hand, the zero point amplitude of vibration is approximately given by $E_0 = \frac{1}{2}kA^2$ or

$A = \pm\sqrt{\frac{2E_0}{k}}$ with an uncertainty $\Delta x \approx 2\sqrt{\frac{2E_0}{k}}$, and thus $\Delta p \Delta x \approx 4\hbar$ which is in accordance with the uncertainty principle $\Delta p \Delta x \leq \hbar$. Thus the existence of zero-point effects is real. These effects are most marked for the particles of low mass and at very low temperatures when the system is essentially in the ground state.

The wave function of the quantum harmonic oscillator is given by

$$\Psi_n(x) = \left[\frac{\alpha}{2^n n! \sqrt{\pi}}\right]^{1/2} e^{-\frac{\alpha^2 x^2}{2}} H_n(\alpha x) \tag{8.52}$$

where

$$\alpha = \left(\frac{mk}{\hbar^2}\right)^{1/4} = \left(\frac{m\omega}{\hbar}\right)^{1/2}$$

The first four wave functions for the quantum harmonic oscillator are given below and illustrated in Figure 8.12.

$$\Psi_0 = \left[\frac{\alpha}{\sqrt{\pi}}\right]^{1/2} e^{-\frac{\alpha^2 x^2}{2}} \tag{8.53}$$

$$\Psi_1 = \left[\frac{\alpha}{2\sqrt{\pi}}\right]^{1/2} 2\alpha x e^{-\frac{\alpha^2 x^2}{2}} \tag{8.54}$$

$$\Psi_2 = \left[\frac{\alpha}{8\sqrt{\pi}}\right]^{1/2} \left(4\alpha^2 x^2 - 2\right) e^{-\frac{\alpha^2 x^2}{2}} \tag{8.55}$$

$$\Psi_3 = \left[\frac{\alpha}{48\sqrt{\pi}}\right]^{1/2} \left(8\alpha^3 x^3 - 12\alpha x\right) e^{-\frac{\alpha^2 x^2}{2}} \tag{8.56}$$

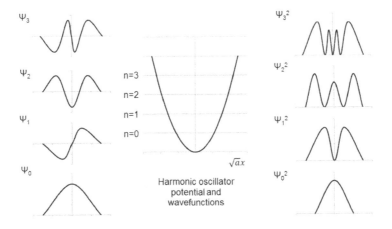

Figure 8.12: The wave functions for the Quantum Harmonic Oscillator.

## 8.3   Exercises

**8.1**. (a) State the following principles and explai the physical significance of each:
(i) Complementary principle
(ii) Uncertainty principle
(iii) Correspondence principle

(b) A quantum harmonic oscillator has zero point energy. Show that the existence of zero point oscillations is consistent with the uncertainty principle.

**8.2**. Show that quantum particles with a wave function $\Psi$ and its complex conjugate, $\Psi^*$, satify the continuity equation

$$\nabla \cdot \mathbf{j} + \frac{\partial \rho}{\partial t} = 0$$

where the probability current  $\mathbf{j} = \frac{\hbar}{2mi}[\Psi^*(\nabla\Psi) - \Psi(\nabla\Psi^*)]$

and the probability density function  $\rho = \Psi^*\Psi = |\Psi|^2$

**8.3**. (a) State the normalisation condition for a wave function.
(b) Given a wave function $\Psi(x)$ of the form

$$\Psi(x) = Ce^{-\frac{(x-x_0)^2}{4\delta x^2}} e^{ik_0 x}$$

Determine $C$.

**8.4**. (a) Quantum particles, each of mass $m$, energy $E$ are moving such that they encounter a potential step, defined by a potential $V(x)$ varying as:

$$V(x) = \begin{cases} 0, & x < 0 \\ V_0, & x \geq 0 \end{cases}$$

(i) Illustrate, with a well labelled diagram, the potential step.
(ii) Write the Schrödinger equation in both regions of the potential step.

(b) Show that for $E > V_0$, the reflection coefficient $R$ and transmission coefficient $T$ are given by

$$R = \frac{[1 - (1 - V_0/E)^{1/2}]^2}{[1 + (1 - V_0/E)^{1/2}]^2}$$

$$T = \frac{4(1 - V_0/E)^{1/2}}{[1 + (1 - V_0/E)^{1/2}]^2}$$

**8.5**. (a) Quantum particles, each of mass $m$, energy $E$ are moving such that they encounter a potential barrier, defined by a potential $V(x)$ varying as:

$$V(x) = \begin{cases} 0, & x < 0 \\ V_0, & 0 < x < a \\ 0, & x > a \end{cases} \qquad (8.57)$$

(i) Illustrate, with a well labelled diagram, the potential barrier.

(ii) Write the Schrödinger equation in all the three regions of the potential barrier.

(b) Show that for the case $E > V_0$ reflection coefficient $R$ and transmission coefficient $T$ are given by

$$R = [1 + \frac{4(E/V_0)(E/V_0 - 1)}{\sin^2(k_2 a)}]^{-1}$$

$$T = [1 + \frac{\sin^2(k_2 a)}{4(E/V_0)(E/V_0 - 1)}]^{-1}$$

with

$$k_2 a = \sqrt{\frac{2mV_0 a^2}{\hbar^2}(\frac{E}{V_0} - 1)}$$

and note that $R + T = 1$.

(c) Show that for the case $E < V_0$ reflection coefficient $R$ and transmission coefficient $T$ are given by

$$R = [1 + \frac{4(E/V_0)(1 - V_0/E)}{\sinh^2 \beta a}]^{-1}$$

$$T = [1 + \frac{\sinh^2 \beta a}{4(E/V_0)(1 - V_0/E)}]^{-1}$$

with

$$\beta a = \sqrt{\frac{2mV_0 a^2}{\hbar^2}(1 - \frac{E}{V_0})}$$

**8.6.** (a) (a) Quantum particles, each of mass $m$, energy $E$ are moving such that they encounter a potential well, defined by a potential $V(x)$ varying as:

$$V(x) = \begin{cases} 0, & x < -a \\ -V_0, & -a < x < a \\ 0, & x > a \end{cases}$$

(i) Illustrate, with a well labelled diagram, the potential well.

(ii) Write the Schrödinger equation in all the three regions of the potential well.

(b) Show that for the case $E > V_0$ reflection coefficient $R$ and transmission coefficient $T$ are given by

$$R = [1 + \frac{4(E/V_0)(1 + E/V_0)}{\sin^2(2k_2a)}]^{-1}$$

$$T = [1 + \frac{\sin^2(2k_2a)}{4(E/V_0)(1 + E/V_0)}]^{-1}$$

with

$$2k_2a = \sqrt{4\frac{2mV_0a^2}{\hbar^2}(1 + \frac{E}{V_0})} = k_2L$$

**8.7.** (a) Quantum particles, each of mass $m$, energy $E$ moving confined in an infinite potential well, whose potential function $V(x)$ is such that

$$V(x) = \begin{cases} 0, & 0 < x < L \\ \infty, & x < 0 \text{ and } x > L \end{cases} \qquad (8.58)$$

(i) Illustrate, with a well labelled diagram, the infinite potential well.
(ii) Write the Schrödinger equation in the infinite potential well.

(b) (i) Show that the energies of the quantum particle in an infinite potential well are quantised given by

$$E_n = \frac{n^2\hbar^2\pi^2}{2mL^2}$$

(ii) Calculate the energies (in eV) in the infinite potential well for $n = 1, 2$ and 3, given $L = 12$nm and $m = 0.067m_e$.

**8.8.** (a) Quantum particle, each of mass $m$, energy $E$ moving confined in a cubical box of side $L$, whose potential function is such that

$$V(x, y, z) = \begin{cases} 0, & \text{inside the box} \\ \infty, & \text{outside the box} \end{cases}$$

(i) Illustrate, with a well labelled diagram, the configuration of particle in a box.
(ii) Write the Schrödinger equation for the particle in a box.

(b) (i) Show that the energies of the quantum particle in an infinite potential well are quantised given by

$$E_n = \frac{\pi^2\hbar^2}{2mL^2}(n_x^2 + n_y^2 + n_z^2)$$

where $n_x, n_y, n_z$ take integral values, and the total energy of the particle in a box for state $n$ is quantised.

**8.9.** A quantum harmonic oscillator has the Hamiltonian of the form $H = T + V$, where $T$ is the kinetic energy operator and $V$ is the potential energy operator, given by

$$H = -\frac{\hbar^2}{2m}\frac{\partial^2}{\partial x^2} + \frac{1}{2}kx^2$$

(a) Write down the Schrödinger Equation for the quantum harmonic oscillator.

(b) Show that by making certain approximations which you must state and explain the wave function for the quantum harmonic oscillator takes the form

$$\Psi(y) = H(y)e^{-\frac{y^2}{2}}$$

where $H(y)$ satisfies the *Hermite second order differential equation* given by

$$\frac{\partial^2 H(y)}{\partial y^2} - 2y\frac{\partial H(y)}{\partial y} + (\lambda - 1)H(y) = 0$$

(c) Show that the energies of a quantum harmonic oscillator are quantised of the form

$$E_n = (n + \frac{1}{2})\hbar\omega$$

**8.10**. The Hamiltonian of the quantum harmonic oscillator for unit mass can be written in the form

$$\hat{H} = \frac{1}{2}\left(\hat{p}^2 + \omega^2\hat{q}^2\right)$$

where the commutator $[\hat{q}, \hat{p}] = i\hbar$

(a) Show that Operators $\hat{a}$ and $\hat{a}^+$ can be defined such that

$$\hat{a} = \frac{1}{(2\hbar\omega)^{1/2}}(\omega\hat{q} + i\hat{p})$$

$$\hat{a}^+ = \frac{1}{(2\hbar\omega)^{1/2}}(\omega\hat{q} - i\hat{p})$$

(b) Show that

$$\text{(i)} \qquad \hat{q} = \left(\frac{\hbar}{2\omega}\right)^{1/2}(\hat{a} + \hat{a}^+)$$

$$\text{(ii)} \qquad \hat{p} = -\left(\frac{\hbar}{2\omega}\right)^{1/2}(\hat{a} - \hat{a}^+)$$

$$\text{(iii)} \qquad [\hat{a}, \hat{a}^+] = 1$$

$$\text{(iv)} \qquad [\hat{a}, (\hat{a}^+)^2] = 2\hat{a}^+$$

$$\text{(v)} \qquad [\hat{a}^2, \hat{a}^+] = 2\hat{a}$$

# Chapter 9

# Introduction to the Structure of the Atom

## 9.1  Thomson's Atomic Model

Thomson's model, also known as "Plum pudding model" of the atom was introduced around 1904 and discarded in 1911 after experimental evidence, and is thus obsolete. In this model, the atom is composed of "electrons" which are negatively charged, embedded in a positive charge, as illustrated in Figure 9.1.

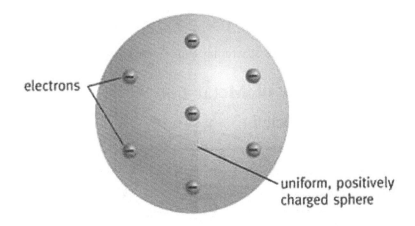

Figure 9.1: Thomson's model of the atom.

## 9.2  Rutherford Atomic Model

Rutherford model of the atom was introduced in 1911 after an interpretation of a scattering experiment where $\alpha$-particles were interacting with gold atoms in a gold foil as illustrated in Figure 9.2. It is, however, obsolete, after further experimentation and theory. The results of the Rutherford experiment are illustrated in Figure 9.3, and were interpreted as follows:

1. The majority of the $\alpha$-particles were undeflected, *implying* that the atom is largely empty space, comprising electron which do not influence the scattering.
2. A few of the $\alpha$-particles were partially deflected, implying they encountered a positive charge.
3. Very few of the $\alpha$-particles were deflected backwards, implying they encountered a positive charge and hence a repulsive force.

The Rutherford scattering problem involves a central force due to the Coulomb interaction between

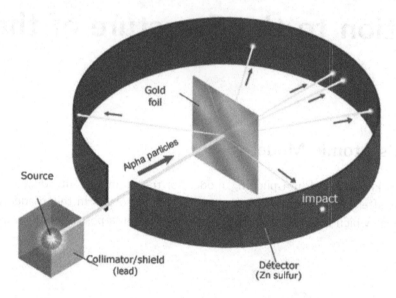

Figure 9.2: Rutherford scattering experimental set up.

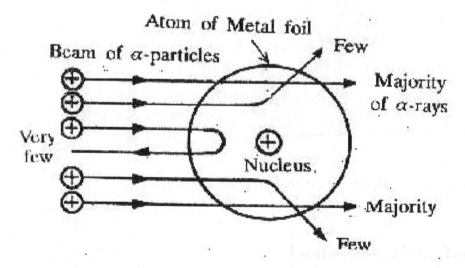

Figure 9.3: Interpretation of Rutherford scattering experiment.

the nucleus and the $\alpha$-particles. A **central force** is a force whose line of action passes through a defined center and whose magnitude depends on the distance from that center. The central force in this case is the Coulomb force, $F$.

$$
\begin{aligned}
F &= \frac{1}{4\pi\epsilon_0}\frac{Z_1 Z_2 e^2}{r^2} \\
&= f(r)\hat{\mathbf{r}} \\
\text{where} \quad f(r) &= \frac{k}{r^2} \\
\text{where} \quad k &= \frac{Z_1 Z_2 e^2}{4\pi\epsilon_0}
\end{aligned}
$$

Consider the **position** of the $\alpha$-particle given by polar coordinates $(\mathbf{r}, \theta)$, where $\mathbf{r} = r\hat{\mathbf{r}}$

The **velocity** of the $\alpha$-particle is given by **v**, where

$$
\begin{aligned}
\mathbf{v} &= \frac{d\mathbf{r}}{dt} \\
&= \frac{d(r\hat{\mathbf{r}})}{dt} \\
&= \dot{r}\hat{\mathbf{r}} + r\frac{d\hat{\mathbf{r}}}{dt} \\
\mathbf{v} &= \dot{r}\hat{\mathbf{r}} + r\dot{\theta}\hat{\theta} \tag{9.1}
\end{aligned}
$$

The **acceleration** of the the $\alpha$-particle is given by **a**, where

$$
\begin{aligned}
\mathbf{a} &= \frac{d\mathbf{v}}{dt} \\
&= \frac{d\dot{r}\hat{\mathbf{r}} + r\dot{\theta}\hat{\theta}}{dt} \\
\mathbf{a} &= (\ddot{r} - r\dot{\theta}^2)\hat{\mathbf{r}} + (r\ddot{\theta} + 2\dot{r}\dot{\theta})\hat{\theta} \tag{9.2} \\
\mathbf{a} &= a_r\hat{\mathbf{r}} + a_\theta\hat{\theta} \tag{9.3}
\end{aligned}
$$

where

$$
\begin{aligned}
a_r &= (\ddot{r} - r\dot{\theta}^2) \quad \text{is the Radial component of the acceleration} \tag{9.4} \\
a_\theta &= r\ddot{\theta} + 2\dot{r}\dot{\theta} \tag{9.5} \\
&= \frac{1}{r}\frac{d(r^2\dot{\theta})}{dt} \quad \text{is the Normal (or Transverse) component of the acceleration} \tag{9.6}
\end{aligned}
$$

The **Angular momentum** of the $\alpha$-particle is given by **L**, which has magnitude $L$, given by

$$
\begin{aligned}
L &= |\mathbf{r} \wedge m\mathbf{v}| \\
&= |\mathbf{r} \wedge m(\dot{r}\hat{\mathbf{r}} + r\dot{\theta}\hat{\theta})| \\
L &= mr^2\dot{\theta} \quad \text{is the magnitude of the Angular momentum} \tag{9.7}
\end{aligned}
$$

The **Equation of motion** for the $\alpha$-particle is given by $m\mathbf{a} = \mathbf{F}$, where

$$m\mathbf{a} = \mathbf{F}$$

$$m\left[(\ddot{r} - r\dot{\theta}^2)\hat{\mathbf{r}} + (r\ddot{\theta} + 2\dot{r}\dot{\theta})\hat{\theta}\right] = f(r)\hat{\mathbf{r}}$$

On equating components,

$$m\left((\ddot{r} - r\dot{\theta}^2)\right) = f(r)\hat{\mathbf{r}} \quad \text{is the Radial equation} \tag{9.8}$$

$$m\left((r\ddot{\theta} + 2\dot{r}\dot{\theta})\right) = 0 \quad \text{is the Transverse equation} \tag{9.9}$$

which on equating components, gives the Radial equation and Transverse equation shown above, and are discussed below.

**Transverse equation**

$$m\left[r\ddot{\theta} + 2\dot{r}\dot{\theta}\right] = 0$$

$$\frac{m}{r}\frac{d(r^2\dot{\theta})}{dt} = 0$$

$$\frac{d(r^2\dot{\theta})}{dt} = 0$$

$$r^2\dot{\theta} = \text{constant} = h = \frac{L}{m}$$

where $h = \frac{L}{m}$ is the angular momentum per unit mass.

**Radial equation**

$$m\left[\ddot{r} - r\dot{\theta}^2\right] = f(r)$$

$$\text{Let} \quad r = \frac{1}{u} \quad \Rightarrow \quad r = u^{-1}$$

$$\dot{r} = -u^{-2}\dot{u} = -\frac{1}{u^2}\frac{du}{dt} = -\frac{1}{u^2}\frac{du}{d\theta}\frac{d\theta}{dt}$$

$$= -r^2\dot{\theta}\frac{du}{d\theta}$$

$$= -h\frac{du}{d\theta}$$

Differentiating again,

$$\ddot{r} = -h\frac{d}{dt}\left(\frac{du}{d\theta}\right)$$

$$= -h\frac{d\theta}{dt}\frac{d}{d\theta}\left(\frac{du}{d\theta}\right)$$

$$= -h\dot{\theta}\frac{d^2u}{d\theta^2}$$

$$= -h^2u^2\frac{d^2u}{d\theta^2}$$

Note that from the transverse equation, we obtained $r^2\dot{\theta} = h, \Rightarrow r\dot{\theta}^2 = \frac{1}{u}h^2u^4$

The Radial equation can be cast in the form

$$m\left[-h^2u^2\frac{d^2u}{d\theta^2} - \frac{1}{u}h^2u^4\right] = f(u^{-1})$$

$$\frac{d^2u}{d\theta^2} + u = -\frac{f(u^{-1})}{mh^2u^2} \quad \text{where} \quad f(u^{-1}) = -ku^2 \qquad (9.10)$$

$$\frac{d^2u}{d\theta^2} + u = \frac{k}{mh^2} \qquad (9.11)$$

which can be identified as a Second Order Inhomogeneous Differential Equation of the orbit, a form of the Binet equation. The solution of this equation is the sum of the complementary function plus a particular integral.

$$u = A\cos(\theta - \theta_0) + \frac{k}{mh^2}$$

The boundary conditions are such that, initially, $\theta_0 = 0$,

$$u = A\cos\theta + \frac{k}{mh^2}$$

$$r = \frac{1}{u} = \frac{1}{A\cos\theta + \frac{k}{mh^2}}$$

$$= \frac{mh^2}{k\left(1 + \frac{Amh^2}{k}\cos\theta\right)}$$

$$= \frac{mh^2}{k\left(1 + e\cos\theta\right)} \qquad (9.12)$$

where $e = \dfrac{Amh^2}{k}$ is the eccentricity of the orbit, and for a hyperbolic orbit, $e > 1$. (9.13)

$$(9.14)$$

From Figure 9.4, it can be seen that the change in momentum associated with the scattering is

$$\Delta\mathbf{p} = \mathbf{p}_f - \mathbf{p}_i$$

$$\Delta p = 2mv_0\sin\frac{\theta}{2}$$

But

$$\Delta p = \int F_{\Delta p}dt$$

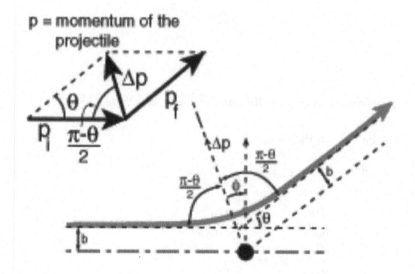

Figure 9.4: Rutherford Scattering

$$= \int F \cos \phi \, dt$$

$$= \int \frac{Z_\alpha Z_2 e^2}{4\pi\epsilon_0 r^2} \cos \phi \, dt$$

$$\text{hence} \quad \Delta p = \frac{Z_\alpha Z_2 e^2}{4\pi\epsilon_0} \int \frac{\cos \phi}{r^2} dt = 2mv_0 \sin \frac{\theta}{2}$$

Using conservation of angular momentum, $L$

$$L = mv_0 b = mr^2 \frac{d\phi}{dt} \quad \text{where } b \text{ is the impact parameter}$$

$$r^2 = \frac{v_0 b}{d\phi/dt}$$

$$\frac{1}{r^2} = \frac{1}{v_0 b} \frac{d\phi}{dt}$$

We can evaluate the integral using boundary conditions,

$$\phi_i = -\frac{1}{2}(\pi - \theta), \quad \phi_f = +\frac{1}{2}(\pi - \theta)$$

$$2mv_0 \sin \frac{\theta}{2} = \frac{Z_\alpha Z_2 e^2}{4\pi\epsilon_0} \int_{\phi_i}^{\phi_f} \frac{\cos \phi}{v_0 b} \frac{d\phi}{dt} dt$$

$$= \frac{Z_\alpha Z_2 e^2}{4\pi\epsilon_0 v_0 b} \int_{\phi_i}^{\phi_f} \cos \phi \, d\phi$$

$$= \frac{Z_\alpha Z_2 e^2}{4\pi\epsilon_0 v_0 b} \left[ \sin \frac{1}{2}(\pi - \theta) - \sin \left\{ -\frac{1}{2}(\pi - \theta) \right\} \right]$$

$$2mv_0 \sin\frac{\theta}{2} = \frac{Z_\alpha Z_2 e^2}{4\pi\epsilon_0 v_0 b} 2\cos\frac{\theta}{2}$$

$$\text{Hence} \quad b = \frac{Z_\alpha Z_2 e^2}{4\pi\epsilon_0 mv_0^2}\cot\frac{\theta}{2} \tag{9.15}$$

is the expression for the impact parameter.

The differential scattering cross section, $\frac{d\sigma}{d\Omega}$ is given by

$$\frac{d\sigma}{d\Omega} = \frac{b}{\sin\theta}\mid\frac{db}{d\theta}\mid \tag{9.16}$$

$$\frac{d\sigma}{d\Omega} = \left(\frac{Z_1 Z_2 e^2}{8\pi\epsilon_0 mv_0^2}\right)^2 \frac{1}{\sin^4(\frac{\theta}{2})} \tag{9.17}$$

or equivalently

$$\frac{d\sigma}{d\Omega} = \left(\frac{Z_1 Z_2 e^2}{8\pi\epsilon_0 mv_0^2}\right)^2 \csc^4(\frac{\theta}{2}) \tag{9.18}$$

or yet in another form

$$\frac{d\sigma}{d\Omega} = \left(\frac{Z_1 Z_2 e^2}{8\pi\epsilon_0 mv_0^2}\right)^2 \frac{4}{(1-\cos\theta)^2} \tag{9.19}$$

$$\tag{9.20}$$

The scattering cross section is illustrated graphically in Figure 9.5.

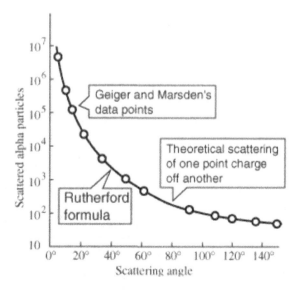

Figure 9.5: Rutherford scattering cross section vs scattering angle.

## 9.3   Bohr's theory of the atom

Bohr's model of the atom was introduced in 1913. In *Bohr's theory of the atom*, the following basic assumptions are made:

(i) In an atom, an electron of mass $m_e$ moves with a velocity $v$ around a nucleus in orbits of radii $r$ such that the angular momentum $L$ of the electron is *quantised* in multiples of $\hbar$. While in these orbits, the electron does not radiate.

$$L = m_e v r = n\hbar \text{ where } n = 1, 2, 3, \cdots \tag{9.21}$$

and hence

$$r = \frac{n\hbar}{m_e v} \tag{9.22}$$

(ii) When an electron makes a transition from a higher energy level to a lower energy level, it is

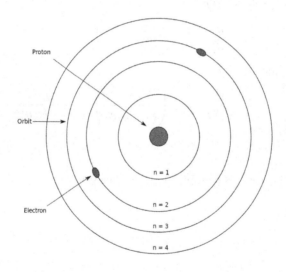

Figure 9.6: Bohr's model of the atom

accompanied by *emission* of radiation, and when an electron *absorbs* a photon it moves from a lower energy level to a higher energy level. The change in energy between the initial energy level $E_i$ and the final energy level $E_f$ is related to the frequency $\nu$ of the associated photon as:

$$E_i - E_f = h\nu \tag{9.23}$$

Although the dynamics of an electron obey quantum mechanical laws, it is possible to get a reasonable understanding if we approximate the dynamics using a classical equation of motion (from Newton's second law) as below. The electron moving in a circular orbit experiences an acceleration $a = v^2/r$. The required centripetal force for the circular motion is supplied by the coulomb interaction between the electron and the nucleus. Thus we have:

$$m_e a \;=\; \sum F$$

$$m_e \frac{v^2}{r} = \frac{kZe^2}{r^2}$$

$$v^2 = \frac{Ze^2}{4\pi\epsilon_0 m_e r}$$

where $Z$ is the atomic number of the atom, $k = 1/(4\pi\epsilon_o)$ is the Coulomb's force constant, and $\epsilon_o = 8.854 \times 10^{-12} c^2 N^{-1} m^{-2}$ is the permittivity of free space. From Bohr's theory, the radius $r_n$ for an electron in an orbit $n$ can be obtained as below.

$$v = \frac{n\hbar}{m_e r} \tag{9.24}$$

Using the equations above,

$$\frac{Ze^2}{4\pi\epsilon_0 m_e r} = \frac{n^2 \hbar^2}{m_e^2 r^2}$$

$$v^2 = \frac{Ze^2}{4\pi\epsilon_0 m_e r} \tag{9.25}$$

and hence the radius of the $n^{th}$ orbit is given as

$$r_n = \frac{4\pi\epsilon_0 n^2 \hbar^2}{m_e Z e^2} \tag{9.26}$$

where for $n = 1$ and $Z = 1$, we have

$$r_1(\text{hydrogen}) = \frac{4\pi\epsilon_0 \hbar^2}{m_e e^2} = a_0 \tag{9.27}$$

known as the *Bohr's radius*, and $r_n = (a_0/Z)n^2$. From above the velocity $v_n$ for an electron in an orbit $n$ can be obtained as below.

$$v_n^2 = \left( \frac{Ze^2}{4\pi\epsilon_0 n\hbar} \right)^2$$

$$v_n = \frac{Ze^2}{4\pi\epsilon_0 n\hbar} \tag{9.28}$$

where for $n = 1$ and $Z = 1$, we have

$$v_1(\text{hydrogen}) = \frac{e^2}{4\pi\epsilon_0 \hbar} = v_0 \tag{9.29}$$

is the speed of the electron in the first Bohr's orbit of the hydrogen atom, and $v_n = v_0 Z/n$.

Thus the speed of the electron in the $n^{th}$ atomic orbit is inversely proportional to the quantum number $n$, and decreases in the ratio, $1 : \frac{1}{2} : \frac{1}{3} : \frac{1}{4} : \frac{1}{5}...$ for $n = 1, 2, 3, 4, 5, ....$

The total energy $E_n$ for an electron in the $n^{th}$ orbit is the sum of its kinetic energy (KE) and potential energy (PE), and can be obtained as below.

$$
\begin{aligned}
E_n &= KE + PE \\
&= \frac{1}{2}m_e v_n^2 - \frac{1}{4\pi\epsilon_0}\frac{Ze^2}{r_n} \\
&= -\frac{m_e Z^2 e^4}{2\hbar^2(4\pi\epsilon_0)^2}\frac{1}{n^2}
\end{aligned}
\tag{9.30}
$$

When the above results are applied to the hydrogen atom $(Z = 1)$, the energy levels are obtained as

$$
\begin{aligned}
E_n &= -\frac{m_e e^4}{2\hbar^2(4\pi\epsilon_0)^2}\frac{1}{n^2} = -E_0\frac{1}{n^2} \\
&= -\frac{(9.11\times10^{-31})(1.6\times10^{-19})^4(9\times10^9)^2}{2(1.05\times10^{-34})^2}\frac{1}{n^2} \\
&= -(2.19\times10^{-18})\frac{1}{n^2}\quad \text{J} \\
&= -\frac{13.6}{n^2}\quad \text{eV}
\end{aligned}
\tag{9.31}
$$

where $E_0 = -13.6$ eV $= -2.19\times10^{-18}$ J is the *ionization energy* of the hydrogen atom..

Spectral wavelengths for the hydrogen spectrum can be obtained from the expression of the energy levels of the hydrogen atom. Let $E_i$ and $E_f$ be the initial and final energy levels respectively in a hydrogen atom. Then

$$
\begin{aligned}
E_i &= -\frac{13.6}{n_i^2}\quad \text{eV} \\
E_f &= -\frac{13.6}{n_f^2}\quad \text{eV} \\
E_i - E_f &= h\nu = -13.6\left(\frac{1}{n_i^2} - \frac{1}{n_f^2}\right)\times1.6\times10^{-19}\quad \text{J} \\
\frac{hc}{\lambda} &= 13.6\left(\frac{1}{n_f^2} - \frac{1}{n_i^2}\right)\times1.6\times10^{-19} \\
\frac{1}{\lambda} &= 1.097\times10^7\left(\frac{1}{n_f^2} - \frac{1}{n_i^2}\right) \\
\frac{1}{\lambda} &= R\left(\frac{1}{n_f^2} - \frac{1}{n_i^2}\right)\quad \text{m}^{-1}
\end{aligned}
\tag{9.32}
$$

where

$$
\begin{aligned}
R &= \frac{13.6\times1.6\times10^{-19}}{hc} \\
&= 1.097\times10^7\quad \text{m}^{-1}
\end{aligned}
\tag{9.33}
$$

is the Rydberg constant. If $n_i < n_f$ there is absorption, and for $n_i > n_f$ there is emission. An energy level diagram for the hydrogen is shown in Figure 9.7, where the Lyman series, Balmer series, Paschen series are illustrated. The series also includes the Bracket series and Pfund series as defined below.

$$\frac{1}{\lambda} = R\left(\frac{1}{1^2} - \frac{1}{n_i^2}\right) \text{ for Lyman series, } n_f = 1, n_i = 2,3,4,\cdots \tag{9.34}$$

$$\frac{1}{\lambda} = R\left(\frac{1}{2^2} - \frac{1}{n_i^2}\right) \text{ for Balmer series, } n_f = 2, n_i = 3,4,5\cdots \tag{9.35}$$

$$\frac{1}{\lambda} = R\left(\frac{1}{3^2} - \frac{1}{n_i^2}\right) \text{ for Paschen series, } n_f = 3, n_i = 4,5,6\cdots \tag{9.36}$$

$$\frac{1}{\lambda} = R\left(\frac{1}{4^2} - \frac{1}{n_i^2}\right) \text{ for Bracket series, } n_f = 4, n_i = 5,6,7\cdots \tag{9.37}$$

$$\frac{1}{\lambda} = R\left(\frac{1}{5^2} - \frac{1}{n_i^2}\right) \text{ for Pfund series, } n_f = 5, n_i = 6,7,8,\cdots \tag{9.38}$$

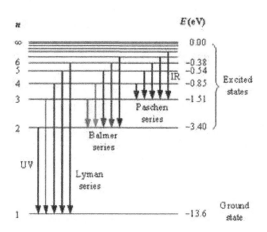

Figure 9.7: Energy levels of the hydrogen spectrum.

## 9.4 Quantum theory of the hydrogen atom

The Schrödinger equation was published in 1926 to describe how a quantum state evolves with time. In this section the Schrödinger equation will be applied to study the hydrogen atom which has spherical symmetry. An understanding of the structure and behavior of the hydrogen atom is essential in physics, chemistry and technology since it is the simplest atomic system, consisting of one electron and one proton.

### 9.4.1  The hydrogen wave functions and energies

Suppose the proton in the hydrogen atom is of mass $m_1$ and an electron of mass $m_2$, which has positions $r_1$ and $r_2$ respectively. This is a *two body problem*, with a Hamiltonian given by

$$H = \frac{p_1^2}{2m_1} + \frac{p_2^2}{2m_2} + V(r_1, r_2) \tag{9.39}$$

where $V(r_1, r_2) = V(r)$, and $r = r_1 - r_2$.

Hence, the Schrödinger Equation for the hydrogen atom is

$$\left[ -\frac{\hbar^2}{2m_1}\nabla_1^2 - \frac{\hbar^2}{2m_2}\nabla_2^2 + V(r) \right] \Psi(r, R) = E_{TOT}\Psi(r, R) \tag{9.40}$$

The two body can be reduced to a *one body problem*

$$\left[ -\frac{\hbar^2}{2\mu}\nabla_r^2 + V(r) \right] \Psi = E\Psi \tag{9.41}$$

where $\mu$ is the reduced mass, satisfying

$$\frac{1}{\mu} = \frac{1}{m_1} + \frac{1}{m_2}$$

$$E = E_{TOT} - \frac{p^2}{2(m_1 + m_2)}$$

### Introduction of the spherical coordinate system

The symmetry of the problem of an electron moving around a proton with a spherically symmetric potential requires that we solve the Schrödinger equation using spherical coordinates. Therefore, the Cartesian coordinates $(x, y, z)$ of the electron are replaced by the spherical polar coordinates $(r, \theta, \phi)$ and the Laplacian operator is transformed into spherical polar coordinates by using the following relations. The spherical coordinate system is illustrated in Figure 9.8.

$$x = r\sin\theta\cos\phi \tag{9.42}$$
$$y = r\sin\theta\sin\phi \tag{9.43}$$
$$z = r\cos\theta \tag{9.44}$$

The Laplacian Operator, $\nabla^2$, in spherical coordinates is given by

$$\nabla^2 = \frac{1}{r^2}\frac{\partial}{\partial r}\left( r^2\frac{\partial}{\partial r} \right) + \frac{1}{r^2\sin\theta}\frac{\partial}{\partial\theta}\left( \sin\theta\frac{\partial}{\partial\theta} \right) + \frac{1}{r^2\sin^2\theta}\frac{\partial^2}{\partial\phi^2} \tag{9.45}$$

$$= \frac{\partial^2}{\partial r^2} + \frac{2}{r}\frac{\partial}{\partial r} + \frac{1}{r^2\sin\theta}\frac{\partial}{\partial\theta}\left( \sin\theta\frac{\partial}{\partial\theta} \right) + \frac{1}{r^2\sin^2\theta}\frac{\partial^2}{\partial\phi^2} \tag{9.46}$$

Hence, the Schrödinger equation in spherical polar coordinates becomes

$$-\frac{\hbar^2}{2\mu}\left\{ \frac{\partial^2}{\partial r^2} + \frac{2}{r}\frac{\partial}{\partial r} + \frac{1}{r^2\sin\theta}\frac{\partial}{\partial\theta}\left( \sin\theta\frac{\partial}{\partial\theta} \right) + \frac{1}{r^2\sin^2\theta}\frac{\partial^2}{\partial\phi^2} \right\}\Psi(r, \theta, \phi) + V(r)\Psi(r, \theta, \phi) = E\Psi(r, \theta, \phi)$$

$$\tag{9.47}$$

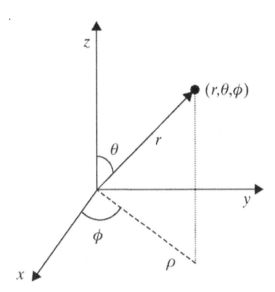

Figure 9.8: Spherical coordinate system

## Separation of Variables

By separation of variables, the wave function $\Psi(r, \theta, \phi)$ can be separated into three functions, each involving a single independent variable, that is, the wave function can be cast as a product of three functions,

$$\Psi(r, \theta, \phi = R(r)P(\theta)F(\phi) = RPF \tag{9.48}$$

where $R(r) = R$ is a function of $r$ only, $P(\theta) = P$ is a function of $\theta$ only, and $F(\phi) = F$ is a function of $\phi$ only. The function $R(r)$ will show the variation of the wave function along a radius vector from the nucleus, when $\theta$ and $\phi$ are constant. The function $P(\theta)$ will show the variation of the wave function with the zenith angle along a circle of on a sphere with nucleus at the center, when $r$ and $\phi$ are constant, and the function $F(\phi)$ will show the variation of with azimuthal angle when $r$ and $\theta$ are constant. Substituting $\Psi$ by $RPF$, and using $V(r) = Ze^2/4\pi\epsilon_0 r$ for hydrogen or hydrogen like atoms, we obtain

$$\frac{PF}{r^2}\frac{d}{dr}\left(r^2\frac{dR}{dr}\right) + \frac{RF}{r^2\sin\theta}\frac{d}{d\theta}\left(\sin\theta\frac{dP}{d\theta}\right) + \frac{RP}{r^2\sin^2\theta}\frac{d^2F}{d\phi^2} + \frac{2\mu}{\hbar^2}\left(E + \frac{Ze^2}{4\pi\epsilon_0 r}\right)RPF = 0 \tag{9.49}$$

In the above equation we have used ordinary derivatives instead of partial derivatives because each function depends on only one variable. Now multiplying the above equation by $\frac{r^2\sin^2\theta}{RPF}$, we obtain

$$\frac{\sin^2\theta}{R}\frac{d}{dr}\left(r^2\frac{dR}{dr}\right) + \frac{\sin\theta}{P}\frac{d}{d\theta}\left(\sin\theta\frac{dP}{d\theta}\right) + \frac{1}{F}\frac{d^2F}{d\phi^2} + \frac{2\mu}{\hbar^2}\left(E + \frac{Ze^2}{4\pi\epsilon_0 r}\right)r^2\sin^2\theta = 0 \tag{9.50}$$

Transposing the third term to the right-hand side, we obtain

$$\frac{\sin^2\theta}{R}\frac{d}{dr}\left(r^2\frac{dR}{dr}\right) + \frac{\sin\theta}{P}\frac{d}{d\theta}\left(\sin\theta\frac{dP}{d\theta}\right) + \frac{2\mu}{\hbar^2}\left(\left(E + \frac{Ze^2}{4\pi\epsilon_0 r}\right)r^2\sin^2\theta\right) = -\frac{1}{F}\frac{d^2F}{d\phi^2} \tag{9.51}$$

## The Azimuthal Equation

The left hand side of the above equation is a function of $r$ and $\theta$ only, while the right hand side is a function of $\phi$ only. Thus each side can be varied independently of the other. Therefore, the equation will be true only if both sides are equal to the same constant, a separation constant, denoted by $m^2$, where $m$ is called the *magnetic quantum number*. Thus we have,

$$-\frac{1}{F}\frac{d^2F}{d\phi^2} = m^2 \qquad (9.52)$$

or

$$\frac{d^2F}{d\phi^2} + m^2F = 0 \qquad (9.53)$$

This is the first of the three differential equations, and is known as *The Azimuthal wave equation*. In the solution of this equation, the solutions are such that $m$ must be a positive or negative integer or zero. The solution of the Azimuthal wave quation is

$$F_m(\phi) = \frac{1}{\sqrt{2\pi}}e^{im\phi} \qquad (9.54)$$

where $m$ is the magnetic quantum number with values

$$m = 0, \pm 1, \pm 2, \cdots, \pm l \qquad (9.55)$$

## The Radial Equation

Equating the left hand side of the above equation containing $R$ and $P$ to $m^2$ and dividing through out by $\sin^2\theta$, we obtain

$$\frac{1}{R}\frac{d}{dr}\left(r^2\frac{dR}{dr}\right) + \frac{1}{P\sin\theta}\frac{d}{d\theta}\left(\sin\theta\frac{dP}{d\theta}\right) + \frac{2\mu r^2}{\hbar^2}\left(E + \frac{Ze^2}{4\pi\epsilon_0 r}\right) = \frac{m^2}{\sin^2\theta} \qquad (9.56)$$

Transposing the second term on the left side to the right hand side, we obtain

$$\frac{1}{R}\frac{d}{dr}\left(r^2\frac{dR}{dr}\right) + \frac{2\mu r^2}{\hbar^2}\left(E + \frac{Ze^2}{4\pi\epsilon_0 r}\right) = \frac{m^2}{\sin^2\theta} - \frac{1}{P\sin\theta}\frac{d}{d\theta}\left(\sin\theta\frac{dP}{d\theta}\right) \qquad (9.57)$$

The left hand side of this equation is a function of $r$ only, while the right hand side is a function of $\theta$ only, and hence both sides of the equation must be equal to the same constant, a separation constant, which is $l(l+1)$, where $l$ is known as the *orbital quantum number*. Hence, we have

$$\frac{1}{R}\frac{d}{dr}\left(r^2\frac{dR}{dr}\right) + \frac{2\mu r^2}{\hbar^2}\left(E + \frac{Ze^2}{4\pi\epsilon_0 r}\right) = l(l+1) \qquad (9.58)$$

Multiplying both sides by $R/r^2$ and rearranging, we obtain

$$\frac{1}{r^2}\frac{d}{dr}\left(r^2\frac{dR}{dr}\right) + \left[\frac{2\mu}{\hbar^2}\left(E + \frac{Ze^2}{4\pi\epsilon_0 r}\right) - \frac{l(l+1)}{r^2}\right]R = 0 \qquad (9.59)$$

or

$$\frac{d^2R}{dr^2} + \frac{2}{r}\frac{dR}{dr} + \left[\frac{2\mu}{\hbar^2}\left(E + \frac{Ze^2}{4\pi\epsilon_0 r}\right) - \frac{l(l+1)}{r^2}\right]R = 0 \tag{9.60}$$

which is known as the *Radial wave equation*. A simplified solution of the radial equation can be obtained by introducing

$$\alpha = \sqrt{-\frac{8\mu E}{\hbar^2}}$$

$$\lambda = \frac{e^2}{4\pi\epsilon_0\hbar}\sqrt{-\frac{\mu}{2E}}$$

which for $n = \lambda$ gives the Energy Eigenvalues as

$$E_n = -\left(\frac{\mu e^4}{32\pi^2\epsilon_0^2\hbar^2}\right)\frac{1}{n^2} \tag{9.61}$$

or

$$E_n = -\left(\frac{\mu e^4}{8\epsilon_0^2 h^2}\right)\frac{1}{n^2} \tag{9.62}$$

which is identical with the expression for energy values obtained by Bohr's theory for the hydrogen atom. Also, one obtains

$$\frac{2}{\alpha} = \frac{4\pi\epsilon_0\hbar^2 n}{\mu e^2} \tag{9.63}$$

which for $n = 1$ gives the first Bohr orbit radius given by $a_0$

$$a_0 = \frac{4\pi\epsilon_0\hbar^2 n}{\mu e^2} \tag{9.64}$$

The solution of the Radial wave equation for the radial wave function for a given value of the principal quantum number $n$ and the orbital quantum number $l$ involves long mathematical details,which we shall not get into, only to show that the expression for the normalised radial wave function is given by

$$R_{nl}(r) = -\left\{\frac{(n-l-1)!}{2n\left[(n+l)!\right]^3}\right\}^{1/2} e^{-Zr/na_0}\left(\frac{2Zr}{na_0}\right)^l L_{n+l}^{2l+1}\left(\frac{2Zr}{na_0}\right) \tag{9.65}$$

where $L_{n+l}^{2l+1}\left(\frac{2Zr}{na_o}\right)$ are the *generalized Laguerre polynomials*. Some hydrogen radial wave functions are given below:

$$R_{10} = 2\left(\frac{Z}{a_0}\right)^{3/2} e^{-Zr/a_0}$$

$$R_{21} = \frac{1}{\sqrt{3}}\left(\frac{Z}{2a_0}\right)^{3/2}\left(\frac{Zr}{a_0}\right) e^{-Zr/2a_0}$$

$$R_{20} = 2\left(\frac{Z}{2a_0}\right)^{3/2}\left(1 - \frac{Zr}{2a_0}\right) e^{-Zr/2a_0}$$

$$R_{32} = \frac{2\sqrt{2}}{27\sqrt{5}}\left(\frac{Z}{3a_0}\right)^{3/2}\left(\frac{Zr}{a_0}\right)^2 e^{-Zr/3a_0}$$

$$R_{31} = \frac{4\sqrt{2}}{3}\left(\frac{Z}{3a_0}\right)^{3/2}\left(\frac{Zr}{a_0}\right)\left(1-\frac{Zr}{6a_0}\right)e^{-Zr/3a_0}$$

$$R_{30} = 2\left(\frac{Z}{3a_0}\right)^{3/2}\left(1-\frac{2Zr}{3a_0}+\frac{2(Zr)^2}{27a_0^2}\right)e^{-Zr/3a_0}$$

The above hydrogen radial wave functions are illustrated graphically in Figure 9.9, and the hydrogen radial probability density are illustrated in Figure 9.10.

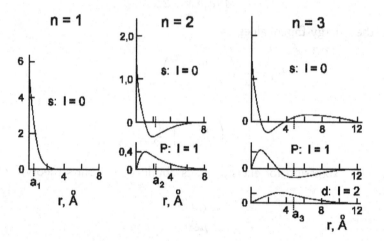

Figure 9.9: Hydrogen radial wave functions

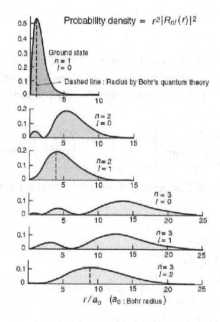

Figure 9.10: Hydrogen Radial Probability density

**The Polar Equation**

Equating $l(l+1)$ to the right hand side of the equation for $P$, we obtain

$$\frac{m^2}{\sin^2\theta} - \frac{1}{P\sin\theta}\frac{d}{d\theta}\left(\sin\theta\frac{dP}{d\theta}\right) = l(l+1) \tag{9.66}$$

Multiplying both sides of this equation by $P$ and rearranging, we obtain

$$\frac{1}{\sin\theta}\frac{d}{d\theta}\left(\sin\theta\frac{dP}{d\theta}\right) + \left[l(l+1) - \frac{m^2}{\sin^2\theta}\right]P = 0 \tag{9.67}$$

This is the *Polar equation*. It is a special form of differential equation known as the *Associated Legendre Equation*. This is often written in the following form, which we can get by changing variables from $\theta$ to $x$.
Let $x = \cos\theta$, which implies $\sin\theta = \sqrt{1-x^2}$, and hence

$$(1-x^2)\frac{d^2P(x)}{dx^2} - 2x\frac{dP(x)}{dx} + \left[l(l+1) - \frac{m^2}{1-x^2}\right]P(x) = 0 \tag{9.68}$$

When $m = 0$, the Associated Legendre Equation reduces to the *Legendre Equation* given by

$$(1-x^2)\frac{d^2P(x)}{dx^2} - 2x\frac{dP(x)}{dx} + l(l+1)P(x) = 0 \tag{9.69}$$

The solution of the Associated Legendre Equation are the Associated Legendre Polynomials, $P_{lm}(\cos\theta)$.

## 9.4.2    Some consequences of the quantum theory of the hydrogen atom

### 1. Quantum numbers $n, l, m_l, m_s$

Principal quantum number, $n = 1, 2, 3, 4, \cdots$
Angular momentum quantum number, $l = 0, 2, \cdots, (n-1)$
Magnetic quantum number, $m_l = -l, -l+1, \cdots, -1, 0, 1, \cdots, l$
Spin quantum number, $m_s = +\frac{1}{2}, -\frac{1}{2}$

**2. Pauli's Exclusion Principle**: states that, in an atom or molecule, no two electrons can have the same set of the four quantum numbers, $n, l, m_l, m_s$.

As a result, we have in an atom:

$$
\begin{aligned}
n = 1, &\quad \text{K shell has 2 electrons,} &\Rightarrow 2(1^2) = 2n^2 \\
n = 2, &\quad \text{L shell has 8 electrons,} &\Rightarrow 2(2^2) = 2n^2 \\
n = 3, &\quad \text{M shell has 18 electrons,} &\Rightarrow 2(3^2) = 2n^2 \\
n = 4, &\quad \text{N shell has 32 electrons,} &\Rightarrow 2(4^2) = 2n^2
\end{aligned}
$$

| $n$ | $l$ | $m_l$ | Orbital | Elements | Shell |
|---|---|---|---|---|---|
| $n=1$ | 0 | 0 | $1s$ | 2 } 2 | $K$ |
| $n=2$ | 0 | 0 | $2s$ | 2 } 8 | $L$ |
|  | 1 | -1, 0, 1 | $2p$ | 6 |  |
| $n=3$ | 0 | 0 | $3s$ | 2 } 18 | $M$ |
|  | 1 | -1, 0, 1 | $3p$ | 6 |  |
|  | 2 | -2, -1, 0, 1, 2 | $3d$ | 10 |  |
| $n=4$ | 0 | 0 | $4s$ | 2 } 32 | $N$ |
|  | 1 | -1, 0, 1 | $4p$ | 6 |  |
|  | 2 | -2, -1, 0, 1, 2 | $4d$ | 10 |  |
|  | 3 | -3, -2, -1, 0, 1, 2, 3 | $4f$ | 14 |  |

Figure 9.11: Quantum Numbers

## 3. Orbitals s, p, d and f

The orbitals s, p, d and f refer states with angular momentum $l = 0, 1, 2, 3$ respectively, and are illustrated in Figure 9.12. The orbitals are generated by the $P(\theta)$ and $F(\phi)$ solutions of the Schrödinger equation in terms of the spherical harmonics $Y_{lm}$

$$Y_{lm}(\theta, \phi) = P_l^m(\cos\theta)F(m\phi) \tag{9.70}$$

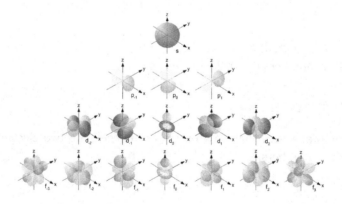

Figure 9.12: Orbitals: s, p, d and f

## 4. Periodic Table of the elements

It has been seen that in each atom there is the principal quantum number $n$ which can have a maximum number of $2n^2$ electrons. As the $n^{th}$ shell is filled by different atoms of different atomic numbers, a table of chemical elements knwn asthe *Periodic Table* is generated as illustrated in Figure 9.13.

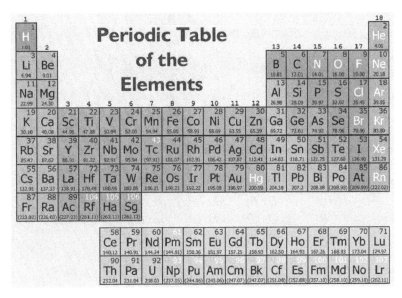

Figure 9.13: The periodic table of elements

### 9.4.3  Quantised Angular Momentum

In discussing the angular momentum, is convenient to use spherical coordinates $(r, \theta, \phi)$ which are related to the rectangular Cartesian coordinates $(x, y, z)$ by the following equations as as discussed earlier.

$$
\begin{aligned}
x &= r \sin\theta \cos\phi \\
y &= r \sin\theta \sin\phi \\
x &= r \cos\theta
\end{aligned}
$$

In *classical mechanics*, angular momentum $\mathbf{L}$ is defined as the cross product of the position vector $\mathbf{r}$ and the linear momentum $\mathbf{p}$, that is, angular momentum is the moment of linear momentum.

$$\mathbf{L} = \mathbf{r} \times \mathbf{p} \tag{9.71}$$

which has components

$$
\begin{aligned}
L_x &= y p_z - z p_y \\
L_y &= z p_x - x p_z \\
L_z &= x p_y - y p_x
\end{aligned}
$$

In *quantum mechanics*, observables are represented by operators, that is

$$\hat{x} \ \rightarrow \ x$$
$$\hat{p}_i \ \rightarrow \ -i\hbar \frac{\partial}{\partial x}$$

Thus the angular momentum components in quantum physics are

$$L_x \ = \ \hat{y}\hat{p}_z - \hat{z}\hat{p}_y \rightarrow -i\hbar \left( y\frac{\partial}{\partial z} - z\frac{\partial}{\partial y} \right) \tag{9.72}$$

$$L_y \ = \ \hat{z}\hat{p}_x - \hat{x}\hat{p}_z \rightarrow -i\hbar \left( z\frac{\partial}{\partial x} - x\frac{\partial}{\partial z} \right) \tag{9.73}$$

$$L_z \ = \ \hat{x}\hat{p}_y - \hat{y}\hat{p}_x \rightarrow -i\hbar \left( x\frac{\partial}{\partial y} - y\frac{\partial}{\partial x} \right) \tag{9.74}$$

Now, from he above equations, the following identities follow

$$\frac{\partial}{\partial x} \ = \ \frac{\partial r}{\partial x}\frac{\partial}{\partial r} + \frac{\partial \theta}{\partial x}\frac{\partial}{\partial \theta} + \frac{\partial \phi}{\partial x}\frac{\partial}{\partial \phi} \tag{9.75}$$

$$\frac{\partial}{\partial y} \ = \ \frac{\partial r}{\partial y}\frac{\partial}{\partial r} + \frac{\partial \theta}{\partial y}\frac{\partial}{\partial \theta} + \frac{\partial \phi}{\partial y}\frac{\partial}{\partial \phi} \tag{9.76}$$

$$\frac{\partial}{\partial z} \ = \ \frac{\partial r}{\partial z}\frac{\partial}{\partial r} + \frac{\partial \theta}{\partial z}\frac{\partial}{\partial \theta} + \frac{\partial \phi}{\partial z}\frac{\partial}{\partial \phi} \tag{9.77}$$

Using the above equations, the angular momentum operators are obtained as

$$L_x \ = \ i\hbar \left( \sin\phi \frac{\partial}{\partial \phi} + \cot\theta \cos\phi \frac{\partial}{\partial \phi} \right) \tag{9.78}$$

$$L_y \ = \ i\hbar \left( -\cos\phi \frac{\partial}{\partial \phi} + \cot\theta \sin\phi \frac{\partial}{\partial \phi} \right) \tag{9.79}$$

$$L_z \ = \ -i\hbar \frac{\partial}{\partial \phi} \tag{9.80}$$

From the expressions of the operators, $L_x, L_y, L_z$ it can be proved that,

$$[L_x, L_y] \ = \ i\hbar L_z \tag{9.81}$$
$$[L_y, L_z] \ = \ i\hbar L_x \tag{9.82}$$
$$[L_z, L_x] \ = \ i\hbar L_y \tag{9.83}$$

The implication of these results is that since the components do not commute with each other, the eigenstate of one component is not an eigenstate of the other components. It is thus not possible to have an exact knowledge of more than one component due to the mutual disturbances of the observations.

## The $z$-component of Angular Momentum

As was shown earlier, the $z$-component of angular momentum is given by

$$L_z = -i\hbar \frac{\partial}{\partial \phi} \tag{9.84}$$

The possible values of the the $z$-component of angular momentum are given by eigenvalue equation

$$-i\hbar \frac{\partial}{\partial \phi} u_m(\phi) = m\hbar u_m(\phi) \tag{9.85}$$

which has a solution

$$u_m(\phi) = e^{im\phi} \tag{9.86}$$

where

$$m = 0, \pm 1, \pm 2, \cdots \tag{9.87}$$

Note that the result that the eigenvalues of $L_z$ are $m\hbar$ is just one of Bohr's assumptions which was imposed on a purely adhoc basis so as to acount for the discrete energy levels of the hydrogen atom.

## The Total Angular Momentum Squared: Eigenfunctions and Eigenvalues

The total angular momentum $\mathbf{L}$ can be found from the components $L_x, L_y, L_z$, and hence the total angular momentum squared is given by

$$\begin{aligned}
\mathbf{L}^2 &= L_x^2 + L_y^2 + L_z^2 \\
&\rightarrow -\hbar^2 \left[ \frac{1}{\sin\theta} \frac{\partial}{\partial\theta} \left( \sin\theta \frac{\partial}{\partial\theta} \right) + \frac{1}{\sin^2\theta} \frac{\partial^2}{\partial\phi^2} \right]
\end{aligned}$$

Before getting the eigenfunctions for $\mathbf{L}^2$, the following observation is important. It can be checked that

$$[L^2, L_z] = 0 \tag{9.88}$$

Since the above two operators are commuting, they must share the same eigenfunction, $Y_{lm}(\theta, \phi)$, that is

$$\begin{aligned}
L_z Y_{lm}(\theta, \phi) &= m\hbar Y_{lm}(\theta, \phi) \\
\mathbf{L}^2 Y_{lm}(\theta, \phi) &= \hbar^2 l(l+1) Y_{lm}(\theta, \phi)
\end{aligned} \tag{9.89}$$

where the possible values of the the angular momentum squared, $\mathbf{L}^2$ can be found by solving the above eigenvalue equation.

Since $L_z$ depends on $\phi$ only, it is possible to write

$$Y_{lm}(\theta, \phi) = P_{lm}(\theta)^{im\phi} \tag{9.90}$$

Thus, we obtain

$$\left[ \frac{1}{\sin\theta} \frac{\partial}{\partial\theta} \left( \sin\theta \frac{\partial}{\partial\theta} \right) - \frac{m^2}{\sin^2\theta} \frac{\partial^2}{\partial\phi^2} \right] P_{lm}(\theta) = -l P_{lm}(\theta) \tag{9.91}$$

Let

$$x = \cos\theta \Rightarrow \frac{1}{\sin\theta} \frac{d}{d\theta} = -\frac{d}{dx} \tag{9.92}$$

and hence,

$$\frac{d}{dx}(1 - x^2)\frac{dP}{dx} + \left[ l(l+1) - \frac{m^2}{1-x^2} \right] P = 0$$

$$\frac{d^2 P}{dx^2} - \frac{2x}{1-x^2} \frac{dP}{dx} + \left[ \frac{l(l+1)}{1-x^2} - \frac{m^2}{1-x^2} \right] P = 0$$

which has singularities at $x = \pm 1$.

### 9.4.4   Spin of an electron

It is well known that an electron has an electronic charge. In addition to an electronic charge, an electron has an intrinsic quantum mechanical property of spin. The evidence of spin was provided by the Stern-Gerlach Experiment of 1921, illustrated in Figure 9.14. Ag atoms were evaporised in an oven and passed through a collimator, after which a beam is passed through a non-uniform magnetic field. The beam splits in two opposite directions, with spin up and spin down. Spin is an intrinsic property, illustrated in Figure 9.15.

The current electronics devices have largely based on the property of an electronic charge. Some electronic devices of the future will be based on the property of "Spin", which forms a new branch of electronics known as *Spintronics*. An example of such devices is shown in Figure 9.16.

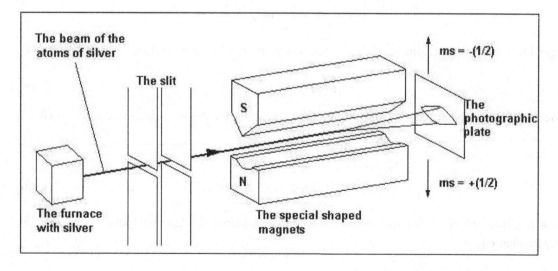

Figure 9.14: the Stern-Gerlach Experiment. On the photographic plate, there are two clear tracks.

Figure 9.15: Spin up and spin down of an electron

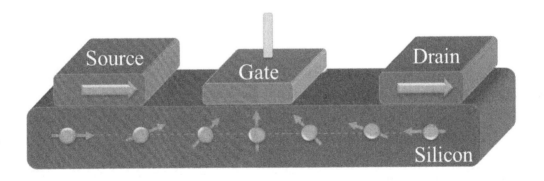

Figure 9.16: Spintronics based devices

## 9.5   Exercises

**9.1.**(a) (i) Explain what is meant by a central force.
(ii) Is the force between the nucleus and the $\alpha$-particle in the Rutherford scattering experiment a central force? Explain.

(b) Show that the motion of $\alpha$-particle in the Rutherford scattering experiment satisfies a second order differential equation of the form

$$\frac{d^2u}{d\theta^2} + u = \frac{k}{mh^2}$$

where $u = 1/r$ and other symbols are in the usual notation.

(c) Give a solution to the second order differential equation in (b) and show that it leads to an equation for the $\alpha$-particle in the Rutherford scattering experiment with a hyperbolic orbit of the form

$$r = \frac{mh^2}{k(1 + e\cos\theta)}$$

where $e > 1$ is the eccentricity for a hyperbola.

**9.2.** (a) Describe the Rutherford scattering experiment and explain the conclusions with respect to the structure of the atom.
(b) (i) An $\alpha$-particle of mass $m$, charge $Z_\alpha e$ and an initial velocity $v_0$ is scatterred by a nucleus with charge $Ze$ and impact parameter $b$. Show that

$$b = \frac{Z_\alpha Z_2 e^2}{4\pi\epsilon_0 m v_0^2} \cot\frac{\theta}{2}$$

(ii) Calculate $b$ given that $m = 6.64 \times 10^{-27}$, $v_0 = 2.0 \times 10^7$ m/s, $Z_\alpha = 2$, $Z = 79$ for gold and $e = 1.6 \times 10^{-19}$.

**9.3.** (a) State an expression for the differential scattering cross section, $\frac{d\sigma}{d\Omega}$
(b) Show that the differential scattering cross section is given by

$$\frac{d\sigma}{d\Omega} = \left(\frac{Z_1 Z_2 e^2}{8\pi\epsilon_0 m v_0^2}\right)^2 \frac{1}{\sin^4(\frac{\theta}{2})}$$

(c) (i) Plot schematically a graph of $\frac{d\sigma}{d\Omega}$ vs $\theta$.
(ii) Calculate $\frac{d\sigma}{d\Omega}$ given that $m = 6.64 \times 10^{-27}$, $v_0 = 2.0 \times 10^7$ m/s, $Z_\alpha = 2$, $Z = 79$ for gold and $e = 1.6 \times 10^{-19}$ at $\theta = \pi/4$.

**9.4.** Using Bohr's theory of the hydrogen atom, calculate the following:
(i) the radius of an electron in the second Bohr orbit?
(ii) the velocity of the electron in the second orbit.

(iii) the frequency of the photon emitted when an electron makes a transition from the energy level $n = 2$ to the energy level $n = 1$, assuming that the energy for the $n^{th}$ level is given by

$$E_n = -\frac{13.6}{n^2} \text{ eV}$$

**9.5.** Calculate the wavelengths and frequencies of the least energetic and the most energetic spectral lines in the Lyman. Balmer, Paschen, Bracket and Pfund series of the hydrogen spectrum. In what regions of electromagnetic spectrum do these lines lie?

**9.6.** (a) Show that the Bohr radius $(= 4\pi\epsilon_0\hbar^2/m_e e^2)$, the Compton wavelength $(h/m_e c)$ divided by $2\pi$, and the classical electron radius $(= e^2/4\pi\epsilon_0 m_e c^2)$ are in the ratio $1 : \alpha : \alpha^2$, where $\alpha = e^2/4\pi\epsilon_0\hbar c$. The quantity $\alpha$ is known as the *fine-structure constant*.

(b) Calculate the value of $\alpha$ and $1/\alpha$.

**9.7.** An electron moving around the nucleus in a hydrogen atom can be reducd to a one body problem described by the Shrodinger equation of the form

$$\left[ -\frac{\hbar^2}{2\mu}\nabla_r^2 + V(r) \right] \Psi = E\Psi$$

where $\mu$ is the reduced mass.

(a) Write down the Schrödinger Equation for the elctron in a hydrogen atom using spherical coordinates. where

$$\nabla^2 = \frac{\partial^2}{\partial r^2} + \frac{2}{r}\frac{\partial}{\partial r} + \frac{1}{r^2 \sin\theta}\frac{\partial}{\partial\theta}\left(\sin\theta\frac{\partial}{\partial\theta}\right) + \frac{1}{r^2\sin^2\theta}\frac{\partial^2}{\partial\phi^2}$$

(b) Assuming the method of separation of variables, where the wave function can be written as

$$\Psi(r, \theta, \phi) = R(r)P(\theta)F(\phi) = RPF$$

Show that the functions $F, R$ and $P$ satisfy

(i) The Azimuthal equation

$$\frac{d^2 F}{d\phi^2} + m^2 F = 0$$

(ii) The Radial Equation

$$\frac{d^2 R}{dr^2} + \frac{2}{r}\frac{dR}{dr} + \left[ \frac{2\mu}{\hbar^2}\left(E + \frac{Ze^2}{4\pi\epsilon_0 r}\right) - \frac{l(l+1)}{r^2} \right] R = 0$$

(iii) The Polar equation

$$\frac{1}{\sin\theta}\frac{d}{d\theta}\left(\sin\theta\frac{dP}{d\theta}\right) + \left[ l(l+1) - \frac{m^2}{\sin^2\theta} \right] P = 0$$

**9.8**. (a) (i) Explain the meaning of each of the four quantum numbers $n, l, m_l, m_s$.

(ii) State Pauli's Exclusion Principle.

(b) Show that the function

$$F_m(\phi) = \frac{1}{\sqrt{2\pi}} e^{im\phi}$$

is a solution of the Azimuthal equation

$$\frac{d^2 F}{d\phi^2} + m^2 F = 0$$

**9.9**. The Radial Equation is given by

$$\frac{d^2 R}{dr^2} + \frac{2}{r}\frac{dR}{dr} + \left[\frac{2\mu}{\hbar^2}\left(E + \frac{Ze^2}{4\pi\epsilon_0 r}\right) - \frac{l(l+1)}{r^2}\right] R = 0$$

can be transformed into a Laguerre equation, whose solution is the normalised radial wave function is given by

$$R_{nl}(r) = -\left\{\frac{(n-l-1)!}{2n\left[(n+l)!\right]^3}\right\}^{1/2} e^{-Zr/na_0} \left(\frac{2Zr}{na_0}\right)^l L_{n+l}^{2l+1}\left(\frac{2Zr}{na_o}\right)$$

where $L_{n+l}^{2l+1}\left(\frac{2Zr}{na_o}\right)$ are the *generalised Laguerre polynomials*.

From tables of functions, you are given the following Generalised Laguerre Polynomials, write down the expressions for Hydrogen Radial Wavefunctions $R_{10}, R_{21}, R_{20}, R_{32}, R_{31}, R_{30}$

$$
\begin{aligned}
L_1^1(x) &= -1 \\
L_2^1(x) &= 2x - 4 \\
L_2^2(x) &= 2 \\
L_3^1(x) &= -3x62 + 18x - 18 \\
L_3^2(x) &= -6x + 18 \\
L_3^3(x) &= -6
\end{aligned}
$$

**9.10**. By defining $x = \cos\theta$, show that the Polar equation

$$\frac{1}{\sin\theta}\frac{d}{d\theta}\left(\sin\theta\frac{dP}{d\theta}\right) + \left[l(l+1) - \frac{m^2}{\sin^2\theta}\right] P = 0$$

can be written in the form

$$(1 - x^2)\frac{d^2 P(x)}{dx^2} - 2x\frac{dP(x)}{dx} + \left[l(l+1) - \frac{m^2}{1 - x^2}\right] P(x) = 0$$

which is the *Associated Legendre Equation*, whose solutions are Assocated Legendre Polynomials, $P_l^m(x)$.

(a) From tables of functions, write down the first few functions $P_0^0(x), P_1^0(x), P_2^0(x), P_2^1(x), P_2^2(x)$.

(b) Hence, write the same functions in terms of $P_l^m(\cos\theta)$, which are,
$P_0^0(\cos\theta), P_1^0(\cos\theta), P_2^0(\cos\theta), P_2^1(\cos\theta), P_2^2(\cos\theta)$.

# Chapter 10

# Introduction to Materials Science I: Crystal Structure

## 10.1 Classification of crystals in terms of symmetry

**Bravais lattices and Crystal systems**

Crystals are 3-dimensional periodic array of atoms. A few solids show low-dimensional behaviour. Crystalline solids show long range order as opposed to amorphous solids which have short range order. Experimental evidence exists to justify the crystalline model, for example X-ray diffraction, electron diffraction etc.

There are over 20000 crystalline substances in the world and all exhibit structural periodicity, and one can use symmetry to classify these substances. Bravais, a Russian mathematician, showed in 1848 that in 3d space there are only 14 different types of lattices known as BRAVAIS LATTICES. This can be shown by using Group Theory, a branch of mathematics concerned with symmetries of systems under operations of translation, rotation and iversions. A change in symmetry may result in dramatic changes in the physical properties of a crystal.

**Unit cells and Fundamental vectors**

Consider a unit cell of a crystal axes $\mathbf{a}, \mathbf{b}$ and $\mathbf{c}$ or $\mathbf{a_1}, \mathbf{a_2}$ and $\mathbf{a_3}$, depending on the notation used. The fundamental vectors are illustrated in Figure 10.1. A Bravais lattice consists of all points with position vectors $\mathbf{R}$ such that

$$\mathbf{R} = n_1\mathbf{a_1} + n_2\mathbf{a_2} + n_3\mathbf{a_3} \qquad (10.1)$$

or

$$\mathbf{R} = n_1\mathbf{a} + n_2\mathbf{b} + n_3\mathbf{c} \qquad (10.2)$$

where $n_1, n_2. n_3$ are integers.

The 14 Bravais lattices are grouped into 7 crystal systems as can be seen in Figure 10.2.

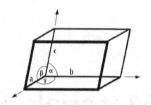

Figure 10.1: Unit cell and fundamental vectors

| Bravais lattice | Parameters | Simple (P) | Volume centered (I) | Base centered (C) | Face centered (F) |
|---|---|---|---|---|---|
| Triclinic | $a_1 \neq a_2 \neq a_3$ $\alpha_{12} \neq \alpha_{23} \neq \alpha_{31}$ | | | | |
| Monoclinic | $a_1 \neq a_2 \neq a_3$ $\alpha_{23} = \alpha_{31} = 90°$ $\alpha_{12} \neq 90°$ | | | | |
| Orthorhombic | $a_1 \neq a_2 \neq a_3$ $\alpha_{12} = \alpha_{23} = \alpha_{31} = 90°$ | | | | |
| Tetragonal | $a_1 = a_2 \neq a_3$ $\alpha_{12} = \alpha_{23} = \alpha_{31} = 90°$ | | | | |
| Trigonal | $a_1 = a_2 = a_3$ $\alpha_{12} = \alpha_{23} = \alpha_{31} < 120°$ | | | | |
| Cubic | $a_1 = a_2 = a_3$ $\alpha_{12} = \alpha_{23} = \alpha_{31} = 90°$ | | | | |
| Hexagonal | $a_1 = a_2 \neq a_3$ $\alpha_{12} = 120°$ $\alpha_{23} = \alpha_{31} = 90°$ | | | | |

Figure 10.2: Crystal Structure:  The 14 Bravais Lattices in 3d

**Packing fraction, $f$**

The packing fraction is the ratio of the volume occupied by atoms to the volume of the unit cell of the lattice structure. It is defined as

$$f = \frac{V_{atoms}}{V_{unit\ cell}} \qquad (10.3)$$

$$f = \frac{(\text{Number of atoms per unit cell}) \times (\text{Volume of an atom})}{V_{unit\ cell}} \qquad (10.4)$$

SOME EXAMPLES OF CALCULATIONS FOR PACKING FRACTION, $f$

Let us illustrate the calculation of the packing fraction, $f$, of the cubic system.

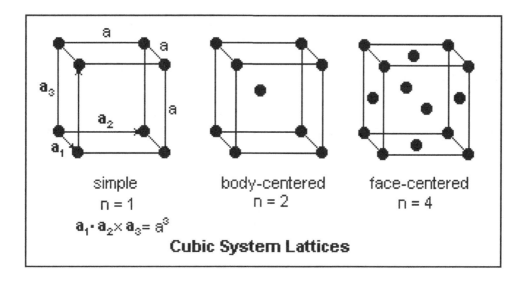

Figure 10.3: Cubic system showing sc, bcc and fcc

**1. Simple cubic (sc)**

In a simple cubic structure, there are eight atoms centred at the corners, each of radius $r$, in a cube of side $a$, and because of close packing, $r = a/2$. There is one eighth of an atom at each corner (See Figures 10.3 and 10.4).

$$\text{Radius of an atom, } r = \frac{a}{2}$$

$$\text{Number of atoms per unit cell} = \frac{1}{8} \text{ of an atom at each corner} \times 8 \text{ corners} = 1$$

$$\text{Packing fraction for sc, } f_{sc} = \frac{(\text{Number of atoms per unit cell}) \times (\text{Volume of an atom})}{V_{unit\ cell}}$$

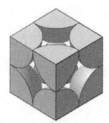

Figure 10.4: Close packing in a simple cubic structure.

$$
\begin{aligned}
&= \frac{1 \times \frac{4}{3}\pi \left(\frac{a}{2}\right)^3}{a^3} \\
&= \frac{4}{3} \frac{\pi a^3}{\times 8a^3} \\
&= \frac{\pi}{6} \\
f_{sc} &= 0.52
\end{aligned}
$$

## 2. Body centred cubic (bcc)

In a body centred cubic structure, there are eight atoms at the corner, each of radius $r$, centred at the corners of a cube of side $a$, and because of close packing (see Figure 10.5), the body diagonal must satisfy

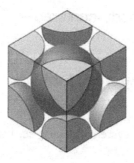

Figure 10.5: Close packing in a bod centred cubic, bcc.

$$
\begin{aligned}
\text{Body diagonal} &= \sqrt{a^2 + a^2 + a^2} = \sqrt{3a^2} \\
&= r + 2r + r = 4r \\
\text{Radius of an atom, } r &= \frac{1}{4}\sqrt{3a^2} = \frac{a}{4}\sqrt{3} \\
\text{Number of atoms per unit cell} &= 1 \text{ atom from the centre} + \frac{1}{8} \text{ of an atom at each corner} \times 8 \text{ corners} = 2 \\
\text{Packing fraction for bcc, } f_{bcc} &= \frac{(\text{Number of atoms per unit cell}) \times (\text{Volume of an atom})}{V_{unit\ cell}}
\end{aligned}
$$

$$= \frac{2 \times \frac{4}{3}\pi \left(\frac{a}{4}\sqrt{3}\right)^3}{a^3}$$

$$= \frac{\sqrt{3}\pi}{8}$$

$$f_{bcc} = 0.68$$

### 3. Face centred cubic (fcc)

In a body centred cubic structure, there are eight atoms at the corners, each of radius $r$, centred at the corners of a cube of side $a$, and half atom from each of the six faces, hence 3 atoms from the faces. Because of close packing (see figure 10.6), the body diagonal must satisfy

Figure 10.6: Close packing in a face centred cubic, fcc.

$$\text{Face diagonal} = \sqrt{a^2 + a^2} = \sqrt{2a^2}$$

$$= r + 2r + r = 4r$$

$$\text{Radius of an atom, } r = \frac{a}{4}\sqrt{2}$$

$$\text{Number of atoms per unit cell} = \text{3 atoms from each face} + \frac{1}{8} \text{ of an atom at each corner} \times \text{8 corners} = 4$$

$$\text{Packing fraction for fcc, } f_{fcc} = \frac{(\text{Number of atoms per unit cell}) \times (\text{Volume of an atom})}{V_{unit\ cell}}$$

$$= \frac{4 \times \frac{4}{3}\pi \left(\frac{a}{4}\sqrt{2}\right)^3}{a^3}$$

$$= \frac{\sqrt{2}\pi}{6}$$

$$f_{fcc} = 0.74$$

### 4. Hexagonal close packed (hcp)

In a hexagonal close packed (hcp) structure, the atoms are arranged in Figure 10.7. Although both the hcp and fcc structures are close pacing of spheres, they differ in the stacking sequence, while the hcp structure is ABABAB squence, the fcc is ABCABC sequence, as illustrated in Figure 10.8. The packing fraction of the hcp structure is 0.74. Prove this as asked in Exercise 10.3.

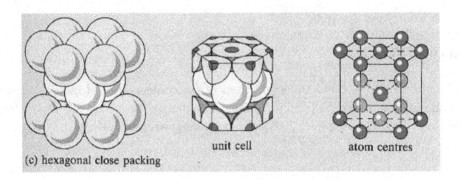

Figure 10.7: Hexagonal Close Packed (hcp) structure.

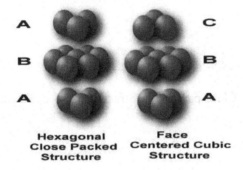

Figure 10.8: hcp and fcc stacking.

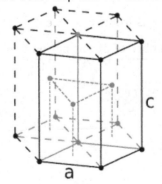

Figure 10.9: Hexagonal close packed (hcp) structure

Examples of crystals with a hcp structure are shown in Figure 10.10.

## Hexagonal Close Packed (HCP) Lattice

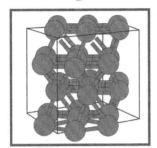 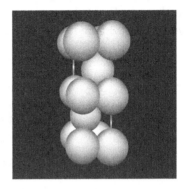

*Hexagonal Close Packed
and related structures*

| Crystal | c/a | Crystal | c/a | Crystal | c/a |
|---------|-------|---------|-------|---------|-------|
| He | 1.633 | Zu | 1.861 | Zr | 1.594 |
| Be | 1.581 | Cd | 1.886 | Gd | 1.592 |
| Mg | 1.623 | Co | 1.622 | Lu | 1.586 |
| Ti | 1.586 | Y | 1.570 | | |

Figure 10.10: Examples of crystals with hcp structure.

## COORDINATION NUMBER, $z$

The coordination number is the total number of nearest neighbouring atoms to a given atom in a lattice. In Figure 10.11, the coordination numbers for the simple cubic, body centred cubic and face centred cubic are shown to be 6, 8 and 12 respectively.

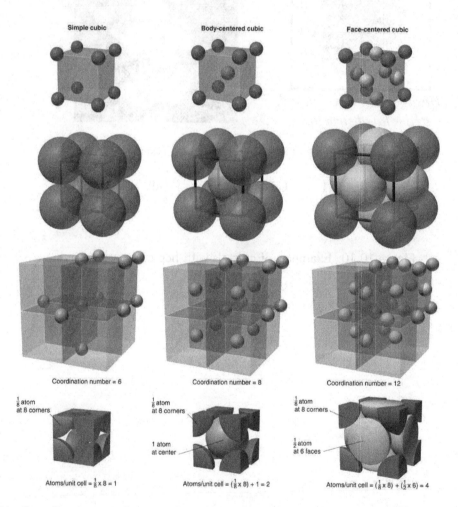

Figure 10.11: Coordination number, z, of sc, bcc and fcc (Source: www.physics.smu.edu/Randal J. Scaliness)

## PRIMITIVE CELLS AND PRIMITIVE VECTORS

### 1. Simple cubic

The primitive vectors of a simple cubic lattice are illustrated in Figure 10.12, given by

$$
\begin{aligned}
\mathbf{a}' &= a\hat{\mathbf{x}} \\
\mathbf{b}' &= a\hat{\mathbf{y}} \\
\mathbf{c}' &= a\hat{\mathbf{z}}
\end{aligned}
$$

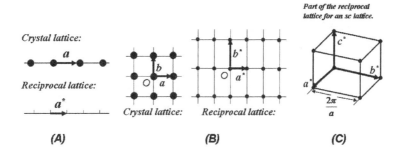

Figure 10.12: Primitive vectors for 1d, 2d and sc.

### 2. Body centred cubic

The primitive vectors of a body centred cubic lattice are illustrated in Figure 10.13.

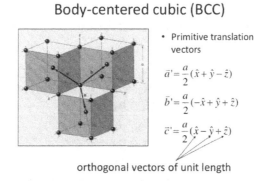

Body-centered cubic (BCC)

- Primitive translation vectors

$$\bar{a}' = \frac{a}{2}(\hat{x} + \hat{y} - \hat{z})$$

$$\bar{b}' = \frac{a}{2}(-\hat{x} + \hat{y} + \hat{z})$$

$$\bar{c}' = \frac{a}{2}(\hat{x} - \hat{y} + \hat{z})$$

orthogonal vectors of unit length

Figure 10.13: Primitive vectors for a bcc.

### 3. Face centred cubic

The primitive vectors of a face centred cubic lattice are illustrated in Figure10.14.

## Face-centered cubic (FCC)

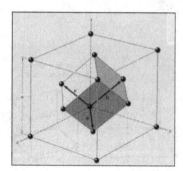

- Primitive translation vectors

$$\vec{a}' = \frac{a}{2}(\hat{x} + \hat{y});$$

$$\vec{b}' = \frac{a}{2}(\hat{y} + \hat{z})$$

$$\vec{c}' = \frac{a}{2}(\hat{z} + \hat{x}).$$

Figure 10.14: Primitive vectors for a fcc.

## RECIPROCAL LATTICE AND DIRECT LATTICE

Every lattice has two important lattices associated with it, namely *the direct lattice* and *the reciprocal lattice*. For example, experimentally, X-ray diffraction maps the reciprocal lattice while a microscope maps the direct lattice.

If $a_1, a_2$ and $a_3$ are a set of primitive vectors in the *direct lattice*, then the primitive vectors $b_1, b_2$ and $b_3$ in the *reciprocal lattice* are given by

$$b_1 = \frac{2\pi a_2 \wedge a_3}{a_1 \cdot a_2 \wedge a_3} \tag{10.5}$$

$$b_2 = \frac{2\pi a_3 \wedge a_1}{a_1 \cdot a_2 \wedge a_3} \tag{10.6}$$

$$b_3 = \frac{2\pi a_1 \wedge a_2}{a_1 \cdot a_2 \wedge a_3} \tag{10.7}$$

## SOME PROPERTIES OF RECIPROCAL AND DIRECT LATTICES

**1.** If $a_j$ is a primitive vector in the direct lattice, and $b_i$ is a primitive vector in the reciprocal lattice, then

$$b_i \cdot a_j = 2\pi \delta_{ij} \tag{10.8}$$

$$\text{where } \delta ij = \begin{cases} 1 \text{ for } i = j \\ 0 \text{ for } i \neq j \end{cases}$$

and $\delta_{ij}$ is known as the KRONECKER-DELTA symbol

**2.** The volume of a primitive cell in the reciprocal lattice is inversely proportional to the volume of the primitive cell in the direct lattice.

$$\text{Volume of the primitive cell in the reciprocal lattice, } V_r = \mathbf{b_1} \cdot \mathbf{b_2} \wedge \mathbf{b_3}$$
$$\text{Volume of the primitive cell in the direct lattice, } V_d = \mathbf{a_1} \cdot \mathbf{a_2} \wedge \mathbf{a_3}$$
$$V_r = \frac{(2\pi)^3}{V_d} \qquad (10.9)$$

which gives the relationship between the volume of a primitive cell in the reciprocal lattice is $V_r$, and the volume of the primitive cell in the direct lattice, $V_d$. Prove this as asked in Exercise 10.4.

**3.** The reciprocal lattices for the sc, bcc and fcc are sc, fcc and bcc respectively, as summarised in the table below.

| Direct lattice | Reciprocal lattice |
|---|---|
| sc | sc |
| bcc | fcc |
| fcc | bcc |

Prove this as asked in Exercise 10.5.

**4.** If $\mathbf{g}$ and $\mathbf{l}$ are a linear combination of vectors in the reciprocal and direct lattices respectively, then

$$e^{i\mathbf{g}\cdot\mathbf{l}} = 1 \qquad (10.10)$$

**Proof**

If $\mathbf{g}$ is a linear combination of vectors in the reciprocal lattice, then

$$\mathbf{g} = g_1\mathbf{b_1} + g_2\mathbf{b_2} + g_3\mathbf{b_3}$$

where $g_1, g_2, g_3$ are integers.

If $\mathbf{l}$ is a linear combination of vectors in the direct lattice, then

$$\mathbf{l} = l_1\mathbf{a_1} + l_2\mathbf{a_2} + l_3\mathbf{a_3} \qquad (10.11)$$

where $l_1, l_2, l_3$ are integers. Hence

$$\begin{aligned} \mathbf{g}\cdot\mathbf{l} &= (g_1\mathbf{b_1} + g_2\mathbf{b_2} + g_3\mathbf{b_3}) \cdot (l_1\mathbf{a_1} + l_2\mathbf{a_2} + l_3\mathbf{a_3}) \\ &= 2\pi\,(g_1l_1 + g_2l_2 + g_3l_3) \\ &= 2\pi N \end{aligned}$$

where $N$ is an integer.
But

$$
\begin{aligned}
e^{i2\pi N} &= \left(e^{i2\pi}\right)^N \\
\cos 2\pi N + i\sin 2\pi N &= (\cos 2\pi + i\sin 2\pi)^N \\
&= 1 \\
\text{Hence } e^{ig\cdot l} &= e^{i2\pi N} = 1 \\
e^{ig\cdot l} &= 1 \text{ , Proved !}
\end{aligned}
$$

**5.** A function $f(\mathbf{r})$ which is periodic with the period of the lattice may be expanded as a sum of Fourier series in the reciprocal lattice vectors $\mathbf{g}$.

$$
f(\mathbf{r}) = \sum_g A(\mathbf{g})e^{i\mathbf{g}\cdot\mathbf{r}} \tag{10.12}
$$

$$
\text{where } A(\mathbf{g}) = \frac{1}{V_{cell}} \int e^{-i\mathbf{g}\cdot\mathbf{r}} f(\mathbf{r})d\mathbf{r} \tag{10.13}
$$

ie. $f(\mathbf{r})$ is a Fourier transform of $A(\mathbf{g})$.

$$
\begin{aligned}
\text{Let } \mathbf{r} &\rightarrow \mathbf{r} + \mathbf{l} \text{ (Periodic)}
\end{aligned}
$$

where $\mathbf{l} = l_1\mathbf{a_1} + l_2\mathbf{a_2} + l_3\mathbf{a_3}$
and $l_1, l_2, l_3$ are integers.

Hence

$$
\begin{aligned}
f(\mathbf{r}+\mathbf{l}) &= \sum_g A(\mathbf{g})e^{i\mathbf{g}\cdot(\mathbf{r}+\mathbf{l})} \\
&= \sum_g A(\mathbf{g})e^{i\mathbf{g}\cdot\mathbf{r}}e^{i\mathbf{g}\cdot\mathbf{l}} \\
&= \sum_g A(\mathbf{g})e^{i\mathbf{g}\cdot\mathbf{r}} \text{ since } e^{i\mathbf{g}\cdot\mathbf{l}} = 1 \\
f(\mathbf{r}+\mathbf{l}) &= f(\mathbf{r})
\end{aligned}
$$

Hence, we have proved that a periodic function can be expanded as a sum of Fourier series.

**Examples of periodic functions in crystals.**

**1. Potential**

The potential, is periodic

$$
V(\mathbf{r}) = V(\mathbf{r}+\mathbf{l}) \tag{10.14}
$$

is illustrated in figure 10.15.

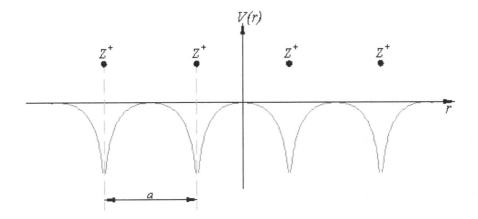

Figure 10.15: Crystal Periodic Potential

## 2. Charge Density

The charge density is periodic

$$\rho(\mathbf{r}) = \rho(\mathbf{r} + \mathbf{l}) \tag{10.15}$$

is illustrated in Figure 10.16.

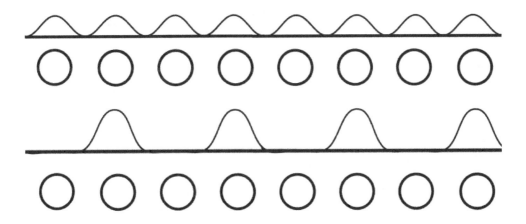

Figure 10.16: Periodic Charge Density

## MILLER INDICES

Miller indices are used to specify lattice planes, and the notation $(hkl)$ is used to denote the orientation of the planes relative to the crystal axes.

### How to find Miller indices (hkl)

(i) Find the intercepts along the $x, y, z$ axes and assign as $(\frac{1}{h}, \frac{1}{k}, \frac{1}{l}$ respectively.
(ii) Take the reciprocals of the intercepts, $(h, k, l)$
(iii) Reduce to integers and write in the form $(hkl)$ as Miller indices.

Applying the above prescription for the the above planes,

| Intercepts | $(1, \infty, \infty)$ | $(1, 1, \infty)$ | $(1, 1, 1)$ |
|---|---|---|---|
| Reciprocals | $(1, 0, 0)$ | $(1, 1, 0)$ | $(1, 1, 1)$ |
| Miller indices | $(100)$ | $(110)$ | $(111)$ |

Hence, the planes in Figure 10.17, have Miller indices (100), (110) and (111), and are referred as the (100), (110) and (111) planes respectively.

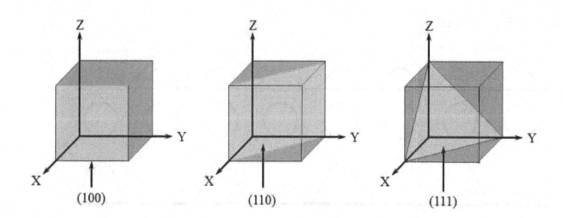

Figure 10.17: Miller indices (100), (110) and (111).

## RECIPROCAL LATTICE AND X-RAY DIFFRACTION

Since X-rays have wavelengths of the same order of magnitude as the crystal lattice spacing, the crystal behaves as a 3d diffraction grating to the X-rays, as illustrated in Figure 10.18. The path difference must equal the integral multiple of a wavelength $\lambda$, and this gives the Bragg condition.

$$2d \sin \theta = n\lambda \qquad (10.16)$$

Consider incident X-rays of frequency $\omega$ and wave vector $\mathbf{k}$ incident on a crystal and diffracted X-rays with wavevector $\mathbf{k}'$.

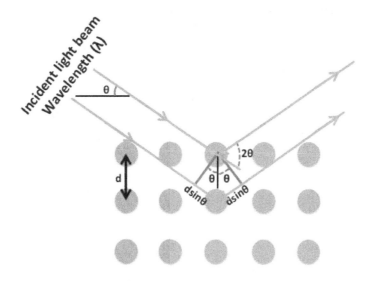

Figure 10.18: Xray Diffraction

It can be shown that $\mathbf{k}' - \mathbf{k}$ is a vector in the reciprocal space, say equal to $\mathbf{G}$. Hence

$$
\begin{aligned}
\mathbf{k}' - \mathbf{k} &= \mathbf{G} \\
\mathbf{k}' &= \mathbf{k} + \mathbf{G} \\
\mathbf{k}' \cdot \mathbf{k}' &= (\mathbf{k} + \mathbf{G}) \cdot (\mathbf{k} + \mathbf{G}) \\
\mathbf{k}'^2 &= \mathbf{k^2} + 2\mathbf{k} \cdot \mathbf{G} + G^2
\end{aligned}
$$

But $\omega = ck$ and $\omega = ck'$, hence $k = k'$, and from above we obtain

$$2\mathbf{k} \cdot \mathbf{G} + G^2 = 0 \tag{10.17}$$

which is also Bragg's condition, which is also satified by the reciprocal lattic vector $-\mathbf{G}$, and hence

$$-2\mathbf{k} \cdot \mathbf{G} + G^2 = 0 \tag{10.18}$$

or

$$2\mathbf{k} \cdot \mathbf{G} = G^2 \tag{10.19}$$

Note that since $\mathbf{k} = k\hat{\mathbf{n}}$ and $\mathbf{G} = \frac{2\pi n}{d}$, then

$$
\begin{aligned}
2\mathbf{k} \cdot \mathbf{G} &= 2kG\sin\theta = G^2 \\
2\frac{2\pi}{\lambda}\sin\theta &= \frac{2\pi n}{d} \\
&\quad\text{or} \\
2d\sin\theta &= n\lambda
\end{aligned}
\tag{10.20}
$$

which recovers the Bragg's condition.

## APPLICATION OF BRAGG'S EQUATION TO A 2D LATTICE

Consider a square lattice, where we have *direct lattice primitive vectors* $\mathbf{a}, \mathbf{b}$ and *reciprocal lattice primitive vectors* $\mathbf{a}^*, \mathbf{b}^*$ given by

$$
\begin{aligned}
\mathbf{a} &= a\mathbf{i} \\
\mathbf{b} &= a\mathbf{j} \\
\mathbf{a}^* &= \frac{2\pi}{a}\mathbf{i} \\
\mathbf{b}^* &= \frac{2\pi}{a}\mathbf{j}
\end{aligned}
$$

which satify

$$
\begin{aligned}
\mathbf{a} \cdot \mathbf{a}^* &= 2\pi \\
\mathbf{b} \cdot \mathbf{b}^* &= 2\pi \\
\mathbf{a} \cdot \mathbf{b}^* &= 0
\end{aligned}
$$

Reciprocal lattice vector $\mathbf{G}$ is a linear combination of primitive vectors $\mathbf{a}^*, \mathbf{b}^*$. Hence, we have

$$
\begin{aligned}
\mathbf{G} &= n_1\mathbf{a}^* + n_2\mathbf{b}^* \\
&= \frac{2\pi}{a}(n_1\mathbf{i} + n_2\mathbf{j}) \\
\mathbf{k} &= (k_x\mathbf{i} + k_y\mathbf{j})
\end{aligned}
$$

Using Bragg equation $2\mathbf{k} \cdot \mathbf{G} + G^2 = 0$ in the square lattice, we obtain

$$
\begin{aligned}
2(k_x\mathbf{i} + k_y\mathbf{j}) \cdot \frac{2\pi}{a}(n_1\mathbf{i} + n_2\mathbf{j}) + \frac{4\pi^2}{a^2}\frac{2\pi}{a}(n_1\mathbf{i} + n_2\mathbf{j}) \cdot \frac{2\pi}{a}(n_1\mathbf{i} + n_2\mathbf{j}) &= 0 \\
\frac{4\pi}{a}(n_1 k_x + n_2 k_y) + \frac{4\pi^2}{a^2}(n_1^2 + n_2^2) &= 0 \\
(n_1 k_x + n_2 k_y) + \frac{\pi}{a}(n_1^2 + n_2^2) &= 0
\end{aligned}
$$

which is the Bragg Equation for a square lattice, giving the condition for diffraction in a square lattice. the integer values for $n_1, n_2$ determines the *Brillouin Zone*, as illustrated below.

$1^{st}$ **Brillouin Zone**

(i) For $n_1 = \pm 1, n_2 = 0 \Rightarrow k_x = n_1\frac{\pi}{a} = \pm\frac{\pi}{a}$
(ii) For $n_1 = 0, n_2 = \pm 1 \Rightarrow k_y = n_2\frac{\pi}{a} = \pm\frac{\pi}{a}$

$2^{nd}$ **Brillouin Zone**

(i) For $n_1 = \pm 1, n_2 = \pm 1$,

$$
\pm k_x \pm k_y + \frac{\pi}{a}(1 + 1) = 0
$$

$$\pm k_y \;=\; \pm k_x - \frac{2\pi}{a}$$

which gives four equations, taking $++,+ \,-,-+,-$ terms

$$k_y \;=\; +k_x - \frac{2\pi}{a}$$
$$k_y \;=\; -k_x - \frac{2\pi}{a}$$
$$k_y \;=\; -k_x + \frac{2\pi}{a}$$
$$k_y \;=\; +k_x + \frac{2\pi}{a}$$

which gives the equations for four lines which bound the second Brillouin zone.

### $3^{rd}$ **Brillouin Zone**

(i) For $n_1 = 0, n_2 = \pm 2$,

$$\pm 2k_y + \frac{4\pi}{a} \;=\; 0$$
$$k_y \;=\; \pm \frac{2\pi}{a}$$
$$k_x \;=\; \pm \frac{2\pi}{a}$$

(ii) For $n_1 = \pm 2, n_2 = 0, \Rightarrow \pm 2k_x + \frac{\pi}{a}4 = 0, or\, k_x = \pm \frac{2\pi}{a}$ as above.

Note that

(i) The $1^{st}$, $2^{nd}$ and $3^{rd}$ Brillouin Zones are illustrated in Figure 10.19.
(ii) Any wave whose wave vector $\mathbf{k}$ drawn from the origin and terminating on the surface of the Brillouin Zone will be diffracted by the crystal.
(iii) Each Brillouin Zone has an area $\left(\frac{2\pi}{a}\right)^2 = \frac{4\pi^2}{a^2}$.

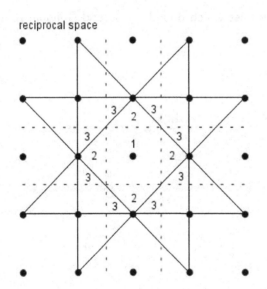

Figure 10.19: Brillouin Zones for a 2D Lattice

## 10.2 Classification of crystals in terms of Binding Mechanisms

Crystalline solids can also be classified on the basis of the interatomic forces binding the atoms together. There are five broad classes.

- Metals

- Covalent crystals

- Ionic crystals

- Molecular crystals

- Hydrogen bonded crystals

### 10.2.1 Metals

When a collection of atoms are brought together to form a metal, the nuclei and inner or core electrons form compact ion core which provide the rigid framework, crystal lattice, through which electrons move freely like particles of a gas. Hence we have an "electron gas", as illustrated in Figure 10.20. The total Hamiltonian of the system is

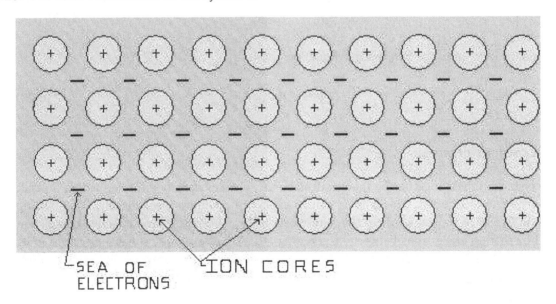

Figure 10.20: Binding Mechanism in Metals

$$H = \sum \frac{\hat{p}_i^2}{2m_i} + \sum \frac{\hat{p}_e^2}{2m_e} + \sum \frac{e^2}{|r_i - r_j|} + \sum_i V_{ii}(R) + \sum U_j(r, R) \qquad (10.21)$$

where
the first term is $KE_{ions}$
the second term is $KE_{electrons}$

the third term is $PE_{ee}$ due electron-electron interaction (Coulomb interaction)

the fourth term is $PE_{ii}$ due ion-ion interaction (Coulomb interaction)

the fifth term is $PE_{ei}$ due electron-ion interaction (Coulomb interaction)

The above Hamiltonian is rather complex, and if we are to make any progress we need to make some *approximations*.

1. *Independent electron approximation*, whereby the $PE_{ee}$ due to electron-electron interaction is considered to be negligible, and hence the third term is negligible.

2. *Free electron approximation*, whereby the $PE_{ei}$ due to electron-ion interaction is considered to be negligible, and hence the fifth term is negligible.

3. *Neglect $KE_{ions}$* since the ions are much heavier compared to electron, hence the first term is negligible.

4. *Neglect $PE_{ii}$* because one ion is shielded from another ion due to electrons in between, hence the fourth term is negligible.

After these approximations, the first, third, fourth and fifth terms are negligible, and only the second term remains, and consider the effect of a single electron

$$H = \frac{\hat{p}_e^2}{2m_e} = -\frac{\hbar^2 \nabla^2}{2m_e} \qquad (10.22)$$

and the problem is to solve the Schrodinger Equation

$$H\Psi = E\Psi \qquad (10.23)$$

## 10.2.2   Covalent crystals

Covalent crystals are held together by the covalent bond. The covalent bond is formed as a result of the sharing of two electrons. This binding can be explained quantum mechanically. This binding is common in Group IV elements, where the outermost shell is incomplete, and effective completion is obtained by sharing of electrons between neigbouring atoms. For example, si has 4 electrons in its outer most shell, the K shell. Four more electrons are required to complete the shell, and this can be achieved by sharing electrons with four neighbours, as illustrated in Figure 10.21.

**Examples of structures of covalent bonding**

**(i) Diamond lattice**

The diamond lattice consists of two interpenetrating fcc lattices displaced by a quarter length along its body diagonal. Examples of the diamond lattice are C, Si, Ge. These are Gp IV elements as can be seen Figure 10.22.

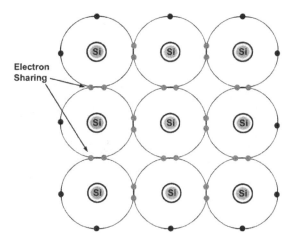

Figure 10.21: Sharing of electrons leading to Covalent Bonding in Si.

| III | IV | V |
|-----|-----|-----|
| B | C | N |
| Al | Si | P |
| Ga | Ge | As |
| In | Sn | Sb |
| Tl | Pb | Bi |

See exercise 10.6 where you are asked to show that the angle between any two covalent bonds in the diamond structure is $109°28'$ and to calculate the packing fraction of the diamond structure.

(b) Calculate the packing fraction of the diamond structure.

### (ii) Zinc blende structure

The Zinc blende lattice consists of two interpenetrating fcc lattices displaced by a quarter length along its body diagonal, with the two fcc lattices consisting of different atoms. Examples of the Zinc blende lattice are shown below, noting hat the midde column is III-V compounds as can be inferred from Figure 10.22.

| Zinc Blende structure | | |
|-----|-----|-----|
| ZnS | GaP | SiC |
| ZnSe | GaAs | CuF |
| ZnTe | GaSb | CuCl |
| CdS | InSb | |

### Low dimensional Materials

There are certain materials which show low dimensional behaviour such as Graphite, where there is strong covalent bonding in a layer of carbon atoms and weak van der Waals bonding between the laters (See Figure 10.23).

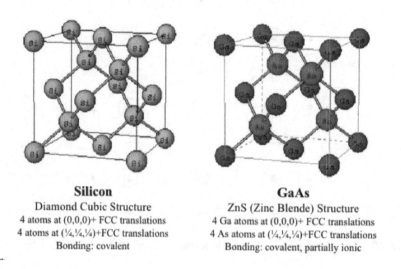

**Silicon**
Diamond Cubic Structure
4 atoms at (0,0,0)+ FCC translations
4 atoms at (¼,¼,¼)+FCC translations
Bonding: covalent

**GaAs**
ZnS (Zinc Blende) Structure
4 Ga atoms at (0,0,0)+ FCC translations
4 As atoms at (¼,¼,¼)+FCC translations
Bonding: covalent, partially ionic

Figure 10.22: Covalent Bonding in Silicon and Zinc Blende structure.

### 10.2.3   Ionic crystals

Ionic crystals are held together by the ionic bond. The ionic bond results from the electrostatic interaction between positive and negative ions. Since the structure does not contain ions of identical size, ionic soilids usually consist of interpenetrating lattices.

**Examples of ionic bonding**

(i) NaCl(2 interpenetrating fcc lattices)
(ii) CsCl (2 interpenetrating sc lattices)
(iii) Perovskite structure (fcc and bcc incorporated together)

**(i) NaCl (2 interpenetrating fcc lattices)**
Examples of the NaCl structure (See Figure 10.24) include most of the alkali halides (I-VII). Some examples are NaCl, PbS, LiH, KCl, KBr, MgO, MnO, AgBr

**(ii) CsCl Structure (2 interpenetrating sc lattices)**
Examples of the CsCl structure are CsCl (see Figure 10.25), CuPd alloy, CuZn alloy, AgMg alloy.

**(iii) Perovskite Structure (fcc and bcc incorporated together)**

Examples of the Perovskite structure are $BaTiO_3, LaCuO_3, SrTiO_3, PbTiO_3$.

$BaTiO_3$ is a ferroelectric crystal (See Figure 10.26). The ions are located as follows.
$Ba^{++}$ ions are at the cube corners
$Ti^{4+}$ ions at the body centre

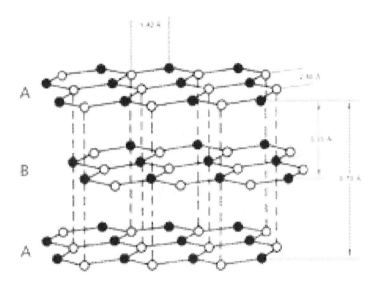

Figure 10.23: Graphite has covalent bonding within a layer and weak Van der Waals bonding between the layers.

$O^{--}$ ions are at the face centres.

Note that the $O^{--}$ form an octahedra with $Ti^{4+}$ at the centre.

### Forces in an Ionic Crystal

There are two major forces between a positive ion and a negative ion in an ionic crystal:

(i) Attractive forces due to Coulomb interaction which is electrostatic in nature.

(ii) Repulsive forces due to overlapping of electronic shellsdue to the Born-Mayer interaction, which is quantum mechanical in nature.

### Potentials in an Ionic Crystal

There are two major potentials in an ionic crystal:

(i) Coulomb potential due to Coulomb interaction of charges at a separation $r_{ij}$.

$$U_{coul} = \pm k \frac{q^2}{r_{ij}} \qquad (10.24)$$

where

$$\lambda = \begin{cases} + \text{ is due to like charges} \\ - \text{ is due to unlike charges} \\ k = \frac{1}{4\pi\epsilon_0} \text{ in SI units} \\ k = 1 \text{ in CGS units} \end{cases} \qquad (10.25)$$

(ii) Born-Mayer potential due to overlapping of the electronic shells of neighbouring charges at a separation, $r_{ij}$.

$$U_{BM} = Ae^{-r_{ij}/b} \qquad (10.26)$$

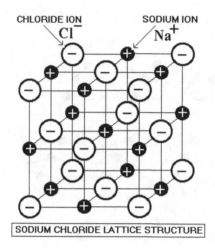

Figure 10.24: Ionic bonding in NaCl structure.

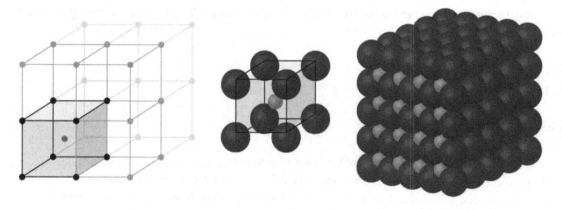

Body-centered simple cubic structure

Figure 10.25: Ionic Bonding in CsCl Structure.

where $A$ and $b$ are constants characteristic of the ionic crystal.

The total potential at an ion $i$ due to another ion at $j$ is the sum of these two contributions, $U_{BM}$ and $U_{Coul}$.

$$U_{ij} = U_{BM} + U_{Coul}$$
$$U_{ij} = Ae^{-r_{ij}/b} + \pm k\frac{q^2}{r_{ij}}$$

The net interaction energy is the sum over interaction energies between ions at $i$ and $j$

$$U_i = \sum_j{}' U_{ij} = \sum_j{}' \left( Ae^{-r_{ij}/b} + \pm k\frac{q^2}{r_{ij}} \right) \tag{10.27}$$

The total potential variation with the interionic separation is illustrated in Figure 10.27.

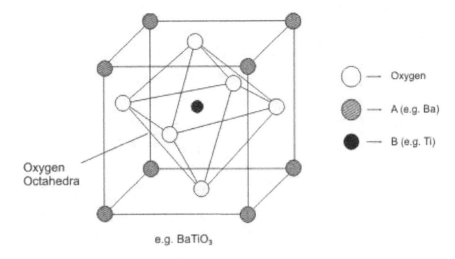

e.g. BaTiO₃

Figure 10.26: Ionic Bonding in Perovskites.

Let us evaluate the sum in a $1 - d$ ionic lattice.

$$U_{ij} = \begin{cases} Ae^{-x/b} - k\dfrac{q^2}{x} & \text{nn interaction only} \\ \pm k\dfrac{q^2}{xp_{ij}} & \text{non nn interaction, +ve for like charges, -ve for unlike charges} \end{cases} \qquad (10.28)$$

where nn denotes "nearest neighbour", and

$$
\begin{aligned}
p_{ij} &= 2, 3, 4, \cdots \text{for non nn} \\
p_{ij} &= 1 \text{for nn} \\
x &= \text{is the equilibrium separation for nn} \\
U_{TOT} &= N U_i \\
&= N \sum_{j}' U_{ij} \\
&= N \sum_{j}' \left( Ae^{-x/b} \pm k\frac{q^2}{xp_{ij}} \right) \\
&= N \left\{ \sum_{j}' Ae^{-x/b} + \sum_{j}' \pm k\frac{q^2}{x_{ij}} \right\}
\end{aligned}
$$

where $x_{ij} = xp_{ij}$

$$
\begin{aligned}
U_{TOT} &= N \left\{ zAe^{-x/b} - (-) \sum_{j}' \pm k\frac{q^2}{x_{ij}} \right\} \\
&= N \left\{ zAe^{-x/b} - \sum_{j}' \pm k\frac{q^2}{x_{ij}} \right\}
\end{aligned}
$$

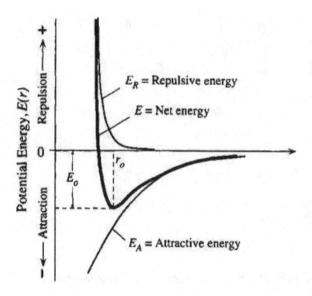

Figure 10.27: Total potential in an Ionic crystal.

$$= N\left\{zAe^{-x/b} - \sum_{j}^{'} \frac{(\pm)q^2}{x_{ij}}\right\}$$

where $z$    is the number of nn

$$U_{TOT} = N\left\{zAe^{-x/b} - \frac{\alpha}{x}kq^2\right\}$$

where $\dfrac{\alpha}{x} = \sum_{j}^{'} \dfrac{(\pm1)}{x_{ij}} = \sum_{j}^{'} \dfrac{(\pm1)}{xp_{ij}}$

and $\alpha$ is known as the MADELUNG CONSTANT.

$$\alpha = \sum_{j}^{'} \frac{(\pm1)}{p_{ij}} \tag{10.29}$$

Evaluating

$$\frac{\alpha}{x} = \sum_{j}^{'} \frac{(\pm1)}{x_{ij}}$$

$$= 2\left\{\frac{1}{x} - \frac{1}{2x} + \frac{1}{3x} - \frac{1}{4x} + \cdots\right\}$$

or

$$\alpha = 2\left\{1 - \frac{1}{2} + \frac{1}{3} - \frac{1}{4} + \cdots\right\}$$

But $\ln(1+x) = x - \dfrac{x^2}{2} + \dfrac{x^3}{3} - \dfrac{x^4}{4} + \cdots$

If $x = 1$

$$\begin{aligned}
\ln 2 &= 1 - \frac{1}{2} + \frac{1}{3} - \frac{1}{4} + \cdots \\
\text{Hence } \alpha &= 2\ln 2 \\
\text{Note that } \ln 2 &= 0.6931 \\
\alpha &= 1.39
\end{aligned}$$

The above analysis has shown that the Madelung Constant, $\alpha$, for a $1-d$ chain of an ionic material comprising alternating positive and negative charges.

The calculation for the Madelung constant for a $3-d$ crystal is very complex, and gives the following results.

| Structure | $\alpha$ |
| --- | --- |
| NaCl | 1.7476 |
| CsCl | 1.7621 |
| ZnS | 1.6381 |

### 10.2.4   Molecular crystals

Molecular crystals are held together by a dipole-dipole interaction, as illustrated in Figure 10.28. The force of interaction between he dipoles is known as the VAN DER WAALS Interaction. Examples of materials which exhibit dipole-dipole interaction are $CO_2$ (soild form), $CH_4$ (solid form),

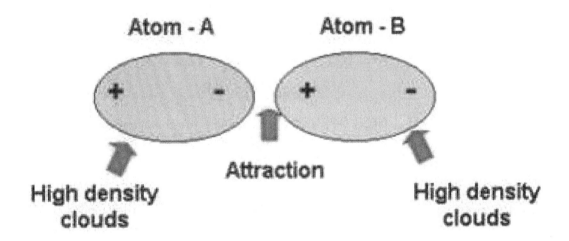

Figure 10.28: Van der Waals interaction is a dipole-dipole interaction between atoms A and B.

Consider two electric dipoles of dipole moments $\mathbf{p_1}$ and $\mathbf{p_2}$ separated by a distance $R$ apart. The interaction energy $U(R)$ is given by

$$U(R) = k \left\{ \frac{\mathbf{p_1} \cdot \mathbf{p_2}}{R^3} - \frac{3(\mathbf{p_1} \cdot \mathbf{R})(\mathbf{p_2} \cdot \mathbf{R})}{R^5} \right\} \qquad (10.30)$$

The above result is set as an exercise to be proved (See problem 10.7).

**Aligned dipoles**: $p_1 \| p_2$

$$
\begin{aligned}
U(R) &= k\left\{\frac{P_1 p_2}{R^3} - \frac{3}{p_1 p_2 R^2}R^5\right\} \\
&= -2k\frac{p_1 p_2}{R^3}
\end{aligned}
$$

If $p_2 = \alpha E$

where $E = \dfrac{2kp_1}{R^3}$

Hence

$$
\begin{aligned}
U(R) &= -\left(\frac{2kp_1}{R^3}\frac{2\alpha k_1 p_1}{R^3}\right) \\
U(R) &= -\frac{C}{R^6}
\end{aligned}
$$

where the negative sign implying an *attractive interaction*, which is identified as $U_{att}$, that is,

$$
U_{att} = -\frac{C}{R^6} \tag{10.31}
$$

There is also a *repulsive interaction*, $U_{rep}$ due to overlapping of electronic shells, which is of the form

$$
U_{rep} = +\frac{B}{R^{12}} \tag{10.32}
$$

Hence, the total interaction, $U_{TOT}$, is given by

$$
\begin{aligned}
U_{TOT}(R) &= U_{rep} + U_{att} \\
&= \frac{B}{R^{12}} - \frac{C}{R^6}
\end{aligned}
$$

which is known as the LENNARD-JONES POTENTIAL, and $B$ and $C$ are characteristic of a particular molecular crystal. The Lennard-Jones potential is illustrated graphically in Figure 10.29.

## 10.2.5   Hydrogen bonded crystals

Hydrogen-bonded crystals are bonded by the Hydrogen bond. The H-bond is electrostatic in character and hence can be understood as a Coulomb interaction between the hydrogen atom and another atom. The hydrogen atom attains an effective positive charge while the other atom attains an effective negative charge.

## Examples of hydrogen bonding
(i) Ice
(ii) DNA (De oxyribo Nucleic Acid)
(iii) KDP (Potassium Dyhydrogen Phosphate, $KH_2PO_3$
(iv) HF (Hydrofluoric Acid)

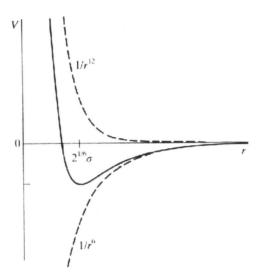

Figure 10.29: Lennard Jones Potential

**Ice**

ice, which is frozen water, is a very open structure where the hydrogen bond is as illustrated in Figure 10.30.

**DNA**

In DNA, there are , A in one strand always pairs with T in the other, while G pairs with C. The nuclei of every cell has at least one DNA molecule to control the production of proteins and carry genetic information from one generation of cells to the next. The DNA molecule is a double helix as illustrated in Figures 10.31 and 10.32.

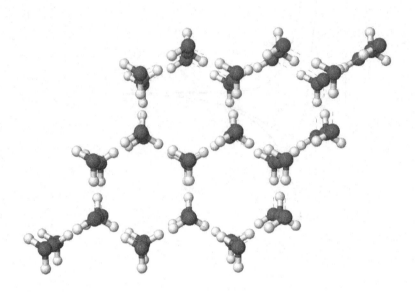

Figure 10.30: Hydrogen Bonding in Ice.

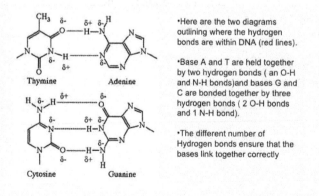

•Here are the two diagrams outlining where the hydrogen bonds are within DNA (red lines).

•Base A and T are held together by two hydrogen bonds ( an O-H and N-H bonds)and bases G and C are bonded together by three hydrogen bonds ( 2 O-H bonds and 1 N-H bond).

•The different number of Hydrogen bonds ensure that the bases link together correctly

Figure 10.31: Hydrogen bonding in DNA

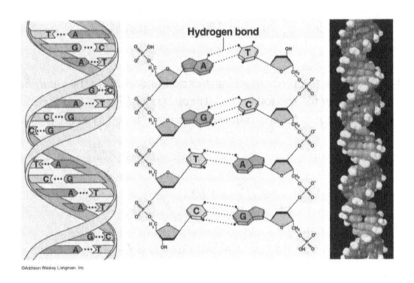

Figure 10.32: Hydrogen Bonding in DNA.

## 10.3   Exercises

**10.1.**(a) Mention the seven crystal systems comprising the 14 Bravais lattices, and for each case define the unit cell in terms of $a_1, a_2, a_3, \alpha, \beta$ and $\gamma$.

(b) Complete the missing information in the dotted lines in the following table.

| Chemical formula | $a_1$ | $a_2$ | $a_3$ | $\alpha$ | $\beta$ | $\gamma$ | Crystal system |
|---|---|---|---|---|---|---|---|
| $Ti_5O_9$ | 5.57 | 7.12 | 8.86 | 97.5° | 112.3° | 108.5° | .......... |
| NaSb | 6.80 | 6.34 | 12.48 | ..... | ...... | 117.7° | monoclinic |
| $CaCl_2$ | 6.24 | 6.43 | 4.20 | 90° | 90° | 90° | .......... |
| $CaC_2$ | .... | 3.87 | 6.37 | 90° | 90° | 90° | tetragonal |
| Ni | 3.52 | 3.52 | 3.52 | 90° | 90° | 90° | .......... |
| $Al_2O_3$ | 5.13 | 5.13 | 5.13 | 55.3° | 55.3° | 55.3° | .......... |
| Zn | 2.66 | .... | 4.95 | 90° | 90° | 120° | hexagonal |

where all the lengths are in Å.

**10.2.**(a) Explain what is meant by the terms *packing fraction* and *coordination number*.
(b) Calculate the packing fractions for the following lattices: (i) sc (ii) bcc (iii) fcc.
(c) Mention the coordination numbers for the following lattices: (i) sc (ii) bcc (iii) fcc.

**10.3.**(a) Show that for an ideal h.c.p

$$\frac{c}{a} = \sqrt{\frac{8}{3}} = 1.633$$

where all symbols are in the usual notation.
(b) Show that the packing fraction of a h.c.p crystal structure is 0.74.

**10.4.**  (a) (i) If $\mathbf{a_1}, \mathbf{a_2}$ and $\mathbf{a_3}$ are primitive vectors for a primitive cell in the *direct lattice*, give expressions for the primitive vectors $\mathbf{b_1}, \mathbf{b_2}$ and $\mathbf{b_3}$ in the *reciprocal lattice*.
(ii) Show that the volume of a primitive cell in the reciprocal lattice, $V_r$, is inversely proportional to the volume of a primitive cell in the direct lattice, $V_d$, in the form

$$V_r = \frac{(2\pi)^3}{V_d}$$

where

$$V_r = \mathbf{b_1} \cdot \mathbf{b_2} \wedge \mathbf{b_3}$$

$$V_d = \mathbf{a_1} \cdot \mathbf{a_2} \wedge \mathbf{a_3}$$

(b) Show that

$$\mathbf{a_1} = \frac{2\pi \mathbf{b_2} \wedge \mathbf{b_3}}{\mathbf{b_1} \cdot \mathbf{b_2} \wedge \mathbf{b_3}}$$

**10.5.**  (a) (i) State the primitive vectors for a s.c structure.

(ii) Show that the reciprocal lattice of a s.c. is a s.c.

(b) (i) State the primitive vectors for a b.c.c structure.
(ii) Show that the reciprocal lattice of a b.c.c. is a f.c.c.

(c) (i) State the primitive vectors for a f.c.c structure.
(ii) Show that the reciprocal lattice of a f.c.c. is a b.c.c.

**10.6**. (a) Show that the angle between any two covalent bonds in the diamond structure is $109°28'$.
(b) Calculate the packing fraction of the diamond structure.

**10.7**. Van der Waals interaction is a dipole-dipole interaction responsible for the binding of molecular crystals. The following model can be used to understand the origin of this interaction. An electric field of a dipole moment $\mathbf{p_1}$ produces an electric field of components $\mathbf{E_r} = E_r \hat{\mathbf{u_r}}$ and $E_\theta = E_\theta \hat{\mathbf{u}}_\theta$ at a distance $R$ away, given by

$$E_r = \frac{2kp_1 \cos\theta}{R^3}$$

$$E_\theta = \frac{kp_1 \sin\theta}{R^3}$$

where $k = 1/4\pi\epsilon_0$ is the electric force constant.
(a) Show that the electric field at $R$ due to the electric dipole $\mathbf{p_1}$ is given by

$$\mathbf{E} = \frac{k}{R^3} \left\{ 3\mathbf{u}_r(\mathbf{p_1} \cdot \mathbf{u}_r) - \mathbf{p_1} \right\}$$

(b) Show that the interaction energy between two electric dipoles of dipole moments $\mathbf{p_1}$ and $\mathbf{p_2}$ at a distance $R$ apart is given by

$$U(R) = k \left\{ \frac{\mathbf{p_1} \cdot \mathbf{p_2}}{R^3} - \frac{3(\mathbf{p_1} \cdot \mathbf{R})(\mathbf{p_2} \cdot \mathbf{R})}{R^5} \right\}$$

**10.8**. (a) Illustrate, diagrammatically, the following planes of a cubic lattice: (100), (110), (111) and (222).
(b) A plane with Miller indices $(hkl)$ has intercepts at $a/h, a/k, a/l$ along the $x, y, z$ axes of a cubic unit cell of cube edge $a$. The spacing $d$ between $(hkl)$ planes is given by $\mathbf{d} = d\hat{\mathbf{d}}$, where

$$\hat{\mathbf{d}} = \cos\alpha \mathbf{i} + \cos\beta \mathbf{j} + \cos\gamma \mathbf{k}$$

where $\cos\alpha, \cos\beta, \cos\gamma$ are direction cosines defined as $d$/intercept along the $x, y, z$ respectively. Hence show that the spacing, $d$, of planes $(hkl)$ in a unit cell of a cubic structure is given by

$$d = \frac{a}{\sqrt{h^2 + k^2 + l^2}}$$

(c) Calculate the spacings $d_{100}, d_{110}, d_{111}$ and $d_{222}$ between the (100),(110), (111), (222) planes, respectively, of a solid with a cubic structure if the cube edge $a$ is $3.16\text{Å}$.

**10.9**. (a) Show that for a bcc (body centred cubic) lattice, the geometrical structure factor is given by

$$F_{hkl} = f\left\{1 + e^{-i\pi(h+k+l)}\right\}$$

where all symbols are in the usual notation, and you are given that the bcc lattice has identical atoms at 000 and $\frac{1}{2}\frac{1}{2}\frac{1}{2}$.

(b) Hence, show that for a bcc $F_{hkl}$ is equal to 0 or $2f$.

**10.10**. (a) Show that for a fcc (face centred cubic) lattice, the geometrical structure factor is given by

$$F_{hkl} = f\left\{1 + e^{-i\pi(k+l)} + e^{-i\pi(h+l)} + e^{-i\pi(h+k)}\right\}$$

where all symbols are in the usual notation, and you are given that fcc lattice has identical atoms at 000; $0\frac{1}{2}\frac{1}{2}$; $\frac{1}{2}0\frac{1}{2}$; $\frac{1}{2}\frac{1}{2}0$.

(b) Hence, show that for a fcc $F_{hkl}$ is equal to 0 or $4f$.

# Chapter 11

# Introduction to Materials Science II: Phonons

## 11.1   What are phonons?

Phonons are quanta of lattice vibrations propagating in a crystal. These vibrations occur at any temperature and they are responsible for several properties, for example, thermal properties of the crystal such as (i) specific heat, (ii) thermal conductivity, (iii) thermal expansion, (iv) melting etc.

## 11.2   Specific Heat Theory

### 11.2.1   Einstein Model of Specific Heat

#### 1. Basic assumptions

In Einstein model, the following *basic assumptions* are made:
(i) Atoms vibrate as harmonic oscillators.
(ii) Each atom vibrates with the same frequency, $\omega_E$.

#### 2. Mean energy per mode, $< E(\omega) >$

From statistical mechanics, the thermodynamic expectation value of an observable $O$, is given by the Gibb's distribution

$$< O > \quad = \quad \frac{\sum_0^\infty O_n e^{-En/k_B T}}{Z} \tag{11.1}$$

$$\text{where } Z \quad = \quad \sum_0^\infty e^{-E_n/k_B T} \text{ is the Partition Function.} \tag{11.2}$$

$$\text{where } E_n \quad = \quad (n + \frac{1}{2})\hbar\omega \text{ is the energy for a quantum harmonic oscillator} \tag{11.3}$$

a result well known from quantum mechanics.

Hence, we obtain the mean energy as

$$
\begin{aligned}
<E(\omega)> &= \frac{\sum_0^\infty E_n e^{-En/k_BT}}{Z} \\[2mm]
&= \frac{\sum_0^\infty (n+\frac{1}{2})\hbar\omega e^{-(n+\frac{1}{2})\hbar\omega/k_BT}}{\sum_0^\infty e^{-(n+\frac{1}{2})\hbar\omega/k_BT}} \\[2mm]
&= \frac{1}{2}\hbar\omega \frac{\sum_0^\infty e^{-(n+\frac{1}{2})\hbar\omega/k_BT}}{\sum_0^\infty e^{-(n+\frac{1}{2})\hbar\omega/k_BT}} + \frac{\sum_0^\infty n\hbar\omega e^{-n\hbar\omega/k_BT}}{\sum_0^\infty e^{-n\hbar\omega/k_BT}}
\end{aligned}
\tag{11.4}
$$

Note that in the second term above,
(i) the denominator is a GP (Geometrical Progression) with the common ratio $e^{-\hbar\omega/k_BT}$
(ii) the numerator is the negative differential of the denominator w.r.t. $(1/k_BT)$

The denominator is

$$
\sum_0^\infty e^{-n\hbar\omega/k_BT} = 1 + e^{-\hbar\omega/k_BT} + e^{-2\hbar\omega/k_BT} + \cdots
$$

$$
\text{But } S_{GP}^\infty = \frac{1}{1-a} \text{ where } a \text{ is a common ratio}
$$

$$
S_{GP}^\infty = \frac{1}{1-e^{-\hbar\omega/k_BT}}
$$

The numerator is

$$
\begin{aligned}
\sum_0^\infty n\hbar\omega e^{-n\hbar\omega/k_BT} &= (-)\frac{d}{d(1/k_BT)} \sum_0^\infty e^{-n\hbar\omega/k_BT} \\[2mm]
&= (-)\frac{d}{d(1/k_BT)} \left\{ \frac{1}{1-e^{-\hbar\omega/k_BT}} \right\} \\[2mm]
&= (-)(-)\left\{ \frac{\hbar\omega e^{-\hbar\omega/k_BT}}{[1-e^{-\hbar\omega/k_BT}]^2} \right\}
\end{aligned}
$$

Hence, the mean energy becomes

$$
\begin{aligned}
<E(\omega)> &= \frac{1}{2}\hbar\omega + \frac{\hbar\omega e^{-\hbar\omega/k_BT}\left[1-e^{-\hbar\omega/k_BT}\right]}{\left[1-e^{-\hbar\omega/k_BT}\right]^2} \\[2mm]
&= \left(\frac{1}{2} + \frac{1}{e^{\hbar\omega/k_BT}-1}\right)\hbar\omega \\[2mm]
<E(\omega)> &= \left(n(\omega)+\frac{1}{2}\right)\hbar\omega
\end{aligned}
\tag{11.5}
$$

$$
\text{where } n(\omega) = \frac{1}{e^{\hbar\omega/k_BT}-1} \text{ is the Bose-Einstein Factor} \tag{11.6}
$$

### 3. Density of states, $D(\omega)$

If the total number of modes is $3N$, then the Density of states funcrion $D(\omega)$ is defined such that

$$3N = \int D(\omega)d\omega \qquad (11.7)$$

So, what is a suitable function for $D(\omega)$ would obey the above equation? One such function is defined as

$$D(\omega) = 3N\delta(\omega - \omega_E) \qquad (11.8)$$

where $\delta(\omega - \omega_E)$ is known as a *delta - function* which has the following properties

$$
\text{where } \delta(\omega - \omega_E) = \left\{ \begin{array}{l} 1 \text{ for } \omega = \omega_E \\ 0 \text{ for } \omega \neq \omega_E \end{array} \right.
$$

$$
\int f(\omega)\delta(\omega - \omega_E)d\omega = f(\omega_E)
$$

Hence

$$
\int D(\omega)d\omega = \int 3N\delta(\omega - \omega_E)d\omega
$$

$$
= 3N, \text{ as required}
$$

### 4. Total thermal energy, $U$

The Total Thermal Energy, $U$, is defined as

$$
\begin{aligned}
U &= \int D(\omega) < E(\omega) > d\omega & (11.9) \\
&= \int 3N\delta(\omega - \omega_E)\left\{\frac{1}{2} + \frac{1}{e^{\hbar\omega/k_BT} - 1}\right\}\hbar\omega d\omega \\
U &= 3N\left\{\frac{1}{2} + \frac{1}{e^{\hbar\omega_E/k_BT} - 1}\right\}\hbar\omega_E & (11.10)
\end{aligned}
$$

### 5. Specific heat, $C_V$

The Specific Heat, $C_V$, is defined as

$$
\begin{aligned}
C_V &= \frac{\partial U}{\partial T} & (11.11) \\
&= 3N\frac{\left\{-\hbar\omega_E e^{\frac{\hbar\omega_E}{k_BT}}\left(\frac{-\hbar\omega_E}{k_b}\right)\left(\frac{1}{T^2}\right)\right\}}{\left[e^{\hbar\omega/k_BT} - 1\right]^2} \\
&= 3Nk_B\left(\frac{\hbar\omega_E}{k_BT}\right)^2\frac{e^{\frac{\hbar\omega_E}{k_BT}}}{\left[e^{\hbar\omega/k_BT} - 1\right]^2} \\
&= \frac{3Nk_By^2e^y}{(e^y - 1)^2} & (11.12)
\end{aligned}
$$

where $y = \frac{\hbar\omega_E}{k_B T} = \frac{\theta_E}{T}$, with $\theta_E = \frac{\hbar\omega_E}{k_B}$.

A graph of $C_V$ against $T$ is illustrated in Figure 11.1.

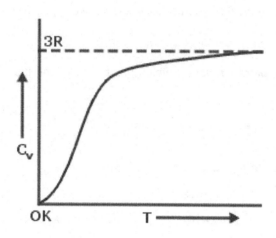

Figure 11.1: Specic Heat Theory in the Einstein Model

There are two limits of physical interest.

**Case (i) High temperature limit, $\hbar\omega_E \ll k_B T$**

$$
\begin{aligned}
C_V &\approx 3Nk_B \left(\frac{\hbar\omega_E}{k_B T}\right)^2 \frac{\left\{1 + \frac{\hbar\omega_E}{k_B T} + \cdots\right\}}{\left\{1 + \frac{\hbar\omega_E}{k_B T} + \cdots - 1\right\}^2} \\
&\approx 3Nk_B \left(\frac{\hbar\omega_E}{k_B T}\right)^2 \left(\frac{k_B T}{\hbar\omega_E}\right)^2 \\
C_V &= 3Nk_B \\
C_V &= 3R = 24.93 \text{ JK}^{-1} = 5.96 \text{ Cal per degree}
\end{aligned}
$$ (11.13)

which is true for all substances, and is known as DULONG AND PETIT'S LAW.

**Case (ii) Low temperature limit, $\hbar\omega_E \gg k_B T$**

$$
\begin{aligned}
C_V &\approx 3Nk_B \left(\frac{\hbar\omega_E}{k_B T}\right)^2 \frac{e^{\frac{\hbar\omega_E}{k_B T}}}{\left[e^{\hbar\omega/k_B T}\right]^2} \\
C_V &\approx 3Nk_B \left(\frac{\hbar\omega_E}{k_B T}\right)^2 e^{-\frac{\hbar\omega_E}{k_B T}} = 3Nk_B y^2 e^{-y}
\end{aligned}
$$ (11.14)

The agreement between Einstein's theory of specific heat with experiment is only good at high temperatures. Experimentally, it is observed that $C_V \propto T^3$ rather than exponential. this is a reflection of the fact that the assumption that all the atoms are vibrating with the same frequency, $\omega_E$, is in fact not correct, and that in reality there is some dispersion present, that is $\omega = \omega(q)$, that is, frequency depends on the wavevector $q$.

### 11.2.2  Debye Model

#### 1. Basic assumptions

In the Debye model, the following *basic assumptions* are made:

(i) Atoms vibrate as harmonic oscillators.
(ii) the atoms do not, in general, vibrate with the same frequency, but with a range of frequencies up to a maximum frequency, $\omega_D$, called the Debye frequency.
(iii) Treat the crystal as a 3-D continuum supporting longitudinal and transverse waves (lattice vibrations).

#### 2. Mean energy per mode, $< E(\omega) >$

The mean energy per mode for harmonic oscillations is, as before, given by

$$< E(\omega) > = \left( \frac{1}{e^{\hbar\omega/k_BT} - 1} + \frac{1}{2} \right) \hbar\omega$$

$$< E(\omega) > = \left( n(\omega) + \frac{1}{2} \right) \hbar\omega$$

$$\text{where } n(\omega) = \frac{1}{e^{\hbar\omega/k_BT} - 1} \text{ is the Bose-Einstein Factor}$$

For the Debye model, we shall consider the Zero Point Energy (ZPE), $\frac{1}{2}\hbar\omega$, to be negligible, and hence the mean energy is taken to be

$$< E(\omega) > \approx \frac{\hbar\omega}{e^{\hbar\omega/k_BT} - 1} \tag{11.15}$$

#### 3. Density of states, $D(\omega)$

We are looking for a Density of states funcrion $D(\omega)$ which satisies

$$3 \int_0^{\omega_D} D(\omega)d\omega = 3N \tag{11.16}$$

where $3N$ is the total number of modes is $3N$, and $\omega_D$ is the Debye frequency. So, what is a suitable function for $D(\omega)$ would obey the above equation? We obtain this by proceeding as follows. Consider a cube of side $L$. The vibrations with displacements $U$ in the crystal satisfy the wave equation,

$$\nabla^2 U = \frac{1}{v_s^2} \frac{\partial^2 U}{\partial t^2} \tag{11.17}$$

This has a solution

$$U = U_0 e^{i(\vec{k}\cdot\vec{r}-\omega t)}$$  (11.18)

$$k^2 U = \frac{\omega^2}{v_s^2}U$$

Hence

$$\omega = v_s k$$

which is the dispersion relation in the continuum limit.

Let us apply **Boundary conditions**

$$x : U(x+L,y,z) = U(x,y,z)$$
$$y : U(x,y+L,z) = U(x,y,z)$$
$$z : U(x,y,z+L) = U(x,y,z)$$

Hence we obtain

$$e^{i(k_x x + k_x L)} = e^{ik_x x} \Rightarrow e^{i(k_x L)} = 1$$
$$e^{i(k_y y + k_y L)} = e^{ik_y y} \Rightarrow e^{i(k_y L)} = 1$$
$$e^{i(k_z z + k_z L)} = e^{ik_z z} \Rightarrow e^{i(k_z L)} = 1$$

which gives

$$k_x L = n_x 2\pi, k_y L = n_y 2\pi, k_z L = n_z 2\pi$$
$$or$$
$$k_x = \frac{n_x 2\pi}{L}, k_y = \frac{n_y 2\pi}{L}, k_z = \frac{n_z 2\pi}{L},$$

In a volume, $(\frac{2\pi}{L})^3$     there is one allowed mode

In a unit volume, there are $\dfrac{L^3}{8\pi^3} = \dfrac{V}{8\pi^3}$ modes

Consider a spherical shell between $k$ and $k + dk$. Its volume $dV_k$ is given by

$$V_k = \frac{4}{3}\pi k^3$$
$$dV_k = 4\pi k^2 dk$$

Total number of modes between $k$ and $k + dk$ is

$$4\pi k^2 dk \frac{V}{8\pi^3} = \frac{Vk^2}{2\pi^2}dk$$
$$= D(\omega)d\omega$$

where $D(\omega)$ is the density of states in the Debye model. Hence we obtain

$$
\begin{aligned}
D(\omega)d\omega &= \frac{Vk^2}{2\pi^2}dk \\
D(\omega) &= \frac{Vk^2}{2\pi^2}\frac{dk}{d\omega} \\
&= \frac{Vk^2}{2\pi^2(d\omega/dk)} \\
\text{But } \omega &= v_s k \\
D(\omega) &= \frac{V\omega^2}{2\pi^2 v_s^2}\frac{1}{v_s} = \frac{V\omega^2}{2\pi^2 v_s^3} \\
D(\omega) &= A\omega^2 \qquad\qquad (11.19)
\end{aligned}
$$

which is a parabolic function, with $A = \frac{V}{2\pi^2 v_s^3}$. The Debye Density of States is illustrated graphically in Figure 11.2.

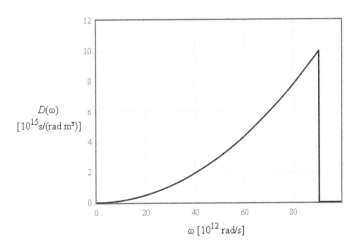

Figure 11.2: Debye Density of States variation with frequency. The maximum frequency is $\omega_D$.

We can obtain an expression for the Debye frequency, $\omega_D$, by evaluating

$$
\begin{aligned}
3\int_0^{\omega_D} \frac{V\omega^2}{2\pi^2 v_s^3}d\omega &= 3N \\
\frac{3V}{2\pi^2 v_s^3}\frac{\omega^3}{3}\Big|_0^{\infty} &= 3N \\
\omega_D^3 &= \frac{6\pi^2 v_s^3 N}{V} \\
\text{or} \\
\omega_D &= \left(\frac{6\pi^2 v_s^3 N}{V}\right)^{1/3} = \left(\frac{6\pi^2 N}{V}\right)^{1/3}v_s
\end{aligned}
$$

$$\text{and } k_D = \frac{\omega_D}{v_s} = \left(\frac{6\pi^2 N}{V}\right)^{1/3}$$

**Exercise**: Calculate an estimate for $\omega_D$ and $k_D$ for Aluminium, given that for aluminium, the Young's modulus, $Y = 7.0 \times 10^{10} \text{N.m}^{-2}$ and $\rho = 2.7 \times 10^3 \text{kg.m}^{-3}$ is the density .

For aluminium, $v_s$ for solids,

$$
\begin{aligned}
v_s &= \sqrt{\frac{Y}{\rho}} \\
&= \sqrt{\frac{7.0 \times 10^{10}}{2.7 \times 10^3}} \\
&= 5.1 \times 10^3 \text{ms}^{-1} \\
&= 5.1 \times 10^5 \text{cms}^{-1}
\end{aligned}
$$

$$\text{but } \frac{N}{V} \approx 10^{23} \text{ atoms/cc}$$

$$
\begin{aligned}
\text{Using }, \omega_D^3 &= \frac{6\pi^2 v_s^3 N}{V} \\
\omega_D^3 &= \left(6\pi^2 \times 5^3 \times 10^{15} \times 10^{23}\right) \\
&= 10^{42} \\
\omega_D &= 10^{14} \text{ Hz or cycles per second} \\
&\text{and} \\
k_D &= \frac{\omega_D}{v_s} \approx 2 \times 10^{10} \text{m}^{-1}
\end{aligned}
$$

## 4. Total thermal energy, $U$

The Total Thermal Energy, $U$, in the Debye model, is defined as

$$U = 3\int_0^{\omega_D} D(\omega) <E(\omega)> d\omega \tag{11.20}$$

$$= 3\int_0^{\omega_D} \left(\frac{V\omega^2}{2\pi^2 v_s^3}\right)\left(\frac{\hbar\omega}{e^{\hbar\omega/k_B T}-1}\right) d\omega$$

$$U = 3\int_0^{\omega_D} \left(\frac{V\hbar}{2\pi^2 v_s^3}\right)\left(\frac{\omega^3}{e^{\hbar\omega/k_B T}-1}\right) d\omega$$

$$\text{Let } x = \frac{\hbar\omega}{k_B T} \Rightarrow d\omega = \frac{k_B T}{\hbar}dx$$

$$x_D = \frac{\hbar\omega_D}{k_B T} = \frac{\theta_D}{T} \text{where } \theta_D = \frac{\hbar\omega_D}{k_B} = \frac{\hbar}{k_B}\left(\frac{6\pi^2 v_s^3 N}{V}\right)^{1/3}$$

$$\text{Hence } U = 3\left(\frac{V\hbar}{2\pi^2 v_s^3}\right)\int_0^{x_D}\left(\frac{k_B T}{\hbar}\right)^3\left(\frac{x^3}{e^x - 1}\right)\left(\frac{k_B T}{\hbar}\right) dx$$

$$
\begin{aligned}
&= \frac{3V(k_BT)(k_BT)^3}{2\pi^2 v_s^3 \hbar^3} \times \frac{3N}{3N} \int_0^{x_D} \frac{x^3}{e^x-1}dx \\
&= \frac{9N(k_BT)(k_BT)^3}{\hbar^3 (\frac{6\pi^2 v_s^3 N}{V})} \int_0^{x_D} \frac{x^3}{e^x-1}dx \\
&= 9N(k_BT)\left(\frac{k_BT}{k_B\theta_D}\right)^3 \int_0^{x_D} \frac{x^3}{e^x-1}dx \\
U &= 9Nk_BT\left(\frac{T}{\theta_D}\right)^3 \int_0^{x_D} \frac{x^3}{e^x-1}dx
\end{aligned}
\tag{11.21}
$$

is the Total thermal energy for a 3D crystal in the Debye model.

## 5. Specific heat, $C_V$

The Specific Heat, $C_V$, is defined as

$$
C_V = \frac{\partial U}{\partial T}
\tag{11.22}
$$

There are two limits of physical interest.

**Case (i) High temperature limit, $k_BT \gg \hbar\omega$ or $T \gg \theta_D$**
In the high temperatute limit, the total thermal energy becomes

$$
\begin{aligned}
U &\approx 9Nk_BT\left(\frac{T}{\theta_D}\right)^3 \int_0^{x_D} \frac{x^3}{1+x+\cdots-1}dx \\
&\approx 9Nk_BT\left(\frac{T}{\theta_D}\right)^3 \int_0^{x_D} x^2 dx \\
&\approx 9Nk_BT\left(\frac{T}{\theta_D}\right)^3 \frac{x^3}{3}\Big|_0^{x_D} \\
&\approx 9Nk_BT\left(\frac{T}{\theta_D}\right)^3 \frac{1}{3}(x_D^3-0) \\
&\approx 9Nk_BT\left(\frac{T}{\theta_D}\right)^3 \frac{1}{3}\left(\frac{\theta_D}{T}\right)^3 \\
U &\approx 3Nk_BT
\end{aligned}
$$

whence

$$
\begin{aligned}
C_V &= \frac{\partial U}{\partial T} = 3Nk_B \\
C_V &= 3R
\end{aligned}
\tag{11.23}
$$

which is DULONG and PETIT'S LAW.

**Case (ii) Low temperature limit, $k_BT \ll \hbar\omega$ or $T \ll \theta_D$**

In the low temperature limit, since $T \to 0$, $x_D = \frac{\theta_D}{T} \to \infty$, and thus the total thermal energy in the low temperature limit becomes

$$U = 9Nk_BT\left(\frac{T}{\theta_D}\right)^3 \int_0^\infty \frac{x^3}{e^x - 1}dx \qquad (11.24)$$

where the integral can be recognized to be a standard integral of the

$$\int_0^\infty \frac{x^{n-1}}{e^x - 1}dx = \Gamma(n)\zeta(n)$$

$$\text{where } \Gamma(n) \text{ is the Gamma function}$$
$$\zeta(n) \text{ is the Riemann-Zeta function}$$
$$\text{But } n - 1 = 3 \Rightarrow n = 4$$
$$\int_0^\infty \frac{x^3}{e^x - 1}dx = \Gamma(4)\zeta(4)$$
$$\text{where } \Gamma(4) = 3! = 6$$
$$\zeta(4) = \frac{\pi^4}{90}$$
$$\int_0^\infty \frac{x^3}{e^x - 1}dx = 6 \times \frac{\pi^4}{90} = \frac{\pi^4}{15}$$
$$\text{Hence}$$
$$U = 9Nk_BT\left(\frac{T}{\theta_D}\right)^3 \frac{\pi^4}{15}$$
$$= \frac{3}{5}\pi^4 Nk_B\frac{T^4}{\theta_D^3}$$
$$\text{Specific Heat, } C_V = \frac{\partial U}{\partial T} = \frac{12\pi^4 Nk_B}{5}\left(\frac{T}{\theta_D}\right)^3$$
$$C_V = 234Nk_B\left(\frac{T}{\theta_D}\right)^3 = BT^3 \qquad (11.25)$$

where $B = \frac{234Nk_B}{\theta_D^3}$.

We have seen the expressions for the specific heat at high and low temperatures. We shall now find a more general expression for the whole temperature range. Starting from the Total Thermal Energy, $U$,

$$U = 3\int_0^{\omega_D} D(\omega) < E(\omega) > d\omega$$
$$= 3\int_0^{\omega_D} \left(\frac{V\omega^2}{2\pi^2 v_s^3}\right)\left(\frac{\hbar\omega}{e^{\hbar\omega/k_BT} - 1}\right)d\omega$$
$$= 3\int_0^{\omega_D} \left(\frac{V\hbar}{2\pi^2 v_s^3}\right)\left(\frac{\omega^3}{e^{\hbar\omega/k_BT} - 1}\right)d\omega$$

Hence, specific heat, $C_V = \frac{\partial U}{\partial T}$

$$= \frac{\partial}{\partial T}\left\{3\int_0^{\omega_D}\left(\frac{V\hbar}{2\pi^2 v_s^3}\right)\left(\frac{\omega^3}{e^{\hbar\omega/k_B T}-1}\right)d\omega\right\}$$

$$= \left(\frac{3V\hbar}{2\pi^2 v_s^3}\right)\int_0^{\omega_D}\frac{\partial}{\partial T}\frac{\omega^3}{e^{\hbar\omega/k_B T}-1}d\omega$$

$$= \left(\frac{3V\hbar}{2\pi^2 v_s^3}\times\frac{\hbar^2 T^2 k_B^3}{\hbar^2 T^2 k_B^3}\right)\int_0^{\omega_D}\frac{(-)\omega^3 e^{\hbar\omega/k_B T}}{[e^{\hbar\omega/k_B T}-1]^2}(-)\frac{\hbar\omega}{k_B T^2}d\omega$$

$$= \left(\frac{3V k_B(k_B T)^2}{2\pi^2 v_s^3\hbar^2}\right)\int_0^{\omega_D}\frac{e^{\hbar\omega/k_B T}}{[e^{\hbar\omega/k_B T}-1]^2}\left(\frac{\hbar^4\omega^4}{k_B^4 T^4}\right)d\omega$$

Let $x = \frac{\hbar\omega}{k_B T}\Rightarrow d\omega = \frac{k_B T}{\hbar}dx$

$$x_D = \frac{\hbar\omega_D}{k_B T} = \frac{\theta_D}{T}\text{where }\theta_D=\frac{\hbar\omega_D}{k_B}=\frac{\hbar}{k_B}\left(\frac{6\pi^2 v_s^3 N}{V}\right)^{1/3}$$

$$= \frac{3k_B(k_B T)^2}{\hbar^2\frac{2\pi^2 v_s^3}{V}}\times\frac{3N}{3N}\int_0^{x_D}\frac{e^x x^4}{[e^x-1]^2}\frac{k_B T}{\hbar}dx$$

$$= \frac{9N k_B(k_B T)^3}{\hbar^3\left(\frac{6\pi^2 v_s^3 N}{V}\right)}\int_0^{x_D}\frac{e^x x^4}{[e^x-1]^2}dx$$

$$= 9N k_B\left(\frac{k_B T}{k_B\theta_D}\right)^3\int_0^{x_D}\frac{e^x x^4}{[e^x-1]^2}dx$$

$$C_V = 9N k_B\left(\frac{T}{\theta_D}\right)^3\int_0^{x_D}\frac{e^x x^4}{[e^x-1]^2}dx \tag{11.26}$$

The above equation is the specific heat dependence on temperature over the whole temperature range. The agreement between Debye theory of specific heat with experiment is good at both high temperatures and low temperatures, as illustrated in Figure 11.3. Debye thory predicts that at low temperatures the specfic heat, $C_V\propto T^3$, which agrees very well with experiment. This is a reflection of the fact the frequency variation with the wave vectors is a fairly reasonable approximation in the continuum limit.

A comparison of specific heat theory predictions due to Einstein model and Debye model is illustrated in Figure 11.4.

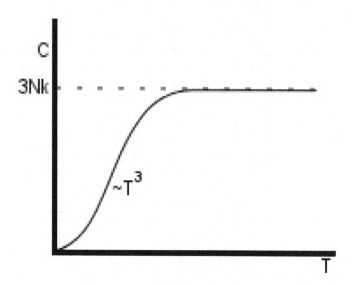

Figure 11.3: Specific Heat variation with temperature in the Debye Theory.

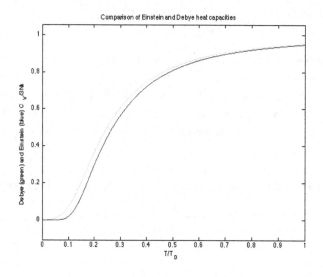

Figure 11.4: Comparison of Einstein model (lower curve) and Debye model (upper curve) of specific heat.

### 11.2.3 Chain model for a monatomic lattice

#### 1. Basic assumptions

The monatomic lattice is modelled as a system of of crystal consisting of a linear chain of identical atoms, each of mass $m$, connected by elastic springs, each of force constant $G$, as illustrated in Figure 11.5.

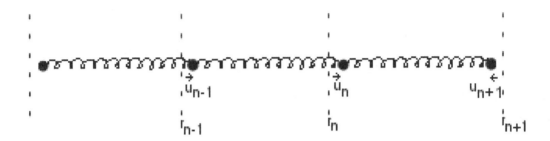

Figure 11.5: A linear chain modelling a monatomic lattice.

#### 2. Equation of Motion ($ma = \sum F$)

This is essentially a force constant problem. For the $n^{th}$ mass, we have

$$
\begin{aligned}
m\frac{d^2 U_n}{dt^2} &= -G(U_n - U_{n-1} + G(U_{n+1} - U_n) \\
&= G(U_{n+1} - 2U_n + U_{n-1})
\end{aligned}
\tag{11.27}
$$

Try solutions of the form

$$
\begin{aligned}
U_n &= A e^{i(qx_n - \omega t)} = A e^{i(qna - \omega t)} \\
U_{n+1} &= A e^{i[qx_{n+1} - \omega t]} = A e^{i[q(n+1)a - \omega t]} \\
U_{n-1} &= A e^{i[qx_{n-1} - \omega t]} = A e^{i[q(n-1)a - \omega t]}
\end{aligned}
$$

Inserting in the equation of motion, we obtain

$$
\begin{aligned}
-m\omega^2 U_n &= GA\left\{ e^{i[q(n+1)a} - 2e^{iqna} + e^{i[q(n-1)a} \right\} e^{-i\omega t} \\
&= GA\left\{ e^{iqa} - 2 + e^{-iqa} \right\} e^{(qna - i\omega t} \\
&= G\left\{ 2\cos qa - 2 \right\} U_n \\
&\quad \text{or} \\
\omega^2 &= \frac{2G}{m}[1 - \cos qa]
\end{aligned}
$$

$$= \frac{4G}{m} \sin^2 \frac{qa}{2}$$

or

$$\omega = 2 \left( \frac{G}{m} \right)^{1/2} \sin \frac{qa}{2} \tag{11.28}$$

which is the monatomic lattice dispersion relation. The dispersion curves are plotted in Figure 11.6.

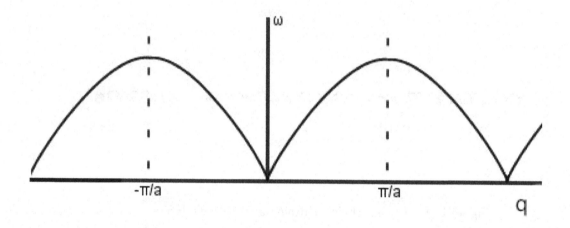

Figure 11.6: Dispersion curves for phonons in a monatomic lattice.

## 3. Mean energy per mode, $< E(\omega) >$

The mean energy per mode for harmonic oscillations is, as before, given by

$$< E(\omega) > = \left( \frac{1}{e^{\hbar\omega/k_BT} - 1} + \frac{1}{2} \right) \hbar\omega$$

$$< E(\omega) > = \left( n(\omega) + \frac{1}{2} \right) \hbar\omega$$

where $n(\omega) = \dfrac{1}{e^{\hbar\omega/k_BT} - 1}$ is the Bose-Einstein Factor

For the monatomic lattice model, we shall consider the Zero Point Energy (ZPE), $\frac{1}{2}\hbar\omega$, to be negligible, and hence the mean energy is taken to be

$$< E(\omega) > \approx \frac{\hbar\omega}{e^{\hbar\omega/k_BT} - 1} \tag{11.29}$$

## 4. Density of states, $D(\omega)$

There is one mode for each interval $\frac{\pi}{L}$ in $q$-space.
In unit interval of $q$-space there are $\frac{L}{\pi}$ modes.

In an interval $= D(\omega)d\omega$ where $D(\omega)$ is the Density of states for a monatomic lattice.

$$
\begin{aligned}
D(\omega)d\omega &= \frac{L}{\pi}dq \\
D(\omega) &= \frac{L}{\pi}\frac{dq}{d\omega} = \frac{L}{\pi(d\omega/dq)}
\end{aligned}
\tag{11.30}
$$

From the dispersion equation

$$
\begin{aligned}
\omega &= 2\left(\frac{G}{m}\right)^{1/2}\sin\frac{qa}{2} = \omega_m\sin\frac{qa}{2} \text{ where } \omega_m = 2\sqrt{\frac{G}{m}} \\
\frac{\omega}{\omega_m} &= \sin\frac{qa}{2} \\
\frac{qa}{2} &= \sin^{-1}\frac{\omega}{\omega_m} \\
q &= \frac{2}{a}\sin^{-1}\frac{\omega}{\omega_m} \\
\frac{dq}{d\omega} &= \frac{2}{a}\frac{1}{(\omega_m^2 - \omega^2)^{1/2}} \\
D(\omega) &= \frac{L}{\pi}\frac{2}{a}\frac{1}{(\omega_m^2 - \omega^2)^{1/2}}
\end{aligned}
\tag{11.31}
$$

which is the Density of states for phonons in a monatomic lattice. note that there is a singularity at $\omega_m$ known as a VAN HOVE SINGULARITY

**5. Total thermal energy, $U$**

The Total Thermal Energy, $U$, is defined as

$$
\begin{aligned}
U &= \int_0^{\omega_m} D(\omega) <E(\omega)> d\omega \\
U &= \frac{2L}{\pi a}\int_0^{\omega_m}\frac{1}{(\omega_m^2 - \omega^2)^{1/2}}\frac{\hbar\omega}{e^{\hbar\omega/k_BT} - 1}d\omega
\end{aligned}
\tag{11.32}
$$

**6. Specific heat, $C_V$**

The Specific Heat, $C_V$, is defined as

$$
C_V = \frac{\partial U}{\partial T}
\tag{11.33}
$$

There are two limits of physical interest.

**Case (i) High temperature limit, $k_BT \gg \hbar\omega$**
In the high temperature limit, the total thermal energy from equation (11.32), becomes

$$
U \approx \frac{2L}{\pi a}\int_0^{\omega_m}\frac{1}{(\omega_m^2 - \omega^2)^{1/2}}\frac{\hbar\omega}{1 + \frac{\hbar\omega}{k_BT} + \cdots - 1}d\omega
$$

$$\approx \frac{2Na}{\pi a} k_B T \int_0^{\omega_m} \frac{1}{(\omega_m^2 - \omega^2)^{1/2}} \frac{\hbar\omega}{\hbar\omega} d\omega$$

$$\approx \frac{2N k_B T}{\pi} \sin^{-1} \frac{\omega}{\omega_m} \Big|_0^{\omega_m}$$

$$\approx \frac{2N k_B T}{\pi} \left(\frac{\pi}{2} - 0\right)$$

$$U \approx N k_B T$$

whence

$$C_V = \frac{\partial U}{\partial T} = \frac{\partial}{\partial T}(N k_B T) = N k_B \qquad (11.34)$$

## Case (ii) Low temperature limit, $k_B T \ll \hbar\omega$

In the low temperature limit, since $T \to 0$, $\frac{\hbar\omega_m}{k_B T} = x_m \to \infty$, and thus the total thermal energy from equation (11.32), in the low temperature limit becomes

$$U \approx \frac{2Na}{\pi a} \int_0^\infty \frac{(k_B T)^2}{\omega_m \hbar} \frac{x}{e^x - 1} dx \qquad (11.35)$$

$$\approx \frac{2N}{\pi} \frac{(k_B T)^2}{\hbar\omega_m} \int_0^\infty \frac{x}{e^x - 1} dx$$

where the integral can be recognized to be a standard integral of the

$$\int_0^\infty \frac{x^{n-1}}{e^x - 1} dx = \Gamma(n)\zeta(n) \qquad (11.36)$$

$$\text{where } \Gamma(n) \quad \text{is} \quad \text{the Gamma function}$$

$$\zeta(n) \quad \text{is} \quad \text{the Riemann-Zeta function}$$

$$\text{But } n - 1 = 1 \Rightarrow n = 2$$

$$\int_0^\infty \frac{x}{e^x - 1} dx = \Gamma(2)\zeta(2)$$

$$\text{where } \Gamma(2) = 1! = 1$$

$$\zeta(2) = \frac{\pi^2}{6}$$

$$\int_0^\infty \frac{x}{e^x - 1} dx = 1 \times \frac{\pi^2}{6} = \frac{\pi^2}{6}$$

$$\text{Hence}$$

$$U = \frac{2N k_B^2 T^2}{\pi \hbar\omega_m} \frac{\pi^2}{6}$$

$$= \frac{\pi}{3} \frac{N k_B^2 T^2}{\hbar\omega_m}$$

$$\text{Specific Heat, } C_V = \frac{\partial U}{\partial T} = \frac{\partial}{\partial T}\left(\frac{\pi}{3} \frac{N k_B^2 T^2}{\hbar\omega_m}\right)$$

$$C_V = \frac{2\pi}{3} \frac{N k_B^2 T}{\hbar\omega_m} \qquad (11.37)$$

### 11.2.4 Chain model for a diatomic lattice

**1. Basic assumptions**

The diatomic lattice is modelled as a system of of crystal consisting of a linear chain of two different atoms per unit cell, of masses $M_1$ and $M_2$, connected by elastic springs, each of force constant $C$, as illustrated in Figure 11.7.

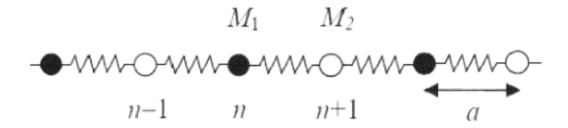

Figure 11.7: A linear chain modelling a diatomic lattice.

**2. Equation of Motion ($M_1 a = \sum F$ and $M_2 a = \sum F$ )**

This is essentially a force constant problem. Suppose that nass $M_1$ is displaced by $U_l$ and $U_{l+2a}$ while $M_2$ is displaced by $V_{l-a}$ and $V_{l+a}$, respectively.

For the mass $M_1$, we have, considering nearest neighbour interaction only,

$$
\begin{aligned}
M_1 \frac{d^2 U_l}{dt^2} &= -C(U_l - V_{l-a} + C(V_{l+a} - U_l) \\
&= C(V_{l-a} - 2U_l + V_{l+a})
\end{aligned}
\tag{11.38}
$$

For the mass $M_2$, we have, considering nearest neighbour interaction only,

$$
\begin{aligned}
M_2 \frac{d^2 V_{l+a}}{dt^2} &= -C(V_{l+a} - U_l) + C(U_{l+2a} - V_{l+a}) \\
&= C(U_l - 2V_{l+a} + U_{l+2a})
\end{aligned}
\tag{11.39}
$$

Try solutions of the form

$$
\begin{aligned}
V_{l-a} &= U_2 e^{i(q(l-a)-\omega t)} \\
V_{l+a} &= U_2 e^{i[q(l+a)-\omega t]}
\end{aligned}
$$

$$U_l = U_1 e^{i[ql-\omega t]}$$

$$U_{l+2a} = U_1 e^{i[q(l+2a)-\omega t]}$$

Hence, we obtain

$$-M_1\omega^2 U_1 e^{iql} = C\left[U_2 e^{iq(l-a)} - 2U_1 e^{iql} + U_2 e^{iq(l+a)}\right]$$

$$-M_1\omega^2 U_1 = C\left[U_2 e^{-iqa} - 2U_1 + U_2 e^{iqa}\right]$$

$$-M_2\omega^2 U_2 e^{i(q+a)} = C\left[U_1 e^{iql} - 2U_2 e^{iq(l+a)} + U_1 e^{iq(l+2a)}\right]$$

$$-M_2\omega^2 U_2 = C\left[U_1 e^{-iqa} - 2U_2 + U_1 e^{iqa}\right]$$

Rearranging, we obtain

$$-M_1\omega^2 U_1 = -2CU_1 + 2CU_2\left[\frac{e^{iqa}+e^{-iqa}}{2}\right] \qquad (11.40)$$

$$-M_2\omega^2 U_2 = -2CU_2 + 2CU_1\left[\frac{e^{iqa}+e^{-iqa}}{2}\right] \qquad (11.41)$$

Hence, we obtain

$$(2C - M_1\omega^2)U_1 - 2C\cos qa\, U_2 = 0$$

$$-2C\cos qa\, U_1 + (2C - M_2\omega^2)U_2 = 0$$

Putting in a matrix form, we obtain

$$\begin{bmatrix} (2C - M_1\omega^2) & -2C\cos qa \\ -2C\cos qa & (2C - M_2\omega^2) \end{bmatrix}\begin{bmatrix} U_1 \\ U_2 \end{bmatrix} = \begin{bmatrix} 0 \\ 0 \end{bmatrix} \qquad (11.42)$$

The equations have a solution if and only if the determinant of the $2 \times 2$ matrix vanishes.

$$(2C - M_1\omega^2)(2C - M_2\omega^2) - 4C^2\cos^2 qa = 0$$

$$4C^2 - 2C(M_1 + M_2)\omega^2 + M_1 M_2\omega^4 - 4C^2\cos^2 qa = 0$$

$$\omega^4 - 2C\left(\frac{1}{M_1} + \frac{1}{M_2}\right)\omega^2 + \frac{4C^2}{M_1 M_2}(1 - \cos^2 qa) = 0$$

$$\omega^4 - 2C\left(\frac{1}{M_1} + \frac{1}{M_2}\right)\omega^2 + \frac{4C^2}{M_1 M_2}\sin^2 qa = 0$$

which has a solution

$$\omega_\pm^2 = \frac{2C}{2}\left(\frac{1}{M_1} + \frac{1}{M_2}\right) \pm \frac{1}{2}\left\{4C^2\left[\frac{1}{M_1} + \frac{1}{M_2}\right]^2 - \frac{16C^2}{M_1 M_2}\sin^2 qa\right\}^{1/2}$$

or

$$\omega_\pm^2 = C\left(\frac{1}{M_1} + \frac{1}{M_2}\right) \pm C\left\{\left[\frac{1}{M_1} + \frac{1}{M_2}\right]^2 - \frac{4}{M_1 M_2}\sin^2 qa\right\}^{1/2} \qquad (11.43)$$

There are two limits of physical interest.

**Case (i)** $q = 0$ limit, $\Rightarrow \sin qa = 0$
the two roots become

$$\omega_+ = \sqrt{2C\frac{1}{M_1} + \frac{1}{M_2}} \text{(Optical branch)} \tag{11.44}$$

$$\omega_- = 0 \text{ (Acoustic branch)} \tag{11.45}$$

**Case (ii)** $q = \frac{\pi}{2a}$ limit, $\Rightarrow \sin\frac{\pi a}{2a} = \sin\frac{\pi}{2} = 1$

$$\omega_\pm^2 = C\left(\frac{1}{M_1} + \frac{1}{M_2}\right) \pm C\left\{\left[\frac{1}{M_1} + \frac{1}{M_2}\right]^2 - \frac{4}{M_1 M_2}\right\}^{1/2}$$

$$= C\left(\frac{1}{M_1} + \frac{1}{M_2}\right) \pm C\left\{\frac{1}{M_1^2} + \frac{1}{M_2^2} + \frac{2}{M_1 M_2} - \frac{4}{M_1 M_2}\right\}^{1/2}$$

$$= C\left(\frac{1}{M_1} + \frac{1}{M_2}\right) \pm \left(\frac{1}{M_1} - \frac{1}{M_2}\right)^{2/2}$$

which gives two roots

$$\omega_+^2 = \frac{2C}{M_1} \text{ and } \omega_-^2 = \frac{2C}{M_2}$$
or
$$\omega_+ = \sqrt{\frac{2C}{M_1}} \text{ and } \omega_- = \sqrt{\frac{2C}{M_2}} \tag{11.46}$$

The dispersion curves are illustrated in Figure 11.8, where $\beta = C, m = M_1, M = M_2$. The two branches are referred to as LONGITUDINAL OPTICAL (LO) PHONONS and LONGITUDINAL ACOUSTIC (LA) PHONONS. By repeating the analysis in a 3d lattice, one obtains extra modes such as 2 Transverse optical phonons and 2 Transverse Acoustical phonons, as illustrated in Figure 11.9.

Let us now look at the **Particle displacements**.

$$\frac{U_1}{U_2} = \frac{2C\cos qa}{2C - M_1\omega^2} \tag{11.47}$$
At $q = 0$
$$\frac{U_1}{U_2} = \frac{2C}{2C - M_1\omega^2}$$
which gives
$$\frac{U_1}{U_2} = -1 \text{ for the Optical mode} \tag{11.48}$$

$$\frac{U_1}{U_2} = +1 \text{ for the Acoustic mode} \tag{11.49}$$

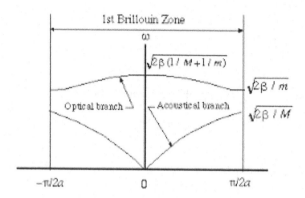

Figure 11.8: Phonons dispersion curves in a diatomic lattice

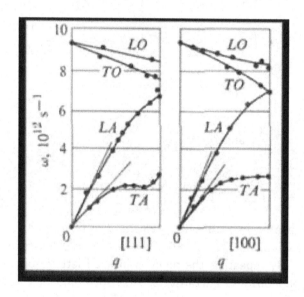

Figure 11.9: Phonons dispersion curves in a diatomic lattice.

The particle displacements are illustrated in Figure 11.10, where it can be noted that $M_1$ and $M_2$ are out of phase for the optic mode, and that $M_1$ and $M_2$ are in phase for the acoustic mode.

## 11.3    Phonon-Photon Interactions

### 11.3.1    Phonon type dielectric function

Consider ions each of mass $m$ in a crystal with displacement **r** under the influence of a force **F**. This is essentially a force constant problem.

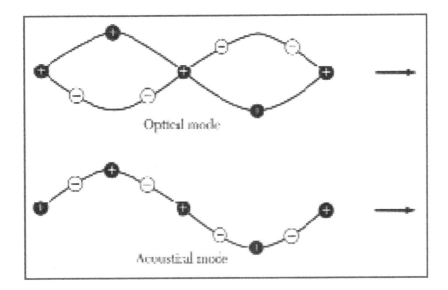

Figure 11.10: Particle displacements in optical phonons annd acoustic phonons.

**Equation of motion**

$$m\frac{d^2\mathbf{r}}{dt^2} = \sum \mathbf{F}$$

$$= -k\mathbf{r} + e\mathbf{E}_{loc}$$

$$\text{where} \quad \mathbf{E}_{loc} = \mathbf{E}_{mac} + \frac{\mathbf{P}}{3\epsilon_0} \tag{11.50}$$

$$\text{where} \quad \mathbf{P} = \mathbf{P}_{elec} + P_{ionic} \tag{11.51}$$

$$\text{But} \quad \mathbf{D} = \epsilon_0 \mathbf{E}_{mac} + \mathbf{P}_{elec} \tag{11.52}$$

$$\mathbf{D} = \epsilon_0 \epsilon(\infty) \mathbf{E}_{mac} \tag{11.53}$$

$$\text{Hence} \quad \epsilon_0 \epsilon(\infty) \mathbf{E}_{mac} = \epsilon_0 \mathbf{E}_{mac} + \mathbf{P}_{elec} \tag{11.54}$$

$$\mathbf{P}_{elec} = \epsilon_0 \left[ \epsilon(\infty) - 1 \right] \mathbf{E}_{mac}$$

$$\text{Hence} \quad \mathbf{E}_{loc} = \mathbf{E}_{mac} + \frac{\mathbf{P_{elec}}}{3\epsilon_0}$$

$$= \mathbf{E}_{mac} + \frac{\epsilon_0 \left[ \epsilon(\infty) - 1 \right] \mathbf{E}_{mac}}{3\epsilon_0}$$

$$= \frac{\mathbf{E}_{mac}}{3} \left[ 3 + \epsilon(\infty) - 1 \right]$$

$$\mathbf{E}_{loc} = \frac{\mathbf{E}_{mac}}{3} \left[ \epsilon(\infty) + 2 \right]$$

$$\text{But} \quad \mathbf{P}_{elec} = \frac{N\alpha_{el}\mathbf{E}_{loc}}{V}$$

$$\frac{N\alpha_{el}}{V} \frac{\mathbf{E}_{mac}}{3} \left[ \epsilon(\infty) + 2 \right] = \epsilon_0 \left[ \epsilon(\infty) - 1 \right] \mathbf{E}_{mac}$$

$$\text{Hence} \quad \frac{N\alpha_{el}}{3V\epsilon_0} \quad = \quad \frac{[\epsilon(\infty) - 1]}{[\epsilon(\infty) + 2]} \tag{11.55}$$

which is the Clausius-Mossotti Equation.

The total polarization $\mathbf{P} = \mathbf{P}_{elec} + P_{ion}$ can be written as

$$\mathbf{P} \quad = \quad \frac{N\alpha_{el}\mathbf{E}_{loc}}{V} + \frac{Ner}{V}$$

Using Clausius-Mosotti Equation,

$$\mathbf{P} \quad = \quad 3\epsilon_0 \frac{[\epsilon(\infty) - 1]}{[\epsilon(\infty) + 2]}\mathbf{E}_{loc} + \frac{Ner}{V}$$

$$\text{Using} \quad \mathbf{E}_{loc} = \mathbf{E}_{mac} + \frac{\mathbf{P}}{3\epsilon_0} \quad \Rightarrow \quad \mathbf{P} = 3\epsilon_0 \left[\mathbf{E}_{loc} - \mathbf{E}_{mac}\right]$$

$$\text{Hence} \quad 3\epsilon_0 \frac{[\epsilon(\infty) - 1]}{[\epsilon(\infty) + 2]}\mathbf{E}_{loc} + \frac{Ner}{V} \quad = \quad 3\epsilon_0 \left[\mathbf{E}_{loc} - \mathbf{E}_{mac}\right]$$

After some algebra,

$$\mathbf{E}_{loc} \quad = \quad \frac{[\epsilon(\infty) + 2]}{3}\mathbf{E}_{mac} + \frac{Ner}{V}\frac{[\epsilon(\infty) + 2]}{9\epsilon_0}$$

$$\mathbf{P} \quad = \quad \frac{Ner}{V}\frac{[\epsilon(\infty) + 2]}{3} + \epsilon_0 \left[\epsilon(\infty) - 1\right]\mathbf{E}_{mac}$$

$$\text{Inserting in} \quad m\ddot{r} \quad = \quad -kr + eE_{loc}, \quad \text{we obtain}$$

$$= \quad -kr + e\left\{\frac{[\epsilon(\infty) + 2]}{3}\mathbf{E}_{mac} + \frac{Ner}{V}\frac{[\epsilon(\infty) + 2]}{9\epsilon_0}\right\}$$

$$m\ddot{r} + \left\{k - \frac{Ne^2}{V}\frac{[\epsilon(\infty) + 2]}{9\epsilon_0}\right\}r \quad = \quad e\frac{[\epsilon(\infty) + 2]}{3}\mathbf{E}_{mac}$$

$$m\ddot{r} + m\omega_T^2 r \quad = \quad e\frac{[\epsilon(\infty) + 2]}{3}\mathbf{E}_{mac}$$

$$\text{where} \quad m\omega_T^2 \quad = \quad \left\{k - \frac{Ne^2}{V}\frac{[\epsilon(\infty) + 2]}{9\epsilon_0}\right\}$$

$$\text{If} \quad r = r_0 e^{-i\omega t} \quad \Rightarrow \quad \ddot{r} = -\omega^2 r$$

$$\left(-m\omega^2 + m\omega_T^2\right)r \quad = \quad e\frac{[\epsilon(\infty) + 2]}{3}\mathbf{E}_{mac}$$

$$\text{Hence} \quad r \quad = \quad \frac{[\epsilon(\infty) + 2]}{3m(\omega_T^2 - \omega^2)}e\mathbf{E}_{mac}$$

$$\mathbf{P} \quad = \quad \frac{Ne}{V}\frac{[\epsilon(\infty) + 2]^2}{3m(\omega_T^2 - \omega^2)}\frac{e\mathbf{E}_{mac}}{3} + \epsilon_0 \left[\epsilon(\infty) - 1\right]\mathbf{E}_{mac}$$

$$= \quad \epsilon_0 \left\{\epsilon(\infty) + \frac{\frac{Ne^2}{\epsilon_0 mV}\left(\frac{\epsilon(\infty)+2}{3}\right)^2}{(\omega_T^2 - \omega^2)} - 1\right\}\mathbf{E}_{mac}$$

$$= \quad \epsilon_0 \left[\epsilon(\omega) - 1\right]\mathbf{E}_{mac}$$

$$\text{where} \quad \epsilon(\omega) = \epsilon(\infty) + \frac{\frac{Ne^2}{\epsilon_0 mV}\left(\frac{\epsilon(\infty)+2}{3}\right)^2}{(\omega_T^2 - \omega^2)}$$

Note that, if $\omega \rightarrow \infty, \epsilon(\infty) = \epsilon_\infty$

$$\omega \rightarrow 0, \epsilon(0) = \epsilon(\infty) + \frac{Ne^2}{\epsilon_0 mV\omega_T^2}\left(\frac{\epsilon(\infty)+2}{3}\right)^2$$

$$\frac{Ne^2}{\epsilon_0 mV}\left(\frac{\epsilon(\infty)+2}{3}\right)^2 = \left[\epsilon(0) - \epsilon(\infty)\right]\omega_T^2$$

$$\epsilon(\omega) = \epsilon(\infty) + \frac{\left[\epsilon(0) - \epsilon(\infty)\right]\omega_T^2}{\omega_T^2 - \omega^2} \tag{11.56}$$

$$\text{Let} \quad S = \epsilon(0) - \epsilon(\infty)$$

$$\epsilon(\omega) = \epsilon(\infty) + \frac{S\omega_T^2}{\omega_T^2 - \omega^2} \tag{11.57}$$

where $\epsilon_{(\omega)}$ is the dielectric function.

When $\epsilon(\omega) = 0$, there are longitudinal oscillations with a frequency $\omega = \omega_L$. How are $\omega_L$ and $\omega_T$ related? Let us find out below.

$$\text{When} \quad \epsilon(\omega) = 0, \rightarrow 0 = \epsilon(\infty) + \frac{S\omega_T^2}{\omega_T^2 - \omega^2}$$

$$0 = \epsilon(\infty) + \frac{\left[\epsilon(0) - \epsilon(\infty)\right]\omega_T^2}{\omega_T^2 - \omega^2}$$

$$0 = \epsilon(\infty)\omega_T^2 - \epsilon(\infty)\omega_L^2 + \epsilon(0)\omega_T^2 - \epsilon(\infty)\omega_T^2$$

$$\omega_L^2 = \frac{\epsilon(0)}{\epsilon(\infty)}\omega_T^2$$

$$\omega_L = \left[\frac{\epsilon(0)}{\epsilon(\infty)}\right]^{1/2}\omega_T \tag{11.58}$$

which is known as the LYDDANE-SACHS-TELLER Relation (or LST Relation).

Generally, the *phonon-type* frequency dependent dielectric function is complex, of the form

$$\epsilon(\omega) = \epsilon(\infty) + \frac{S\omega_T^2}{\omega_T^2 - \omega^2 - i\omega\Gamma} \tag{11.59}$$

There are two cases of physical interest.

**Case 1: Zero damping** $(\Gamma = 0)$
With *zero damping*, the dielectric function reduces to

$$\epsilon(\omega) = \epsilon(\infty) + \frac{S\omega_T^2}{\omega_T^2 - \omega^2}$$

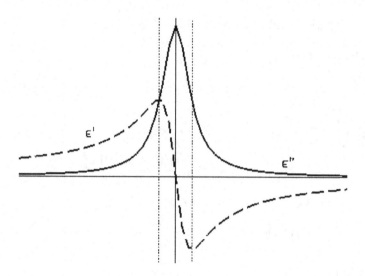

Figure 11.11: Phonon type dielectric function

It can be noted that the dielectric function is infinite at the resonant frequency $\omega_T$.

**Case 2: Finite damping** $(\Gamma \neq 0)$

With *finite damping*, the dielectric function reduces to

$$
\begin{aligned}
\epsilon(\omega) &= \epsilon(\infty) + \frac{S\omega_T^2}{(\omega_T^2 - \omega^2 - i\omega\Gamma)} \\
&= \epsilon(\infty) + \frac{S\omega_T^2}{(\omega_T^2 - \omega^2 - i\omega\Gamma)} \times \frac{(\omega_T^2 - \omega^2 + i\omega\Gamma)}{(\omega_T^2 - \omega^2 + i\omega\Gamma)} \\
&= \epsilon(\infty) + \frac{S\omega_T^2 \left(\omega_T^2 - \omega^2\right)}{\left[\left(\omega_T^2 - \omega^2\right)^2 + \omega^2\Gamma^2\right]} + i\frac{S\omega_T^2\omega\Gamma}{\left[\left(\omega_T^2 - \omega^2\right)^2 + \omega^2\Gamma^2\right]} \\
\epsilon(\omega) &= \epsilon'(\omega) + i\epsilon''(\omega)
\end{aligned}
$$

where

$$
\begin{aligned}
\epsilon'(\omega) &= \epsilon(\infty) + \frac{S\omega_T^2 \left(\omega_T^2 - \omega^2\right)}{\left[\left(\omega_T^2 - \omega^2\right)^2 + \omega^2\Gamma^2\right]} \\
\epsilon''(\omega) &= \frac{S\omega_T^2\omega\Gamma}{\left[\left(\omega_T^2 - \omega^2\right)^2 + \omega^2\Gamma^2\right]}
\end{aligned}
$$

where $\epsilon'(\omega)$ is the *Real Part of the dielectric function* and $\epsilon''(\omega)$ is the *Imaginary Part of the dielectric function*. These are illustrated graphically in Figure 11.11.

## 11.3.2 Bulk Phonon Polaritons

To study phonon-photon iteraction we use Maxwell's equations. Consider a material medium which is described by a dielectric function $\epsilon(\omega)$. The electric field $\mathbf{E}$ and magnetic field $\mathbf{B}$ satisfy Maxwell's equations which are given below in the differential form.

$$
\begin{aligned}
\nabla \cdot \mathbf{E} &= 0 \\
\nabla \cdot \mathbf{B} &= 0 \\
\nabla \wedge \mathbf{E} &= -\frac{\partial \mathbf{B}}{\partial t} \\
\nabla \wedge \mathbf{B} &= \mu_0 \frac{\partial \mathbf{D}}{\partial t}
\end{aligned}
$$

where the displacement vector $\mathbf{D} = \epsilon_0 \mathbf{E} + \mathbf{P} = \epsilon_0 \epsilon(\omega)\mathbf{E}$, which for vacuum is only $\mathbf{D} = \epsilon_0 \mathbf{E}$. It can be seen from Maxwell's equations that

$$
\begin{aligned}
\nabla \wedge (\nabla \wedge \mathbf{E}) &= \nabla \wedge \left(-\frac{\partial \mathbf{B}}{\partial t}\right) \\
&= -\frac{\partial}{\partial t}(\nabla \wedge \mathbf{B}) \\
&= -\frac{\partial}{\partial t}\left(\mu_0 \frac{\partial}{\partial t}\epsilon_0 \epsilon(\omega)\mathbf{E}\right) \\
\nabla(\nabla \cdot \mathbf{E}) - \nabla^2 \mathbf{E} &= -\mu_0 \epsilon_0 \epsilon(\omega)\frac{\partial^2 \mathbf{E}}{\partial t^2}
\end{aligned}
$$

Consider a solution of the electric field of the form

$$
\begin{aligned}
E &= E_0 e^{i(\mathbf{k}\cdot\mathbf{r}-\omega t)} \\
\frac{\partial \mathbf{E}}{\partial t} &= -i\omega \mathbf{E} \\
\frac{\partial^2}{\partial t^2} &= (-i\omega)(-i\omega \mathbf{E}) = -\omega^2 \mathbf{E}
\end{aligned}
$$

It can also be shown that

$$
\nabla^2 \mathbf{E} = -k^2 \mathbf{E}
$$

Hence, we obtain

$$
\begin{aligned}
0 - (-k^2\mathbf{E}) &= -\mu_0 \epsilon_0 \epsilon(\omega)(-)\omega^2 \mathbf{E} \\
k^2 &= \frac{1}{c^2}\epsilon(\omega)\omega^2 \\
\epsilon(\omega) &= \frac{c^2 k^2}{\omega^2} \tag{11.60}
\end{aligned}
$$

The above equation is the *dispersion relations* for *bulk phonon polaritons*, where the dielectric function is of the form

$$
\epsilon(\omega) = \epsilon(\infty) + \frac{S\omega_T^2}{\omega_T^2 - \omega^2}
$$

The bulk phonon polaritons dispersion curves are illustratd in Figure 11.12.

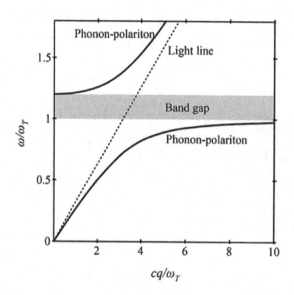

Figure 11.12: Bulk phonon polaritons dispersion curves.

## 11.4 Exercises

**11.1.** Consider a 1-d monatomic lattice of length $L$. Using the Debye model, show that
(a) The total thermal energy due to phonons is given by

$$U = \frac{L}{\pi \hbar v_s}(k_B T)^2 \int_0^{x_D} \frac{x}{e^x - 1}dx$$

where $x = \hbar\omega/k_B T$ and all other symbols are in the usual notation.
(b) The specific heat, $C_V$, is
(i) $Nk_B$ at high temperatures.
(ii) $\frac{\pi^2}{3}Nk_B\frac{T}{\theta_D}$ at low temperatures.
*Hints*:
(i) Density of states for a 1-d lattice, $D_1(\omega) = L/\pi v_s$
(ii) The following standard integral may be useful.

$$\int_0^\infty \frac{x}{e^x - 1}dx = \frac{\pi^2}{6}$$

**11.2.** Consider a 2-d monatomic lattice of area $A$. Using the Debye model, show that
(a) The total thermal energy due to phonons is given by

$$U = \frac{A}{\pi \hbar^2 v_s^2}(k_B T)^3 \int_0^{x_D} \frac{x^2}{e^x - 1}dx$$

where $x = \hbar\omega/k_B T$ and all other symbols are in the usual notation.
(b) The specific heat, $C_V$, is
(i) $2Nk_B$ at high temperatures.
(ii) $28.85 Nk_B(\frac{T}{\theta_D})^2$ at low temperatures.
*Hints*:
(i) Density of states for a 2-d lattice, $D_2(\omega) = A\omega/2\pi v_s^2$
(ii) The following standard integral may be useful.

$$\int_0^\infty \frac{x^2}{e^x - 1}dx = 2.404$$

**11.3.** The Debye density of states for a 3-d lattice is given as

$$D(\omega) = \frac{V}{2\pi^2}\frac{\omega^2}{v_s^3}$$

where all symbols are in the usual notation.
(a) Show that the Debye frequency, $\omega_D$, is given by

$$\omega_D = (\frac{6\pi^2 v_s^3 N}{V})^{1/3}$$

(b) Calculate an approximate value for the the Debye frequency, $\omega_D$, of aluminium, assuming the following:

Young's modulus for aluminium, $Y = 7.0 \times 10^{10}$N.m$^{-2}$

Density of aluminium, $\rho = 2.7 \times 10^3$kg.m$^{-3}$

Velocity, $v_s = \sqrt{\frac{Y}{\rho}}$, and take $N/V \approx 10^{23}$ atoms/cc.

(c) Show that the contribution of zero-point oscillations to the total thermal energy due to phonons in a 3-d lattice is

$$\frac{9N\hbar\omega_D}{8}$$

**11.4.** (a) The dispersion relation for phonons in a 1d *monatomic lattice* is given by

$$\omega = 2(\frac{G}{m})^{1/2}\sin\frac{1}{2}qa$$

where all symbols are in the usual notation of lattice dynamics.

(a) Plot the dispersion curves for phonons in a 1d monatomic lattice.

(b) Show that the total energy $E$ of a vibrational mode in a 1d monatomic lattice is given by

$$E = \frac{1}{2}m\sum_n(\frac{du_n}{dt})^2 + \frac{1}{2}G\sum_n(u_{n+1} - u_n)^2$$

where $u_n, u_{n+1}$ are displacements for the $n^{th}$ and $(n+1)^{th}$ atoms respectively.

(c) (i) Suppose $u_n = \frac{A}{\sqrt{N}}\cos(qna - \omega t)$ and $u_{n+1} = \frac{A}{\sqrt{N}}\cos(q(n+1)a - \omega t)$, show that the time averaged total energy, $<E>$, is given by

$$<E> = \frac{1}{2}m\omega^2 A^2$$

where you may assume that

$$\text{Time average, } <\sin^2(kx - \omega t)> = \frac{1}{2}$$

(ii) At room temperature, say 300 K, the vibrational mode in 1-d will have approximately the classical energy $k_BT$. Calculate the amplitude $A$ if $m = 64$ a.m.u. (1 a.m.u $= 1.66 \times 10^{-27}$ kg.) and $\omega = 2.0 \times 10^{13}$ s$^{-1}$

**11.5.** (a) State the basic assumptions used in the Debye model of specific heat theory.

(b) Assuming an expression for the total thermal energy $U$ for phonons in a 3d lattice given by

$$U = 3\int_0^{\omega_D}(\frac{V}{2\pi^2}\frac{\omega^2}{v_s^3})(\frac{\hbar\omega}{e^{\hbar\omega/k_BT} - 1})d\omega$$

show that the specific heat, $C_V$, is given by

$$C_V = 9Nk_B(\frac{T}{\theta_D})^3\int_0^{x_D}\frac{e^x x^4}{[e^x - 1]^2}dx$$

where $x = \hbar\omega/k_B T$ and all other symbols are in the usual notation.

**11.6**. Low temperature measurements for specific heat of Nickel are obtained as shown below.

| T (K) | | 2 | 4 | 6 | 8 | 10 | 12 | 14 | 16 |
|---|---|---|---|---|---|---|---|---|---|
| $C_V(\times 10^{-3}$Cal. mole $^{-1}$ K$^{-1}$) | | 4.1 | 8.0 | 13.2 | 18.6 | 23.8 | 30.4 | 37.9 | 48.3 |

Plot a suitable graph and deduce the Debye temperature of Nickel.

**11.7**. Show that the local electric field, $\mathbf{E}_{loc}$, is related to the macroscopic electric field, $\mathbf{E}_{mac}$, by the relation

$$\mathbf{E}_{loc} = \mathbf{E}_{mac} + \frac{\mathbf{P}_{elec}}{3\epsilon_0}$$

**11.8** (a) (i) Write down the expression for the *phonon-type* dielectric function, $\epsilon(\omega)$ *in the absence of damping*, defining all symbols used.
(ii) Hence plot schematically $\epsilon(\omega)$ vs $\omega$.
(b) Prove that the LST relation is given by

$$\omega_L = [\frac{\epsilon(0)}{\epsilon(\infty)}]^{1/2}\omega_T$$

where all symbols are in the usual notation.
(c) (i) Write down an expression for the dielectric function *in presence of damping*, and give its real and imaginary parts.
(ii) Hence plot schematically the real and imaginary parts of the dielectric function as a function of $\omega$.

**11.9**. (a) What is meant by a *polariton*?
(b) (i) Show that the dispersion relation for *bulk polaritons* is given by

$$\frac{c^2 q^2}{\omega^2} = \epsilon(\omega)$$

where all symbols are in the usual notation.
(ii) Plot schematically the dispersion curves of bulk phonon-type polaritons.
(c) Show that the group velocity, $v_g = \frac{\partial\omega}{\partial q}$, for bulk polaritons is given by

$$v_g = \frac{c}{\epsilon^{1/2}(\omega)[1 + \frac{\omega}{2\epsilon(\omega)}\frac{\partial\epsilon(\omega)}{\partial\omega}]}$$

**11.10**. (a) From definitions of the electronic polarisation vector $\vec{P}_{elec}$ in the following forms,

$$\vec{P}_{elec} = \epsilon_0[\epsilon(\infty) - 1]\vec{E}_{mac}$$

and

$$\vec{P}_{elec} = \frac{N\alpha_{elec}}{V}\vec{E}_{loc}$$

show that the Clausius-Mossotti relation is given by

$$\frac{[\epsilon(\infty) - 1]}{[\epsilon(\infty) + 2]} = \frac{N\alpha_{elec}}{3V\epsilon_0}$$

where you may assume that $E_{loc}$ and $E_{mac}$ are related by

$$\vec{E}_{loc} = \vec{E}_{mac} + \frac{\vec{P}_{elec}}{3\epsilon_0}$$

(b) A generalised form of the Clausius-Mossotti relation gives the dielectric function $\epsilon$ of a ferroelectric as satisfying

$$\frac{[\epsilon - 1]}{[\epsilon + 2]} = \frac{1}{3\epsilon_0} \sum N_j \alpha_j$$

Show that the dielectric function $\epsilon$ can also be written as

$$\epsilon = \frac{1 + \frac{2}{3\epsilon_0} \sum N_j \alpha_j}{1 - \frac{1}{3\epsilon_0} \sum N_j \alpha_j}$$

# Chapter 12

# Introduction to Materials Science III: Electrons

## 12.1 Introduction

Electrons are involved in all applications in electronics. In this section we shall review some properties of electrons as understood classically, and also quantum mechanically.

## 12.2 Classical theory of electrons in materials

The classical theory of electrons in mateials is also known as *Drudes's Theory*. It is successful in explaining a number of phenomena, although a proper theory must use quantum mchanics. Let us examine the explanation of classical theory of the following phenomena:

(i) Total thermal energy, $U$

$$\text{For one electron, thermal energy} \quad = \quad \frac{1}{2}mv^2 = \frac{3}{2}k_BT$$

$$\text{Total thermal energy for } n \text{ electrons, } U \quad = \quad \frac{3}{2}nk_BT \tag{12.1}$$

(ii) Specific Heat, $C_V$

$$
\begin{aligned}
\text{Specific Heat, } C_V \quad &= \quad \frac{\partial U}{\partial T} \\
&= \quad \frac{\partial}{\partial T}\left(\frac{3}{2}nk_BT\right) \\
C_V \quad &= \quad \frac{3}{2}nk_B
\end{aligned}
\tag{12.2}
$$

This result is *not good* at low temperaturs, that is, it does not agree with experiment, and therefore needs to be modified.

(iii) Thermal conductivity, $k$

Recall the result of thermal conductivity, $k$, obtained from kinetic theory, given by

$$k = \frac{1}{3}C_V v l \qquad (12.3)$$

where    $v$ is velocity

$l$ is mean free path and $l = v\tau$

(iv) Electrical conductivity, $\sigma$, is given by

$$\sigma = \frac{ne^2\tau}{m} \qquad (12.4)$$

$$= \frac{1}{\rho}$$

where    $\rho$ is resistivity

The proof for the expression for electrical conductivity given above is due to Drude's theory which we summarise below.

An electric current is due to electrons in a conductor, usually a metal. Consider a metal of length $l$ and cross sectional $A$ as illustrated in Figure 12.1. Let there be $n$ electrons per unit volume.

Let the conduction move with a drift velocity $v$.

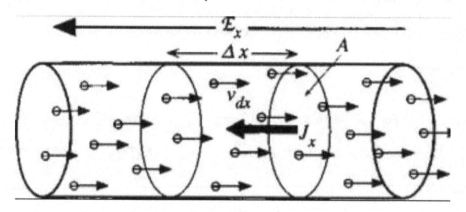

Figure 12.1: An electric current is due electrons in a metal

The number of electrons in a length $l$ is $nAl$. If each electron is of charge $e$, then the total charge, $q$ is given by

$$q = nAle$$

Crossing the length $l$ takes time, $t = \frac{l}{v}$. Then

The electric current, $I = \frac{q}{t}$

$$= \frac{nAle}{t}$$

$$= nAev$$

$$= jA \quad \text{where } j = nev \text{ is the current density, that is } j = \frac{I}{A}$$

Under an electric field $E$, there is a force, $\quad F \;=\; ma = eE$

But acceleration is given by, $\quad a \;=\; \frac{v}{\tau} \quad \text{where } \tau \text{ is relaxation time}$

Hence $\quad m\frac{v}{\tau} \;=\; eE$

$$v \;=\; \frac{\tau eE}{m}$$

Hence, the current density, $\quad j \;=\; nev = ne\left(\frac{eE\tau}{m}\right)$

$$j \;=\; \left(\frac{ne^2\tau}{m}\right)E$$

$$j \;=\; \sigma E \quad \text{which is a form of Ohm's law} \qquad (12.5)$$

where $\quad \sigma \;=\; \frac{ne^2\tau}{m} \quad \text{which is the electrical conductivity}$

$$\rho \;=\; \frac{1}{\sigma} = \frac{m}{ne^2\tau} \quad \text{where } \rho \text{ is the Resistivity} \quad (12.6)$$

The above form of Ohm's law is related to the familiar form of Ohm's as follows.

The current density, $\quad j \;=\; \sigma E = \frac{I}{A}$

The electric field, $\quad E \;=\; \frac{V}{l}$

$$\frac{I}{A} \;=\; \sigma\frac{V}{l}$$

$$\frac{I}{A} \;=\; \frac{1}{\rho}\frac{V}{l}$$

$$V \;=\; I\left(\frac{\rho l}{A}\right)$$

$$V \;=\; IR \quad \text{which is the familiar Ohm's law with } R = \frac{\rho l}{A} \qquad (12.7)$$

and $R$ is identified as Resistance, $\rho$ is the Resistivity. Resistivity is chareteristic of a material and varies from one material to another.

The electrical conductivity of a metal such as Alumin um decreases with temperature, while the conductivity of a semiconductor such as Germanium increases with temperature, as illustrated in Figure 12.2. The electrical resistivity of a metal and superconductor increases with temperature, while the resistivty of a semiconductor decreases with temperature, as illustrated in Figure 12.3.

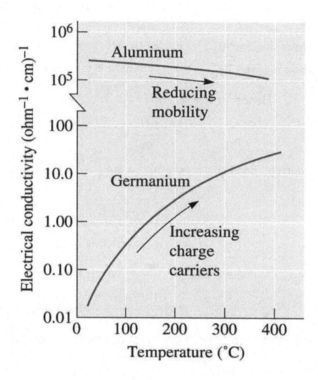

Figure 12.2: Variation of electrical conductivity with temperature for a metal (Aluminum) and a semiconductor (Germanium).

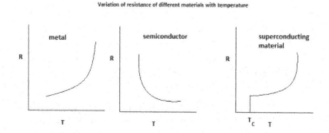

Figure 12.3: Resistivity variation with Temperature for a Metal (Normal and Superconductor) and Semiconductor

(v) Wiedemann-Franz Law: Ratio of thermal conductivity to electrical conductivity

$$\frac{k}{\sigma} = \frac{\frac{1}{3}C_V vl}{ne^2\tau/m}$$
$$= \frac{1}{3}\frac{C_V vlm}{ne^2\tau}$$
$$= \frac{1}{3}\left(\frac{3}{2}nk_B\right)\frac{mv^2}{ne^2}$$
$$= \frac{1}{3}\left(\frac{3}{2}nk_B\right)\frac{3k_BT}{ne^2}$$

$$\frac{k}{\sigma} = \frac{3}{2}\left(\frac{k_B}{e}\right)^2 T \text{ which is known as WIEDEMANN-FRANZ LAW} \tag{12.8}$$

It can be noted that

$$\frac{k}{\sigma T} = \frac{3}{2}\left(\frac{k_B}{e}\right)^2 = 1.11 \times 10^{-8} \text{ W. ohm.K}^{-2} \tag{12.9}$$

is a constant, known as the LORENTZ NUMBER. The experimental value, however, is $2.21 \times 10^{-8}$ W. ohm.K$^{-2}$, which is *twice* the value predicted by this classical theory. There is therefore a problem with the classical theory.

## 12.3  Quantum Theory of electrons in materials

### 12.3.1  Free electron theory

**Basic assumptions in free electron theory**

Let us consider *a particle in a box*. Consider a quantum particle confined in a cubical box of side $L$,whose potential function is such that

$$V(x, y, z) = \begin{cases} 0, & \text{inside the box} \\ \infty, & \text{outside the box} \end{cases} \tag{12.10}$$

as illustrated in Figure 12.4.

**Equation of motion:  The Schrödinger equation**

For the particle in a box, the Schrödinger equation becomes

$$-\frac{\hbar^2}{2m}\nabla^2\Psi(\mathbf{r}) = E\Psi(\mathbf{r}) \tag{12.11}$$

$$-\frac{\hbar^2}{2m}\left(\frac{\partial^2}{\partial x^2} + \frac{\partial^2}{\partial y^2} + \frac{\partial^2}{\partial z^2}\right)\Psi(x, y, z) = E\Psi(x, y, z)$$

Consider a solution

$$\Psi(x) = Ae^{ik_x x} + Be^{-ik_x x} \tag{12.12}$$

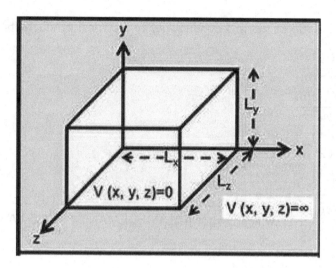

Figure 12.4: Particle in a box.

Applying boundary conditions, $\Psi(x = 0) = \Psi(x = L)$, one obtains $A + B = 0$, or $B = -A$, and hence

$$\Psi(x) = A(e^{ik_x x} - e^{-ik_x x}) = 2iA \sin k_x x = C \sin k_x x \qquad (12.13)$$

which gives $k_x L = n_x \pi$ or $k_x = n_x \pi / L$, and hence $p_x = \hbar k_x = n_x \pi \hbar / L$. Similarly for $\Psi(y)$ and $\Psi(z)$. The wavefunction for a particle in the box is a solution of the Schrödinger equation, and is given by

$$\begin{aligned}
\Psi(x, y, z) &= \sin k_x x \sin k_y y \sin k_z z \\
&= \sin(\frac{n_x \pi x}{L}) \sin(\frac{n_y \pi y}{L}) \sin(\frac{n_z \pi z}{L})
\end{aligned}$$

Using this in the Schrödinger equation, one obtains

$$\frac{\hbar^2}{2m} \left( k_x^2 + k_y^2 + k_z^2 \right) \Psi(x, y, z) = E\Psi(x, y, z) \qquad (12.14)$$

which gives the energies as

$$\begin{aligned}
E &= \frac{\hbar^2 k^2}{2m} \\
&= \frac{\hbar^2}{2m} \left( k_x^2 + k_y^2 + k_z^2 \right) \\
E_n &= \frac{\pi^2 \hbar^2}{2mL^2} (n_x^2 + n_y^2 + n_z^2) \qquad (12.15)
\end{aligned}$$

where $n_x, n_y, n_z$ take integral values, and the total energy of the particle in a box for state $n$ is quantized.

**Electron Density of states, $g(E)$**

The electron density of states $g(E)$ is the number of allowed states between energy interval $E$ and $E + dE$. To find the number of states, $N(E)$, we must find the volume of an octant of volume of radius $k$ (See Figure 12.5), that is

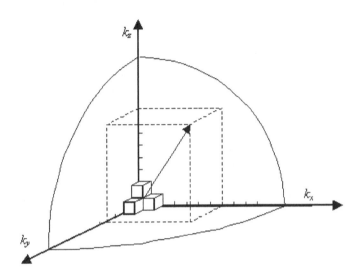

Figure 12.5: An element of spherical shell of radius $k$.

$$V_k = \frac{4}{3}\pi k^3$$

$$\text{Hence, an element has volume, } dV_k = 4\pi k^2 dk$$

$$\text{Considering a shell with one eigth of the volume, } V_s = \frac{1}{8}dV_k$$

$$= \frac{1}{8}4\pi k^2 dk$$

$$= \frac{1}{2}\pi k^2 dk$$

Density of states, $g(E)$, satisfies

$$g(E)dE = (2)\frac{V_s}{(\pi/L)^3}$$

where multiplication by 2 above is due to spin up and spin down

$$g(E)dE = (2)\frac{\frac{1}{2}\pi k^2 dk}{(\pi/L)^3}$$

$$= \frac{L^3}{\pi^2}k^2 dk$$

$$= \frac{Vk^2}{\pi^2}dk, \text{ where } V = L^3$$

$$\text{But for free electrons, } E = \frac{\hbar^2 k^2}{2m}$$

$$\text{Hence, } dE = \frac{\hbar^2 k}{m} dk$$

$$\text{Hence, } g(E)dE = \frac{V}{\pi^2}\left(\frac{2mE}{\hbar^2}\right)\frac{m}{\hbar^2 k} dE$$

$$= \frac{V}{\pi^2}\left(\frac{2mE}{\hbar^2}\right)\left(\frac{m}{\hbar^2}\right)\left(\frac{\hbar^2}{2mE}\right)^{1/2} dE$$

$$= \frac{V}{\pi^2}\frac{2m^2 E}{\hbar^4}\frac{\hbar}{(2mE)^{1/2}} dE$$

$$= \frac{V}{\pi^2}\frac{(2mE)}{\hbar^3}\frac{m}{(2mE)^{1/2}} dE$$

$$\text{Hence, } g(E) = \frac{V}{2\pi^2}\left(\frac{2m}{\hbar^2}\right)^{3/2} E^{1/2} \qquad (12.16)$$

A graph of the density of states, $g(E)$ against energy $E$ is illustrated in Figure 12.6.

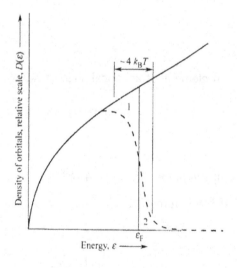

Figure 12.6: Electron Density of States in 3d

## Fermi-Dirac Distribution

Electrons are *fermions* and they satisfy Fermi-Dirac statistics where the Fermi-Dirac distribution function is given by

$$f(E) = \frac{1}{e^{(E-\mu)/k_B T} + 1} \qquad (12.17)$$

where $\mu = \mu(T)$ is the chemical potential, and $\mu(T = 0) = E_F$, and $E_F$ is the Fermi Energy. A graphical variation of $f(E)$ is illustrated in Figure 12.7, where $< n_k >= f(E)$.

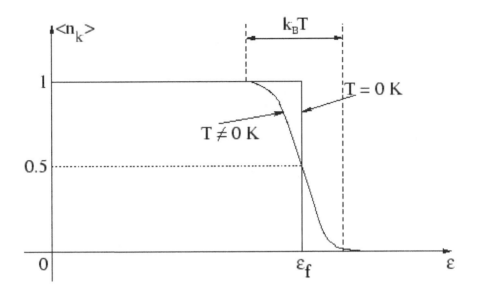

Figure 12.7: Fermi Dirac Statistics Distribution Function

**Number of Particles, $N(E)$**

The number of particles, $N(E)$, is defined as

$$N(E) = \int_0^\infty g(E)f(E)dE$$

where $g(E)$     is the electron density of states

$f(E)$     is the Fermi-Dirac Distribution functions

It is of interest to evaluate the number of particles in $T = 0$ limit.

$$
\begin{aligned}
N(E) &= \int_0^\infty g(E)f(E)dE & (12.18)\\
&= \int_0^\infty g(E)dE \text{ since } f(E) = 1 \text{ for } E < E_F\\
&= \int_0^{E_F} \frac{V}{2\pi^2}\left(\frac{2m}{\hbar^2}\right)^{3/2}E^{1/2}dE\\
&= \frac{V}{2\pi^2}\left(\frac{2m}{\hbar^2}\right)^{3/2}\frac{2E^{3/2}}{3}\Big|_0^{E_F}\\
&= \frac{V}{3\pi^2}\left(\frac{2m}{\hbar^2}\right)^{3/2}E_F^{3/2} & (12.19)
\end{aligned}
$$

which can be rearranged to give $E_F$

$$E_F = \frac{\hbar^2}{2m}\left(\frac{3\pi^2 N}{V}\right)^{2/3} \quad \text{is the FERMI ENERGY} \quad (12.20)$$

or

$$E_F = \frac{\hbar^2 k_F^2}{2m}$$

where =

$$k_F = \left(\frac{3\pi^2 N}{V}\right)^{1/3} \quad \text{is the FERMI WAVEVECTOR} (12.21)$$

**Electronic Total thermal energy, $U_{TOT}$**

The electronic total energy, $U_{TOT}$), is defined as

$$u_{TOT} = \int_0^\infty Eg(E)f(E)dE$$

where $g(E)$    is the electron density of states

$f(E)$    is the Fermi-Dirac Distribution functions

There are two cases of physical interest.

**Case (i) Zero temperature limit, $T = 0$**

$$U_{TOT}^0 = \int_0^\infty Eg(E)f(E)dE \qquad (12.22)$$

$$= \int_0^\infty Eg(E)dE \text{ since } f(E) = 1 \text{ for } E < E_F$$

$$= \int_0^{E_F} E\frac{V}{2\pi^2}\left(\frac{2m}{\hbar^2}\right)^{3/2} E^{1/2}dE$$

$$= \int_0^{E_F} \frac{V}{2\pi^2}\left(\frac{2m}{\hbar^2}\right)^{3/2} E^{3/2}dE$$

$$= \frac{V}{2\pi^2}\left(\frac{2m}{\hbar^2}\right)^{3/2} \int_0^{E_F} E^{3/2}dE$$

$$= \frac{V}{2\pi^2}\left(\frac{2m}{\hbar^2}\right)^{3/2} E_F^{5/2}\frac{2}{5}$$

$$= \frac{3NE_F^{5/2}}{5E_F^{3/2}}$$

$$U_{TOT}^0 = \frac{3}{5}NE_F \qquad (12.23)$$

**Case (ii) Finite temperature limit, $T \neq 0$, but$T \ll T_F$**

$$U_{TOT} = \int_0^\infty Eg(E)f(E)dE$$

$$= \int_0^\infty E \frac{V}{2\pi^2} \left(\frac{2m}{\hbar^2}\right)^{3/2} E^{1/2} \frac{1}{e^{(E-\mu)/k_B T} + 1} dE$$

$$= \int_0^\infty A \frac{E^{3/2}}{e^{(E-\mu)/k_B T} + 1} dE \text{ where } A = \frac{V}{2\pi^2} \left(\frac{2m}{\hbar^2}\right)^{3/2}$$

$$\text{Let } x = \frac{E}{k_B T} \Rightarrow k_B T dx = dE$$

$$U_{TOT} = A \int_0^\infty \frac{(k_B T)^{3/2} (k_B T) x^{3/2}}{e^{(x-\beta\mu)} + 1} dx \text{ where } \beta = \frac{1}{k_B T}$$

$$= A\beta^{-5/2} \int_0^\infty \frac{x^{3/2}}{e^{(x-\beta\mu)} + 1} dx$$

Using tables of integrals, we obtain

$$U_{TOT} = \frac{3}{5} N E_F \left\{ 1 + \frac{5\pi^2}{12} \left(\frac{k_B T}{E_F}\right)^2 \right\} \tag{12.24}$$

**Mean energy, $\overline{E}$**

The mean energy, $\overline{E}$, is defined as

$$\overline{E} = \frac{\int_0^\infty E g(E) f(E) dE}{\int_0^\infty g(E) f(E) dE}$$

$$= \frac{U_{TOT}}{N(E)}$$

There are two cases of physical interest.

**Case (i) Zero temperature limit, $T = 0$**

$$\overline{E_0} = \frac{\int_0^\infty E g(E) f(E) dE}{\int_0^\infty g(E) f(E) dE}$$

$$= \frac{U_{TOT}^0}{N(E}$$

$$= \frac{\frac{3}{5} N E_F}{N}$$

$$\overline{E_0} = \frac{3}{5} E_F \tag{12.25}$$

**Case (ii) Finite temperature limit, $T \neq 0$, but$T \ll T_F$**

$$\overline{E} = \frac{\int_0^\infty E g(E) f(E) dE}{\int_0^\infty g(E) f(E) dE}$$

$$= \frac{U_{TOT}}{N(E)}$$

$$= \frac{3}{5}\frac{NE_F}{N}\left\{1 + \frac{5\pi^2}{12}\left(\frac{k_BT}{E_F}\right)^2\right\}$$

$$= \frac{3}{5}E_F\left\{1 + \frac{5\pi^2}{12}\left(\frac{k_BT}{E_F}\right)^2\right\} \tag{12.26}$$

**Electronic Specific heat, $C_V$**

$$C_V = \frac{\partial U_{TOT}}{\partial T} \tag{12.27}$$

$$= \frac{\partial}{\partial T}\left[\frac{3}{5}E_F\left\{1 + \frac{5\pi^2}{12}\left(\frac{k_BT}{E_F}\right)^2\right\}\right]$$

$$= \frac{\pi^2}{2}\frac{NK_B^2 T}{E_F}$$

$$C_V = \gamma T \text{ where } \gamma = \frac{\pi^2}{2}\frac{NK_B^2}{E_F} \tag{12.28}$$

The total specific heat at $T \ll T_F$ and $T \ll T_D$ in a material consists of the sum of the phonon contribution and the electronic contribution.

$$C_V = \gamma T + BT^3 \tag{12.29}$$

$$\text{where } \gamma = \frac{\pi^2}{2}\frac{NK_B^2}{E_F}$$

$$B = \frac{12\pi^4}{5}Nk_B\left(\frac{1}{\theta_D}\right)$$

where the first term is the electronic contribution linear in $T$, and the second term is the phonon contribution which is cubic in $T$. Rearranging the above equation, we obtain

$$\frac{C_V}{T} = \gamma + BT^2 \tag{12.30}$$

A graph of $\frac{C_V}{T}$ against $T^2$ is a straight line with a slope $B$ and intercept $\gamma$, as illustrated in Figure 12.8. In the same figure, there is an insert showing a plot of $\frac{C_V}{T}$ against $T$ with a discontinuity in the specific heat at very low temperatures.

## 12.3.2  Kronig-Penney Model

Consider an electron moving in a periodic array of rectangular potential wells. This model is referred to as THE KRONIG-PENNEY model. The potential can be defined as follows.

$$V(x) = \begin{cases} 0, & \text{for } 0 < x < a \text{ Region 1 and periodic in such regions} \\ V_0, & \text{for } -b < x < 0 \text{ Region 2 and periodic in such regions} \end{cases} \tag{12.31}$$

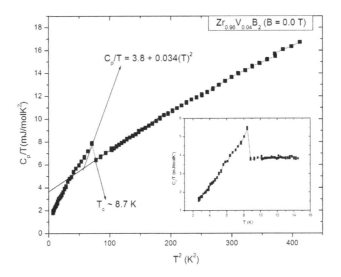

Figure 12.8: Specific Heat contributions by electrons and phonons at low temperatures. The graph in the insert shows a plot of $\frac{C_V}{T}$ against $T$ with a discontinuity in the specific heat at very low temperatures. (source:www.intechopen.com/A.J.S. Machado et.al.)

as illustrated in Figure 12.9. The wavefunction for such an electron moving in a periodic potential in a lattice must be periodic and is known as a BLOCH FUNCTION, satisfying

$$\Psi(x) = U_k(x)e^{ikx} \tag{12.32}$$

or

$$U_k(x) = e^{-ikx}\Psi \tag{12.33}$$

where $U_k(x)$ must also be periodic with the lattice, that is,

$$U_k(a) = U_k(-b) \tag{12.34}$$

**Equation of motion: The Schrödinger equation**

Applying the Schrödinger equation,
In region 1, $0 < x < a$,

$$-\frac{\hbar^2}{2m}\frac{\partial^2}{\partial x^2}\Psi = E\Psi \tag{12.35}$$

In region 2, $-b < x < 0$,

$$-\frac{\hbar^2}{2m}\frac{\partial^2}{\partial x^2}\Psi + V_0\Psi = E\Psi \tag{12.36}$$

These equations can be rearranged to become

$$\frac{\partial^2}{\partial x^2}\Psi + \alpha^2\Psi = 0$$

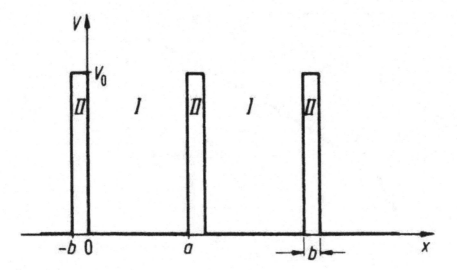

Figure 12.9: Kronig Penney Model

$$\frac{\partial^2}{\partial x^2}\Psi - \beta^2\Psi = 0$$

where

$$\alpha^2 = \frac{2mE}{\hbar^2}$$

$$\beta^2 = \frac{2mE}{\hbar^2}[V_0 - E]$$

Solutions are of the form

$$\Psi_1(x) = A^{i\alpha x} + Be^{-i\alpha x}$$
$$\Psi_2(x) = C^{\beta x} + De^{-\beta x}$$

and

$$U_1(x) = A^{i(\alpha-k)x} + Be^{-i(\alpha+k)x}$$
$$U_2(x) = C^{(\beta-ik)x} + De^{-(\beta+ik)x}$$

To evaluate the four constants $A, B, C$ and $D$, we need four equations. These equations arise from the boundary conditions.

$$U_1(0) = U_2(0)$$
$$U_1(a) = U_2(-b)$$
$$\left(\frac{dU_1}{dU_2}\right)_{x=0} = \left(\frac{dU_2}{dU_2}\right)_{x=0}$$
$$\left(\frac{dU_1}{dU_2}\right)_{x=a} = \left(\frac{dU_2}{dU_2}\right)_{x=-b}$$

From the above equations, a solution exits if

$$\left(\frac{\beta^2 - \alpha^2}{2\alpha\beta}\right) \sinh \beta b \sin \alpha a + \cosh \beta b \cos \alpha a = \cos k(a+b) \qquad (12.37)$$

The above equation is exact. Let us study in the limit of $\delta$-function potentials, which are

$$V_0 \quad \rightarrow \quad \infty \text{ (large)}$$
$$b \quad \rightarrow \quad 0 \text{ (small)}$$
$$\text{such that } V_0 b \quad \text{is finite}$$

Under these conditions, we obtain

$$\beta \quad \rightarrow \quad \infty$$
$$\beta b \quad \text{is finite}$$
$$\sinh \beta b \quad \rightarrow \quad \beta b$$
$$\cosh \beta b \quad \rightarrow \quad 1$$
$$\beta^2 - \alpha^2 \quad = \quad \frac{2m}{\hbar^2}[V_0 - E] - \frac{2mE}{\hbar^2}$$
$$= \quad \frac{2m}{\hbar^2}[V_0 - 2E] \approx \frac{2mV_0}{\hbar^2} \text{ since } V_0 \text{ is very large}$$

Hence, we obtain

$$\frac{2mV_0}{\hbar^2 2\alpha\beta}\beta b \sin \alpha a + \cos \alpha a \quad = \quad \cos ka$$
$$\left(\frac{mV_0 ba}{\hbar^2 \alpha a}\right) \sin \alpha a + \cos \alpha a \quad = \quad \cos ka$$
$$P\frac{\sin \alpha a}{\alpha a} + \cos \alpha a \quad = \quad \cos ka \qquad (12.38)$$

where $P = \frac{mV_0 ba}{\hbar^2}$. We plot the LHS of the above equation against $\alpha a$ for $P = \frac{3\pi}{2}$ in Figure 12.10. Note that the RHS can only take values between -1 and +1.

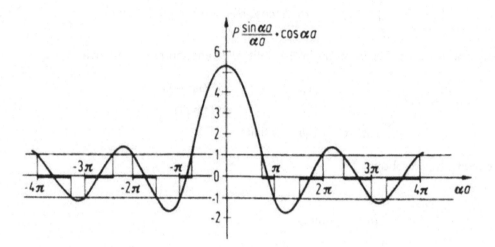

Figure 12.10: Kronig Penney Graph

## REMARKS ON THE KRONIG-PENNEY MODEL

1. Only certain values of $\alpha a$ are allowed, and others are forbidden. This means only certain values of energy $E$ are allowed and some are forbidden i.e ENERGY BANDS and ENERGY GAPS ARISE.

2. The **width** of allowed energies **increases** with increasing values of $\alpha a$.

3. The **width** of allowed energies **decreases** with increasing values of $P$.

4. When $P = 0$ all energies are allowed, that is we recover the free electron model, that is

$$
\begin{aligned}
\cos \alpha a &= \cos ka \\
\alpha &= k \Rightarrow \alpha^2 = k^2 = \frac{2mE}{\hbar^2} \\
&\text{or} \\
E &= \frac{\hbar^2 k^2}{2m}
\end{aligned}
$$

which is the free electron model.

5. **Dispersion curves** ($E$ vs $k$ graph)

Note the $E$ vs $k$ graph (See Figure 12.10 and 12.11) has discontinuities when the wave vector has values

$$k = \frac{n\pi}{a} \tag{12.39}$$

Figure 12.11 is plotted in what is called the EXTENDED ZONE SCHEME. The graph in the extended zone scheme can be compressed into a scheme where the wavevectors are in the region

$$-\frac{\pi}{a} < k < \frac{\pi}{a} \tag{12.40}$$

and this is known as the REDUCED ZONE SCHEME, as illustrated in Figure 12.12.

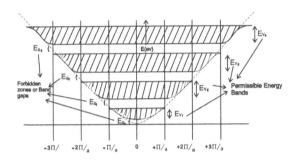

Figure 12.11: E vs k in Extended Zone Scheme

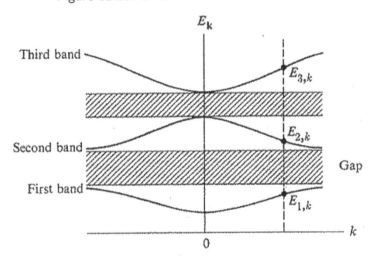

Figure 12.12: E vs k in Reduced Zone Scheme, within $-\frac{\pi}{a} < k < \frac{\pi}{a}$

## 6. Conduction Band (CB) and Valence Band (VB)

Consideration of the highest energy bands results into two important bands known as the

- Conduction band (CB)

- Valence band (VB)

## 12.4   Classification of materials in terms of energy band structures: Metals, Semiconductors and Insulators

Materials are classified into metals, semiconductors and insulators depending on their energy band structures, as illustrtated in Figure 12.13.

- **Metals** are materials in which the conduction band (CB) and valence band overlap.

- **Semiconductors** are materials in which the conduction band (CB) and valence band are separated by an energy gap of the order of 2 eV or less.  These materials are of intermediate character between metals and insulators.

- **Insulators** are materials in which the conduction band (CB) and valence band are separated by an energy gap of the order of 2 eV or larger.

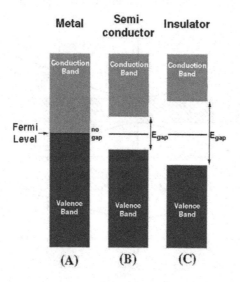

Figure 12.13: Classification of Metals, Semiconductors and Insulators in terms of energy bands.

## 12.5 Metals

### 12.5.1 Plasmon type dielectric function

Plasmons are quanta of a plasma. A plasma is a medium with equal concentration of positive charges and negative charges such that at least one charge type is mobile. Consider an electron in an electric field. If each electron is of mass $m$ in a crystal under the influence of an electric field **E** has a displacement $x$. This is essentially a force constant problem.

**Equation of motion**

$$m\frac{d^2x}{dt^2} = \sum \mathbf{F}$$
$$= -eE$$
$$\text{If} \quad x = x_0 e^{-i\omega t}$$
$$\dot{x} = -i\omega x$$
$$\ddot{x} = (-i\omega)(-i\omega)x = -\omega^2 x$$
$$-m\omega^2 x = -eE$$
$$x = \frac{eE}{m\omega^2}$$
$$\text{Dipole moment of one electron} = -ex$$
$$= -\frac{e^2 E}{m\omega^2}$$
$$\text{Hence, Polarisation,} \quad P = \text{dipole moment per unit volume}$$
$$= -nex$$
$$P = -\frac{ne^2 E}{m\omega^2} \tag{12.41}$$

where $n$ is the concentration, that is, the number of electrons per unit volume.

$$\text{Hence, the displacement vector,} \quad \mathbf{D} = \epsilon_0 \mathbf{E} + \mathbf{P} = \epsilon_0 \epsilon(\omega)\mathbf{E}$$
$$\epsilon(\omega) = 1 + \frac{P}{\epsilon_0 E}$$
$$= 1 - \frac{ne^2}{\epsilon_0 m\omega^2}$$
$$\epsilon(\omega) = 1 - \frac{\omega_p^2}{\omega^2} \tag{12.42}$$
$$\text{where} \quad \omega_p^2 = \frac{ne^2}{\epsilon_0 m} \quad \text{and } \omega_p \text{ is the plasma frequency.} \tag{12.43}$$

The *plasmon-type* frequency dependent dielectric function withdamping is of the form

$$\epsilon(\omega) = 1 - \frac{\omega_p^2}{\omega^2 + i\omega\Gamma}$$

where $\Gamma$ is a damping parameter.

There are two cases of physical interest.

**Case 1: Zero damping** $(\Gamma = 0)$
With *zero damping*, the dielectric function reduces to

$$\epsilon(\omega) = 1 - \frac{\omega_p^2}{\omega^2}$$

It can be noted that the dielectric function is zero at the frequency $\omega_p$, as illustrated in Figure 12.14.

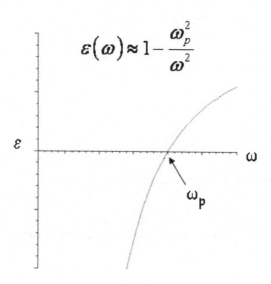

Figure 12.14: Plasmon type dielectric function

**Case 2: Finite damping** $(\Gamma \neq 0)$
With *finite damping*, the plasmon-type dielectric function reduces to

$$\begin{aligned}
\epsilon(\omega) &= 1 - \frac{\omega_p^2}{(\omega^2 + i\omega\Gamma)} \\
&= 1 - \frac{\omega_p^2}{(\omega^2 + i\omega\Gamma)} \times \frac{(\omega^2 - i\omega\Gamma)}{(\omega^2 - i\omega\Gamma)} \\
&= 1 - \frac{\omega_p^2\omega^2}{[\omega^4 + \omega^2\Gamma^2]} + i\frac{\omega_p^2\omega\Gamma}{[\omega^4 + \omega^2\Gamma^2]} \\
\epsilon(\omega) &= \epsilon'(\omega) + i\epsilon''(\omega) \\
\textit{where} \\
\epsilon'(\omega) &= 1 - \frac{\omega_p^2\omega^2}{[\omega^4 + \omega^2\Gamma^2]} \\
\epsilon''(\omega) &= \frac{\omega_p^2\omega\Gamma}{[\omega^4 + \omega^2\Gamma^2]}
\end{aligned}$$

where $\epsilon'(\omega)$ is the *Real Part of the dielectric function* and $\epsilon''(\omega)$ is the *Imaginary Part of the dielectric function*.

### 12.5.2 Plasmon-Photon Interactions: Polaritons

Plasmon-photon interactions are studied by solving Maxwell's equation as was done in chapter 1 (See section 1.5) or chapter 11 (see section 11.3.2), but using the plasmom type dielectric function

$$\epsilon(\omega) = 1 - \frac{\omega_p^2}{\omega^2}$$

The bulk plasmon dispersion equation is given by

$$\frac{c^2 q^2}{\omega^2} = \epsilon(\omega) \tag{12.44}$$

while the surface plasmon dispersion equation is given by

$$\frac{c^2 q_{1x}^2}{\omega^2} = \frac{\epsilon_1 \epsilon(\omega)}{\epsilon_1 + \epsilon(\omega)} \tag{12.45}$$

The dispersion curves for bulk and surface plasmon type polaritons are illustrated in Figure 12.15.

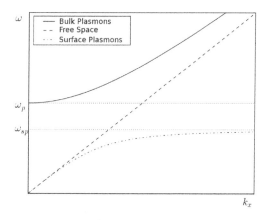

Figure 12.15: Bulk and Surface Plasmon type polaritons.

## 12.6 Semiconductors

### 1. Introduction

Recall a section of the periodic table as shown in Table 1, showing Gp IIB, III, IV, V and VI, where a number of semiconducting elements and constituents of semiconducting compounds occur.

| IIB | III | IV | V  | VI |
|-----|-----|-----|-----|-----|
|     | B   | C  | N  | O  |
|     | Al  | Si | P  | S  |
| Zn  | Ga  | Ge | As | Se |
| Cd  | In  | Sn | Sb | Te |
| Hg  | Tl  | Pb | Bi | Po |

Table 12.1: Some elements in Group IIB, III, IV, V and VI.

## 2. Semiconducting elements and Semiconducting compounds

- Semiconducting elements include Gp IV elements Si, Ge, S (Grey) and Gp VI element Te

- Semiconducting compounds include
  (a) III-V compounds Eg GaP, GaAs, InSb
  (b) IV-IV compounds Eg SiC
  (c) IV-VI compounds Eg PbS, PbSe, PbTe
  (d) II-VI compounds Eg ZnO, ZnS, ZnSE, CdS, CdTe

## 3. Direct gap and Indirect gap semiconductors

The energy gap between the CB and VB can be *direct* or *indirect*. When the CB edge and VB edge have the same $k$-value, then the material is a **DIRECT GAP SEMICONDUCTOR**, while if the $k$-value of the CB edge and VB edge have are different, then the material is an **INDIRECT GAP SEMICONDUCTOR**.

Note that

(a) In a direct gap semiconductor, there can be "direct transition", and conservation of energy requires

$$\hbar\omega = E_g \tag{12.46}$$

where $E_g$ is the energy gap, in either emission as in Figure 12.16 or absorption or emission as in Figure 12.17.

(b) In an indirect gap semiconductor, there can be an "Indirect transition" which is a phonon assisted transition. and conservation of energy requires

$$\hbar\omega + \hbar\Omega = E_g \tag{12.47}$$

where $E_g$ is the energy gap, and $\hbar\Omega$ is the energy of the phonon, in either emission as in Figure 12.16 or absorption or emission as in Figure 12.17.

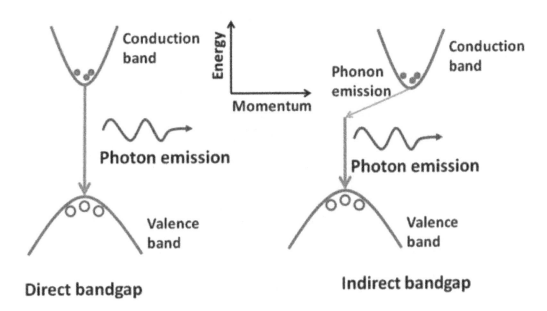

Figure 12.16: Direct gap and Indirect gap Semiconductors.

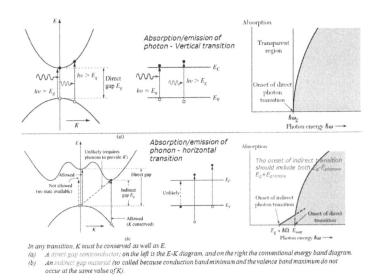

Figure 12.17: Direct gap and Indirect gap Semiconductors Absorption and Emission.

**4. Energy gap,$E_g$**

| Crystal | $d$ or $i$ | $E_g(eV)$ $T = 0K$ | $E_g(eV)$ $T = 300K$ |
|---------|-----------|--------------------|----------------------|
| Si      | i         | 1.17               | 1.14                 |
| Ge      | i         | 0.74i              | 0.67                 |
| InSb    | d         | 0.24               | 0.18                 |
| InP     | d         | 1.42               | 1.33                 |
| GaP     | i         | 2.32               | 2.26                 |

Table 12.2: Some values of Energy gaps for indirect and direct gap semiconductors.

**5. The effective mass of an electron in a CB**

Consider an electron moving in some allowed energy band. The following relation will be satisfied

$$E = \frac{p^2}{2m} = \frac{\hbar^2 k^2}{2m}$$

$$\frac{dE}{dp} = \frac{p}{m} = v$$

$$\text{Hence } v = \frac{dE}{dp} = \frac{1}{\hbar}\frac{dE}{dk}$$

$$\frac{dv}{dt} = \frac{d}{dt}\left(\frac{1}{\hbar}\frac{dE}{dk}\right) = \frac{1}{\hbar}\frac{d}{dk}\left(\frac{dE}{dt}\right)$$

$$= \frac{1}{\hbar}\frac{d}{dk}\left(\frac{dE}{dk}\frac{dk}{dt}\right) = \frac{1}{\hbar}\frac{d^2E}{dk^2}\frac{1}{\hbar}\frac{d(\hbar k)}{dt}$$

$$= \frac{1}{\hbar}\frac{d^2E}{dk^2}\frac{1}{\hbar}\frac{dp}{dt}$$

$$= \frac{1}{\hbar^2}\frac{d^2E}{dk^2}\sum F$$

$$\frac{dv}{dt} = \frac{\sum F}{m^*}$$

$$\text{where } \frac{1}{m^*} = \frac{1}{\hbar^2}\frac{d^2E}{dk^2} \quad \text{where } m^* \text{ is the effective mass}$$

$$\text{or}$$

$$\text{Hence } m^* = \frac{\hbar^2}{(d^2E/dk^2)} = \left(\frac{1}{\hbar^2}\frac{d^2E}{dk^2}\right)^{-1} \tag{12.48}$$

Note that the effective mass $m^*$ is inversely proportional to the second derivatve of the $E$ vs $k$ dispersion curve.

## 6. Electron density of states in a semiconductor and concentration of electrons and holes

To get the density of states for electrons and holes in a semiconductor, we shall start from the free electron theory density of states, $g(E)$, and modify it to suit the semiconductor situation. For unit volume, $g(E)$, is given by

$$g(E) = \frac{1}{2\pi^2}\left(\frac{2m}{\hbar^2}\right)^{3/2} E^{1/2} \tag{12.49}$$

There are two cases of physical interest.

### Case (a) Electrons in the CB.

The following modifications are made for electrons in the CB.
(i) Replace the free electron mass, $m$, by the effective mass of the electron, $m_e^*$, that is, $m \to m_e^*$,
(ii) Replace energy, $E$, by $(E - E_c)$, that is, $E \to (E - E_c)$, where $E_c$ is the conduction band edge (bottom of the CB),
Hence, the electron density of states in the conduction band, $\eta_c(E)$, is given by

$$\eta_c(E) = \frac{1}{2\pi^2}\left(\frac{2m_e^*}{\hbar^2}\right)^{3/2}(E - E_c)^{1/2} \tag{12.50}$$

The concentration of electrons in the CB is given by

$$n_e = \int_{E_c}^{\infty} \eta_c(E) f(E)\, dE$$

$$\text{where} \quad f(E) = \frac{1}{e^{(E-E_F)/k_BT} + 1} \text{ is the Fermi-Dirac distribution}$$

$$\text{If} \quad (E - E_F) \gg k_BT$$

$$f(E) \approx e^{-(E-E_F)/k_BT} \approx e^{(E_F - E)/k_BT}$$

$$\text{Hence, the concentration,} \, n_e = \int_{E_c}^{\infty} \frac{1}{2\pi^2}\left(\frac{2m_e^*}{\hbar^2}\right)^{3/2}(E - E_c)^{1/2}e^{(E_F - E)/k_BT}\, dE$$

$$\text{Let} \quad \frac{E - E_c}{k_BT} = x, \Rightarrow dE = (k_BT)dx \text{ noting when } E = E_c, x = 0$$

$$n_e = \int_0^{\infty} \frac{1}{2\pi^2}\left(\frac{2m_e^*}{\hbar^2}\right)^{3/2} x^{1/2}(k_BT)^{1/2}e^{(E_F - E_c)/k_BT}e^{-(E-E_c)/k_BT}(k_BT)dx$$

$$= N_e e^{(E_F - E_c)/k_BT} \tag{12.51}$$

$$\text{where} \quad N_e = 2\left(\frac{m_e^* k_BT}{2\pi\hbar^2}\right)^{3/2}$$

### Case (b) Holes in the VB.

The following modifications are made for holes in the VB.
(i) Replace the free electron mass, $m$, by the effective mass of the hole, $m_h^*$, that is, $m \to m_h^*$,
(ii) Replace energy, $E$, by $(E_v - E)$, that is, $E \to (E_v - E)$, where $E_v$ is the valence band edge

(top of the VB),

Hence, the hole theory density of states in the valence band, $\eta_v(E)$, is given by

$$\eta_v(E) = \frac{1}{2\pi^2}\left(\frac{2m_e^*}{\hbar^2}\right)^{3/2}(E_v - E)^{1/2} \tag{12.52}$$

The concentration of holes in the VB is given by

$$n_h = \int_{-\infty}^{E_v} \eta_h(E) f_h(E) dE$$

$$\text{where} \quad f_h(E) = 1 - f(E) = 1 - \frac{1}{e^{(E-E_F)/k_BT} + 1} \quad \text{is the Fermi-Dirac distribution}$$

$$= \frac{e^{(E-E_F)/k_BT}}{e^{(E-E_F)/k_BT} + 1}$$

$$\text{If} \quad (E - E_F) \ll k_BT$$

$$f_h(E) \approx e^{(E-E_F)/k_BT} \approx e^{-(E_F-E)/k_BT}$$

Hence, similarly, the hole concentration, $n_h = N_h e^{-(E_F-E_v)/k_BT}$ (12.53

$$\text{where} \quad N_h = 2\left(\frac{m_h^* k_B T}{2\pi\hbar^2}\right)^{3/2}$$

## 7. Fermi level, $E_F$ in a semiconductor

To obtain the Fermi energy, $E_F$, we proceed as follows. In an intrinsic semiconductor, the concentration of electrons, $n_e$ is the same as the concentration of holes, $n_h$, that is

$$\text{From,} \quad n_e = N_e e^{(E_F-E_c)/k_BT}$$

$$n_h = N_h e^{-(E_F-E_v)/k_BT}$$

$$\text{If} \quad n_e = n_h$$

$$N_e e^{(E_F-E_c)/k_BT} = N_h e^{-(E_F-E_v)/k_BT}$$

Rearranging, we obtain

$$e^{2E_F/k_Bt} = e^{(E_c+E_v)/k_BT}\frac{N_h}{N_e}$$

Taking ln on both sides

$$\frac{2E_F}{k_BT} = \frac{(E_c + E_v)}{k_BT} + \ln\frac{N_h}{N_e}$$

$$E_F = \frac{1}{2}(E_c + E_v) + \frac{1}{2}k_BT\ln\left\{\frac{2\left(\frac{m_e^* k_B T}{2\pi\hbar^2}\right)^{3/2}}{2\left(\frac{m_h^* k_B T}{2\pi\hbar^2}\right)^{3/2}}\right\}$$

$$E_F = \frac{1}{2}(E_c + E_v) + \frac{3}{4}\ln\left(\frac{m_h^*}{m_e^*}\right) \tag{12.54}$$

If $m_e^* \approx m_h^*$, then

$$E_F \approx \frac{1}{2}(E_c + E_v)$$

$$\text{Since} \quad E_g = E_c - E_v$$
$$2E_F + E_g = 2E_C$$
$$\text{or}$$
$$\frac{1}{2}E_g = (E_c - E_F)$$

## 8. The intrinsic carrier concentration, $n_i$

Recall that the concentrations of electrons and holes are given by

$$n_e = N_e e^{(E_F-E_c)/k_BT}$$
$$n_h = N_h e^{-(E_F-E_v)/k_BT}$$
$$\text{If} \quad n_e = n_h = n_i$$
$$\text{Then} \quad n_e n_h = n_i^2$$
$$= N_e e^{(E_F-E_c)/k_BT} N_h e^{-(E_F-E_v)/k_BT}$$
$$= N_e N_h e^{-(E_c-E_v)/k_BT}$$
$$= N_e N_h e^{-E_g/k_BT} \text{ where } E_g = E_c - E_v$$
$$n_i^2 = 4\left(\frac{m_e^* k_BT}{2\pi\hbar^2}\right)^{3/2}\left(\frac{m_h^* k_BT}{2\pi\hbar^2}\right)^{3/2} e^{-E_g/k_BT}$$

Takng square root on both sides,

$$n_i = (N_e N_h)^{1/2} e^{-E_g/2k_BT}$$
$$\text{or}$$
$$n_i = 2(m_e^* m_h^*)^{3/4}\left(\frac{k_BT}{2\pi\hbar^2}\right)^{3/2} e^{-E_g/2k_BT} \quad (12.55)$$

is the intrinsic carrier concentration. In Table 12.3, values of $n_i$ for Si and Ge are shown.

| | $E_g$(eV) | $n_i$ at T=273K |
|---|---|---|
| Si | 1.12 | $10^9$ |
| Ge | 0.65 | $10^{12}$ |

Table 12.3: Examples of intrinsic carrier concentrations, $n_i$, for si and Ge at 273K.

## 9. Conductivity, $\sigma$ and Mobilities, $\mu_e$ and $\mu_h$

The conductivity, $\sigma$, of a semiconductor is defined as

$$\sigma = ne(\mu_e + \mu_h)$$
$$\text{where} \quad n = n_i \text{ the intrinsic carrier concentration}$$
$$e \quad \text{is electronic charge}$$
$$\mu_e \quad \text{is the electron mobility}$$
$$\mu_h \quad \text{is the hole mobility}$$

$$\text{Hence } \sigma = 2e(m_e^* m_h^*)^{3/4}\left(\frac{k_B T}{2\pi\hbar^2}\right)^{3/2} e^{-E_g/2k_B T}(\mu_e + \mu_h)$$

$$\sigma = A e^{-E_g/2k_B T} \tag{12.56}$$

where A is slowly varying with temperature, and overall the conductivity of a semiconductor increases with increasing temperature.

### 10. Doped semiconductors: $n$-type and $p$-type

### (a) $n$-type semiconductor

Consider a Si crystal which has a valency of 4 (tetravalent) as a host lattice. Doping occurs when an impurity, say phophorous or antimony, which has a valency of 5 (pentavalent) is substituted for a silicon atom due to a defect, the excess electron will be free since it loosely bound compared to the other valence electrons. Thus, phosphorous or antimony is a *"donor"*, and the impurity electron will have energy levels between the CB and the VB. Since conduction will be due to the *excess electron*, Si is said to have been doped to become an n-type semiconductor, as illustrated in Figure 12.18.

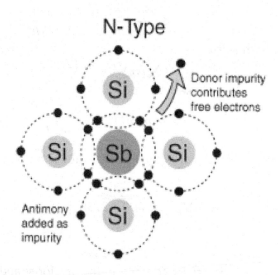

Figure 12.18: n-type semiconductor: Si doped with Gp V

The excess electron will be like in a hydrogen atom ie hydrogenic, satisfying the Schrödinger equation

$$\left[-\frac{\hbar^2}{2m^*}\nabla^2 + \frac{Ze^{*2}}{4\pi\epsilon_0 r}\right]\Psi = E\Psi$$

where $m^*$ is the effective mass
$e^*$ is the effective charge, $= \dfrac{e}{\sqrt{\epsilon(0)}}$

which has a solution for energies as

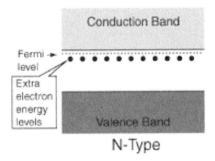

Figure 12.19: n-type with the donor energy level within the energy gap.

$$E_n = -\frac{m^*e^{*4}}{2(4\pi\epsilon_0)^2\hbar^2}\frac{1}{n^2} = -\frac{13.6}{n^2}\left[\frac{m^*}{m\epsilon(0)^2}\right] \text{ eV} \quad (12.57)$$

$$\text{and Bohr orbit radius, } a_n = \frac{4\pi\epsilon_0\hbar^2n^2}{m^*e^{*2}} \quad (12.58)$$

$$\text{where } n = 1, 2, 3, \ldots$$

The donor energy levels appear within the energy gap as illustrated in Figure 12.19.

**(b) $p$-type semiconductor**

Consider a Si crystal which has a valency of 4 (tetravalent) as a host lattice. Doping occurs when an impurity, say boron, which has a valency of 3 (trivalent) is substituted for a silicon atom due to a defect, the excess hole will be free to move in the lattice. Thus, boron is an *"acceptor"*, and the impurity hole will have energy levels between the CB and the VB. Since conduction will be due to the *hole*. Si is said to have been doped to become a p-type semiconductor, as illustrated in Figure 12.20.

The acceptor energy levels appear within the energy gap as illustrated in Figure 12.21.

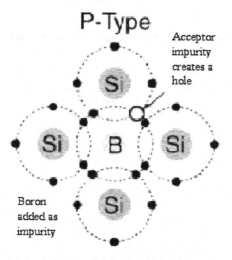

Figure 12.20: p-type seminductor: Si doped with Gp III

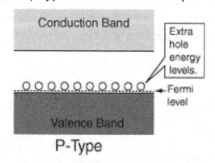

Figure 12.21: p-type semiconductor showing the acceptor energy level within the energy gap.

## 12.7 Exercises

**12.1.** (a) Using classical theory of electrons in solids, show that the thermal conductivity $\kappa$, electrical conductivity $\sigma$ and temperature $T$ satisfy Weidmann-Franz law given by

$$\frac{\kappa}{\sigma T} = \frac{3}{2}(\frac{k_B}{e})^2$$

where all symbols are in the usual notation.

(b) The quantity $\frac{3}{2}(\frac{k_B}{e})^2$ in (a) above is known as the *Lorentz number*. Calculate its value.

(c) In the table below, experimental values of thermal conductivities ($\kappa$) and electrical conductivities ($\sigma$) for selected metals are given. Calculate the Lorentz numbers and complete the following table.

| METAL | T (K) | $\kappa$ (watt-cm$^{-1}$ K$^{-1}$) | $\sigma$ (ohm-cm$^{-1}$ | $\frac{\kappa}{\sigma T}$ (watt-ohm/K$^2$ |
|-------|-------|------|------|------|
| Cu | 373 | 3.82 | $4.47 \times 10^5$ | |
| Mg | 373 | 1.5 | $1.79 \times 10^5$ | |
| Zn | 373 | 1.1 | $1.28 \times 10^5$ | |
| Al | 373 | 2.3 | $2.82 \times 10^5$ | |
| Pb | 373 | 0.35 | $3.71 \times 10^5$ | |

(d) Comment on the values of the Lorentz number obtained from (b) and (c).

**12.2.** (a) Show that electrical conductivity in a metal is given by

$$\sigma = \frac{ne^2\tau}{m}$$

where all symbols are in the usual notation.

(b) (i) Calculate the electrical conductivity, $\sigma$, for copper given that, for copper: Atomic weight = 63.5 g/mol, Density = 8.96 gm.cm$^{-3}$, Avogadro's number, N=$6.022 \times 10^{23}$ mol$^{-1}$, electronic charge, $e = 1.602 \times 10^{-19}$ C, electronic mass, $m = 9.11 \times 10^{-31}$ kg, $\tau = 2.44 \times 10^{-14}$ s.

(ii) Calculate the resistivity of copper from the data in (b)(i).

**12.3.** (a) Show that the electron contribution to the total energy at absolute zero, $U^0_{TOT}$, is given by

$$U^0_{TOT} = \frac{3}{5}NE_F$$

where all symbols are in the usual notation and you may assume the expression for the Fermi distribution function and that of the electron density of states function.

(b) Hence show that pressure $P$, volume $V$ and $U^0_{TOT}$ are related by

$$P = \frac{2}{3}\frac{U^0_{TOT}}{V}$$

where you may assume the thermodynamic result that

$$P = -\frac{\partial U^0_{TOT}}{\partial V}$$

(c) (i) Hence, show that the electron contribution to the bulk modulus at absolute zero, $B$, volume $V$ and $U_{TOT}^0$ are related by

$$B = \frac{10}{9} \frac{U_{TOT}^0}{V}$$

where you may assume the result that

$$B = -V \frac{\partial P}{\partial V}$$

(ii) The metal sodium has density of 1013 kg m$^{-3}$ and its atomic weight is 23. Calculate the electron gas contribution to its bulk modulus at absolute zero.

**12.4.** (a) An electron, each of mass $m$, energy $E$ moving confined in a cubical box of side $L$, whose potential function is such that

$$V(x, y, z) = \begin{cases} 0, & \text{inside the box} \\ \infty, & \text{outside the box} \end{cases}$$

(i) Illustrate, with a well labelled diagram, the configuration of electron in a box.
(ii) Write the Schrödinger equation for the particle in a box.

(b) (i) Show that the energies of the electron n a box are quantised given by

$$E_n = \frac{\pi^2 \hbar^2}{2mL^2}(n_x^2 + n_y^2 + n_z^2)$$

where $n_x, n_y, n_z$ take integral values, and the total energy of the electron in a box for state $n$ is quantised.

**12.5.** (a) Electrons, each of mass $m$, energy $E$ move in a periodic array of a potential $V(x)$ is such that
$$V(x) = \begin{cases} 0, & \text{for } 0 < x < a \text{ Region 1 and periodic in such regions} \\ V_0, & \text{for } -b < x < 0 \text{ Region 2 and periodic in such regions} \end{cases}$$

(i) Illustrate, with a well labelled diagram, the periodic potential.
(ii) Write the Schrödinger equation in the infinite potential well.

(b) (i) Show that the energies of the electron satisfy

$$\left( \frac{\beta^2 - \alpha^2}{2\alpha\beta} \right) \sinh \beta b \sin \alpha a + \cosh \beta b \cos \alpha a = \cos k(a + b)$$

where all symbols are in the usual notation.

(ii) This problem is the Kronig-Penney model. What are the main conclusions of this model?

**12.6.** (a) What is meant by a *plasmon*?

(b) (i) Derive the expression for the *plasmon-type* dielectric function, $\epsilon(\omega)$, given by

$$\epsilon(\omega) = 1 - \frac{\omega_p^2}{\omega^2}$$

and define all symbols used.

(ii) Hence plot schematically $\epsilon(\omega)$ vs $\frac{\omega}{\omega_p}$.

(c) (i) State, without proof, the dispersion relation for bulk plasmon type polaritons.

(ii) Plot schematically the dispersion curves of bulk plasmon-type polaritons.

**7.** (a) Explain what is meant by a *plasmon polariton*.

(b) (i) Show that the dispersion relation for *surface polaritons* at a single interface is given by

$$\frac{c^2 q_{1x}^2}{\omega^2} = \frac{\epsilon_1 \epsilon(\omega)}{\epsilon_1 + \epsilon(\omega)}$$

where all symbols are in the usual notation.

(ii) Plot schematically the dispersion curves for *plasmon-type* surface polaritons.

(c) Show that the group velocity, $v_g = \frac{\partial \omega}{\partial q_{1x}}$, for surface plasmon polaritons is given by

$$v_g = \frac{c\epsilon(\omega)^{1/2}\{\epsilon_1 + \epsilon(\omega)\}^{3/2}}{\epsilon_1^{3/2}\{\frac{\epsilon(\omega)[\epsilon_1+\epsilon(\omega)]}{\epsilon_1} + \frac{\omega}{2}\frac{\partial\epsilon(\omega)}{\partial\omega}\}}$$

**12.8.** (a) Assuming an expression for the density of states for *free electrons*, give arguments to justify that electron density of states in the conduction band, $\eta_c(E)$, is given by

$$\eta_c(E) = \frac{1}{2\pi^2}\left(\frac{2m_e^*}{\hbar^2}\right)^{3/2}(E - E_c)^{1/2}$$

(b) Show that the concentration of electrons in the CB is given by

$$n_e = N_e e^{(E_F - E_c)/k_B T}$$
$$\text{where}\quad N_e = 2\left(\frac{m_e^* k_B T}{2\pi\hbar^2}\right)^{3/2}$$

where all symbols are in the usual notation.

**12.9.** (a) Assuming an expression for the density of states for *free electrons*, give arguments to justify that the hole density of states in the valence band, $\eta_v(E)$, is given by

$$\eta_v(E) = \frac{1}{2\pi^2}\left(\frac{2m_e^*}{\hbar^2}\right)^{3/2}(E_v - E)^{1/2}$$

(b) Show that the concentration of holes in the valence band is given by

$$n_h = N_h e^{-(E_F - E_v)/k_B T}$$
$$\text{where}\quad N_h = 2\left(\frac{m_h^* k_B T}{2\pi\hbar^2}\right)^{3/2}$$

where all symbols are in the usual notation.

**12.10**. (a) Assuming expressions for electron and hole concentrations given by

$$n_e = N_e e^{(E_F - E_c)/k_B T}$$
$$n_h = N_h e^{-(E_F - E_v)/k_B T}$$

Show that the expression for electrical conductivity $\sigma$ of an intrinsic semiconductor is given by

$$\sigma = 2e(m_e^* m_h^*)^{3/4}\left(\frac{k_B T}{2\pi\hbar^2}\right)^{3/2} e^{-E_g/2k_B T}(\mu_e + \mu_h)$$
$$\sigma = A e^{-E_g/2k_B T}$$

where all symbols are in the usual notation.

(b) Estimate, making reasonable assumptions, the energy gap (in eV) of an intrinsic semiconductor if the specimen is observed to have an electrical conductivity at 300 K to be 1000 times its electrical conductivity at liquid nitrogen temperature (77 K).

# Chapter 13

# Introduction to Materials Science IV: Magnetic materials

## 13.1  Classification of magnetic materials

Magnetic materials are classified according to their value of magnetic susceptibility, $\chi_m$, which is defined as

$$\chi_m = \frac{M}{H} \tag{13.1}$$

where    $M$    is the Magnetization

$H$    is the applied magnetic field

There are three types of magnetic materials, depending on the value of $\chi_m$: Diamagnetic materials, Paramagnetic materials and spin ordered systems, and these are studied in the sections below. Magnetic materials can be seen in the periodic table as arranged in Figure 13.1.

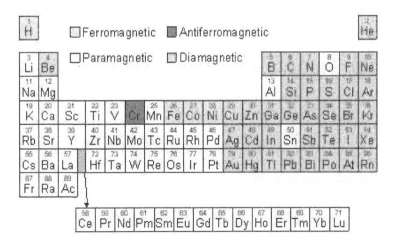

Figure 13.1: Magnetic materials in the Periodic Table.

## 13.2   Diamagnetic materials

Diamagnetic materials are characterized as follows:
(i) The magnetic susceptibility is negative ($\chi_m < 0$)
(ii) There are no permanent magnetic dipoles.
Examples of diamagnetic materials and values of $\chi_m$ are shown in Table 13.1. Other diagmagnetic materials are: water, nitrogen, copper, bismuth etc.

Table 13.1: Examples of diamagnetic materials and values of $\chi_m$.

| Gp 1 Ions | $\chi_m(\times 10^{-6} cm^3/\text{mole})$ | Noble gases | $\chi_m(\times 10^{-6} cm^3/\text{mole})$ |
|-----------|-------------------------------------------|-------------|-------------------------------------------|
| Li$^+$ | -0.1 | He | -1.9 |
| Na$^+$ | -6.1 | Ne | -7.2 |
| K$^+$ | -14.6 | A | -19.4 |
| Rb$^+$ | -22.0 | Kr | -28.0 |
| Cs$^+$ | -35.1 | Xe | -43.0 |

## 13.3   Paramagnetic materials

(i) The magnetic susceptibility is small and positive ($\chi_m > 0$)
(ii) The individual atoms possess a permanent magnetic dipole moment. With no external magnetic field, the dipole moments are randomly oriented, showing "disorder", with no net magnetization. With an applied magnetic field, the dipoles orient themselves parallel to the field since this is a state of low energy in comparison to the antiparallel direction. This is illustrated in Figure 13.2.

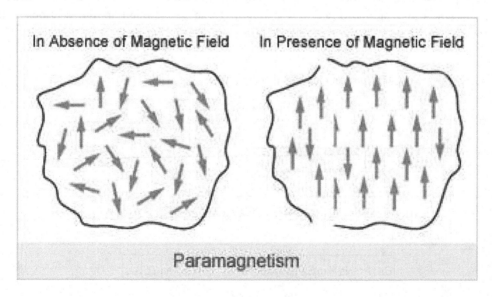

Figure 13.2: Magnetic dipoles in a Paramagnet, in the absence of a magnetic field and in the presence of a magnetic field.

In this chapter, we shall caculate the magnetic susceptibity, $\chi_m$, of a paramagnet by two methods: classically by Langevin theory and quantum mechanically by Brillouin theory.

### 13.3.1 Classical theory of paramagnetism: Langevin theory

Consider a magnetic dipole of magnetic dipole moment, $\mu$, in a magnetic field **B** or **H**, as illustrated in Figure13.3. The potential energy $E$ of a magnetic dipole of magnetic dipole moment, $\mu$, in a

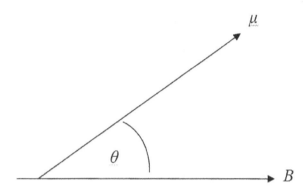

Figure 13.3: A dipole moment in a magnetic field

magnetic field **H** is given by

$$
\begin{aligned}
E &= -\mu \cdot \mathbf{H} & (13.2) \\
&= -\mu H \cos \theta \\
dE &= \mu H \sin \theta d\theta & (13.3)
\end{aligned}
$$

From statistical mechanics, the probability that a magnet dipole has energy $E$ is

$$
f(E) = Ae^{-E/k_B T} \tag{13.4}
$$

The number $dN$ of such magnetic dipoles oriented between $\theta$ and $\theta + d\theta$ is given by

$$
\begin{aligned}
dN &= 2\pi \sin \theta d\theta \times f(E) \\
&= (-2\pi)f(E)d(\cos\theta) \\
\text{Hence,} \quad N &= \int dN \\
&= \int (-2\pi) Ae^{-E/k_B T} d(\cos\theta) & (13.5)
\end{aligned}
$$

Magnetization, **M**, is the net dipole moment per unit volume. If there are $N$ magnetic dipoles per unit volume, then the magnetization is given by

$$
\begin{aligned}
M &= N\overline{\mu} & (13.6) \\
&= \frac{N \int dN \mu \cos \theta}{\int dN}
\end{aligned}
$$

$$= \frac{N \int_{-1}^{+1} \mu \cos\theta (-2\pi) A e^{-E/k_B T} d(\cos\theta)}{\int_{-1}^{+1} (-2\pi) A e^{-E/k_B T} d(\cos\theta)}$$

$$M = \frac{N \int_{-1}^{+1} \mu \cos\theta e^{+(\mu H \cos\theta/k_B T)} d(\cos\theta)}{\int_{-1}^{+1} e^{+\mu H \cos\theta/k_B T} d(\cos\theta)}$$

Let $\cos\theta = y$ and $\dfrac{\mu H}{k_B T} = x$

Hence

$$M = \frac{N\mu \int_{-1}^{+1} y e^{xy} dy}{\int_{-1}^{+1} e^{xy} dy} \qquad (13.7)$$

The integrals in the numerator and denominator of equation (13.7) can be evaluated as follows.

First, the denominator

$$\int_{-1}^{+1} y e^{xy} dy = \frac{e^{xy}}{x} \left\{ y - \frac{1}{x} \right\} \Big|_{-1}^{+1}$$

$$= \frac{e^{+x}}{x} \left\{ 1 - \frac{1}{x} \right\} - \frac{e^{-x}}{+x} \left\{ -1 - \frac{1}{x} \right\}$$

$$= \frac{e^{+x}}{x} + \frac{e^{-x}}{x} - \frac{1}{x} \left\{ \frac{e^{+x}}{x} - \frac{e^{-x}}{x} \right\} \qquad (13.8)$$

Secondly, the numerator

$$\int_{-1}^{+1} e^{xy} dy = \frac{e^{xy}}{x} \Big|_{-1}^{+1}$$

$$= \frac{e^{+x}}{x} - \left( + \frac{e^{-x}}{x} \right)$$

$$= \frac{e^{+x}}{x} - \frac{e^{-x}}{x} \qquad (13.9)$$

Hence, the magnetization becomes

$$M = \frac{N\mu \int_{-1}^{+1} y e^{xy} dy}{\int_{-1}^{+1} e^{xy} dy} = \frac{\frac{e^{+x}}{x} + \frac{e^{-x}}{x} - \frac{1}{x} \left\{ \frac{e^{+x}}{x} - \frac{e^{-x}}{x} \right\}}{\frac{e^{+x}}{x} - \frac{e^{-x}}{x}}$$

$$= N\mu \left[ \coth(x) - \frac{1}{x} \right] \qquad (13.10)$$

$$M = N\mu L(x) \qquad (13.11)$$

where $L(x) = \coth(x) - \dfrac{1}{x} \qquad (13.12)$

is known as the LANGEVIN FUNCTION, which is illustrated graphically in Figure 13.4.  Let us evaluate the magnetization, $M$, in the high temperature limit, where $\mu H \ll k_B T$ (small $x$),

$$\coth x = \frac{1}{x} + \frac{x}{3} - \frac{x^3}{15} + \cdots \qquad (13.13)$$

$$L(x) = \frac{1}{x} + \frac{x}{3} + \cdots - \frac{1}{x}$$

$$\approx \frac{1}{3}x \qquad (13.14)$$

$$M = N\mu\frac{1}{3}\frac{\mu H}{k_B T}$$

$$= \frac{N\mu^2}{3k_B}\frac{H}{T}$$

$$= \chi_m H$$

$$\text{where} \quad \chi_m = \frac{N\mu^2}{3k_B T}$$

$$\chi_m = \frac{C}{T} \qquad (13.15)$$

$$(13.16)$$

which is known of CURIE'S LAW, and $C = \frac{N\mu^2}{3k_B}$ is known as CURIE'S CONSTANT. A graph of $\chi$

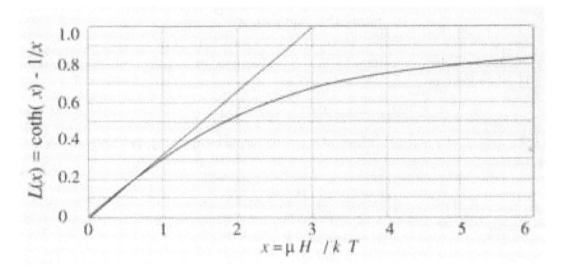

Figure 13.4: Langevin function

against $T$ is illustrated in Figure 13.5.

## 13.3.2   Quantum theory of paramagnetism: Brillouin theory

Let us calculate the magnetization, $M = N\mu$, where the $\mu$ is related to the quantised angular momentum, $J_z$ by

$$\mu_z = -g\mu_B J_z \qquad (13.17)$$

where   $g$     is the Lande factor

   $\mu_B$     is the Bohr magneton   $(\mu_B = \dfrac{e\hbar}{2m_e})$

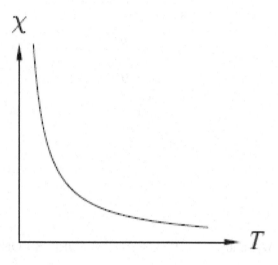

Figure 13.5: Curie's Law

Noting that the total angular momentum $J = L + S$ is quantised, its z-component, $J_z$ is quantised with $(2J+1)$ values given by

$$J_z = -J, -J+1, \cdots, J-1, J \qquad (13.18)$$

The potential energy $E$ of a magnetic dipole of magnetic dipole moment, $\mu_z$, in a magnetic field $\mathbf{H}$ is given by

$$
\begin{aligned}
E &= -\mu_z \cdot \mathbf{H} \\
&= -\mu_z H \cos\theta \\
&= -\mu_z H \quad \text{for} \quad \theta = 0 \\
&= +g\mu_B H J_z \qquad (13.19)
\end{aligned}
$$

which has $(2J+1)$ values, corresponding to the quantised values of $J_z$ and hence of $\mu_z$. Let us now calculate the *magnetization*, $M = N\overline{\mu}_z$, where the average magnetic moment $\overline{\mu}_z$ is calculated using the Gibb's distribution from statistical mechanics.

$$\overline{\mu}_z = \frac{\sum_i \mu_z e^{-E/k_B T}}{\sum_i e^{-E/k_B T}} \qquad (13.20)$$

where the energy $E = +g\mu_B H J_z$. Hence the magnetization is given by

$$
\begin{aligned}
M &= N\overline{\mu}_z \\
&= N\frac{\sum_i \mu_z e^{-E/k_B T}}{\sum_i e^{-E/k_B T}} \\
&= \frac{N \sum_{J_z=-J}^{J} (-g\mu_B J_z) e^{(-g\mu_B H J_z)/k_B T}}{\sum_{J_z=-J}^{J} e^{(-g\mu_B H J_z)/k_B T}}
\end{aligned}
$$

$$\text{Let} \quad a = \frac{g\mu_B H}{k_B T}$$

$$M = \frac{N g\mu_B \sum_{J_z=-J}^{J}(-J_z)e^{-aJ_z}}{\sum_{J_z=-J}^{J}e^{-aJ_z}} = N g\mu_B \frac{N_m}{D_n} \tag{13.21}$$

Note that on examining equation (13.21) above, we find that

(i)The denominator, $D_n$, is a geometrical progression (GP) with a common ratio $r = e^{-a}$, and $(2J+1)$ terms, that is

$$e^{-a(-J)} + e^{-a(-J+1)} + \cdots + e^{-a(+J)}$$

(ii)The numerator, $N_m$, is a differential of the denominator, $D_n$. Let us proceed to evaluate (i) and (i).

(i) Denominator, $D_n$

$$D_n = \sum_{J_z=-J}^{J} e^{-aJ_z}$$

Using an expression for sum of a GP, $\quad S_n^{GP} = \frac{a(1-r^n)}{(1-r)}$

Hence, we obtain

$$D_n = e^{aJ}\left\{\frac{1-e^{-a(2J+1)}}{1-e^{-a}}\right\}$$

$$= e^{a/2}e^{aJ}\left\{\frac{1-e^{-a(2J+1)}}{e^{a/2}-e^{-a/2}}\right\}$$

$$= \frac{e^{\frac{a}{2}(2J+1)}-e^{-\frac{a}{2}(2J+1)}}{e^{a/2}-e^{-a/2}}$$

$$D_n = \frac{\sinh\left(\frac{2J+1}{2a}\right)a}{\sinh(a/2)} \tag{13.22}$$

(ii) Numerator, $N_m$

Noting that the numerator, $N_m$, is the differential of the denominator, $D_n$, we obtain

$$N_n = \sum_{J_z=-J}^{J}(-J_z)e^{-aJ_z}$$

$$= \sum_{J_z=-J}^{J}\frac{\partial}{\partial a}\left\{e^{-aJ_z}\right\}$$

$$= \frac{\partial}{\partial a}\left\{\sum_{J_z=-J}^{J}e^{-aJ_z}\right\}$$

$$= \frac{\partial}{\partial a} D_n$$

$$= \frac{\partial}{\partial a} \left\{ \frac{\sinh\left(\frac{2J+1}{2a}\right) a}{\sinh(a/2)} \right\}$$

$$= \frac{\left(\frac{2J+1}{2}\right) \sinh(a/2) \cosh\left(\frac{2J+1}{2}\right) a - \frac{1}{2} \sinh\left(\frac{2J+1}{2}\right) a \cosh(a/2)}{\sinh^2(a/2)}$$

$$N_m = \left(\frac{2J+1}{2}\right) \frac{\cosh\left(\frac{2J+1}{2}\right) a}{\sinh(a/2)} - \frac{1}{2} \frac{\sinh\left(\frac{2J+1}{2}\right) a \cosh(a/2)}{\sinh^2(a/2)} \qquad (13.23)$$

Hence we can calculate the magnetization, $M = Ng\mu_B \frac{N_m}{D_n}$, where $N_m$ and $D_n$ have been calculated in equations 13.23 and 13.22 respectively.

$$M = Ng\mu_B \frac{N_m}{D_n}$$

$$\text{where} \quad \frac{N_m}{D_n} = \frac{\left(\frac{2J+1}{2}\right) \frac{\cosh\left(\frac{2J+1}{2}\right) a}{\sinh(a/2)} - \frac{1}{2} \frac{\sinh\left(\frac{2J+1}{2}\right) a \cosh(a/2)}{\sinh^2(a/2)}}{\frac{\sinh\left(\frac{2J+1}{2a}\right) a}{\sinh(a/2)}}$$

$$= \left(\frac{2J+1}{2}\right) \coth\left(\frac{2J+1}{2}\right) a - \frac{1}{2} \coth\frac{a}{2}$$

$$\text{Let} \quad aJ = x \quad \text{ie} \quad x = \frac{g\mu_B H J}{k_B T}$$

$$\text{Hence} \quad M = Ng\mu_B J \left\{ \left(\frac{2J+1}{2J}\right) \coth\left(\frac{2J+1}{2J}\right) x - \frac{1}{2J} \coth\frac{x}{2J} \right\} \qquad (13.24)$$

$$M = Ng\mu_B J B_J(x) \qquad (13.25)$$

$$\text{where} \quad B_J(x) = \left\{ \left(\frac{2J+1}{2J}\right) \coth\left(\frac{2J+1}{2J}\right) x - \frac{1}{2J} \coth\frac{x}{2J} \right\} \qquad (13.26)$$

is the BRILLOUIN FUNCTION, which is illustrated graphically in figure 13.6.
On examining the Brillouin function, there are several cases of physical interest.

(i) Case 1, $J \to \infty$

In this limit,

$$\lim_{J \to \infty} \frac{2J+1}{2J} = 1 + \frac{1}{2J} = 1$$

$$B_\infty(x) = \coth x - \frac{1}{2J} \left\{ \frac{1}{x/2J} + \frac{1}{3}\left(\frac{x}{2J}\right) + \cdots \right\}$$

$$= \coth x - \frac{1}{x}$$

$$B_\infty(x) = L(x) \qquad (13.27)$$

which is the Langevin function. In the limit $J \to \infty$, quantum effects are negligible and spatial quantisation is unimportant, and hence the Brillouin function reduces to the Langevin function.

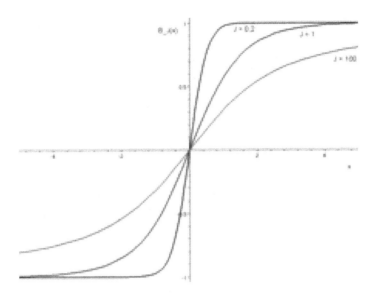

Figure 13.6: Brillouin function, $B_J(x)$ vs $x$.

(ii) Case (ii), $J = \frac{1}{2}$

In this limit,

$$
\begin{aligned}
B_{1/2}(x) &= 2\coth 2x - \coth x \\
&= \frac{2}{\tanh 2x} - \frac{1}{\tanh x} \\
&= \frac{2(1 + \tanh^2 x)}{\tanh 2x} - \frac{1}{\tanh x} \\
&= \frac{1 + \tanh^2 x - 1}{\tanh x} \\
B_{1/2}(x) &= \tanh x
\end{aligned}
\tag{13.28}
$$

$$
\text{Hence,} \quad M = Ng\mu_B J \tanh\left(\frac{g\mu_B H J}{k_B T}\right) \tag{13.29}
$$

$$
M = N\mu_B \tanh\left(\frac{\mu_B H}{k_B T}\right) \quad \text{for} \quad g = 2, J = 1/2 \tag{13.30}
$$

### 13.3.3 Quantum theory of paramagnetism due to conduction electrons: Pauli theory

In this section, we discuss paramagnetism arising from conduction electrons. An electron has an intrinsic spin and magnetic moment, $\mu_m$, which is such that

$$
\mu_m = \begin{cases} -\mu, & \text{spin down} \\ +\mu, & \text{spin up} \end{cases}
\tag{13.31}
$$

Energy of an electron in a magnetic field, **H**, is given by, $E = -\mu_m \cdot \mathbf{H}$, and for the two spins we have

$$E = \begin{cases} +\mu H, & \text{spin down} \\ -\mu H, & \text{spin up} \end{cases} \tag{13.32}$$

The density of states function, $g(E)$, becomes modified in the presence of the magnetic field, as illustrated in Figure 13.7.

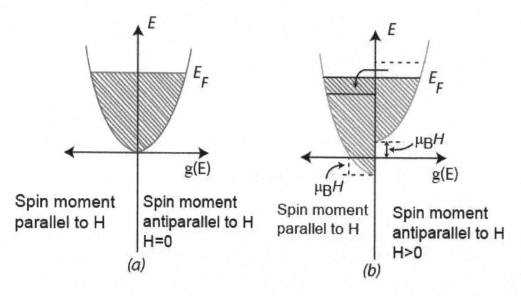

Figure 13.7: Electron Density of states in presence of a magnetic field.

For *spin up electrons*, we have

$$
\begin{aligned}
g(E) & \rightarrow g(E + \mu H) \\
N_+ & = \frac{1}{2} \int_{-\mu H}^{E_F} f(E) g(E + \mu H) dE \\
& = \frac{1}{2} \int_0^{E_F} f(E) g(E) dE + \frac{1}{2} \mu H g(E_F)
\end{aligned}
$$

For *spin down electrons*, we have

$$
\begin{aligned}
g(E) & \rightarrow g(E - \mu H) \\
N_+ & = \frac{1}{2} \int_{-\mu H}^{E_F} f(E) g(E - \mu H) dE \\
& = \frac{1}{2} \int_0^{E_F} f(E) g(E) dE - \frac{1}{2} \mu H g(E_F)
\end{aligned}
$$

Magnetization, $M$, is the net dipole moment per unit volume, is given by

$$M = \mu(N_+ - N_-)$$

$$= \mu \left\{ \frac{1}{2} \int_0^{E_F} f(E)g(E)dE + \frac{1}{2}\mu H g(E_F) - \frac{1}{2} \int_0^{E_F} f(E)g(E)dE + \frac{1}{2}\mu H g(E_F) \right\}$$

$$= \mu^2 H g(E_F) \qquad (13.33)$$

$$\text{But } \quad g(E_F) = \frac{3}{2}\frac{N}{E_F} = \frac{3}{2}\frac{N}{k_B T_F}$$

$$M = \mu^2 H \frac{3}{2}\frac{N}{k_B T_F} = \frac{3}{2}\frac{N\mu^2}{k_B T_F}H \qquad (13.34)$$

is the *Pauli Spin Magnetisation*. From this, we can identify the Magnetic Susceptibility, $\chi$, as

$$\chi = \frac{M}{H} = \mu^2 g(E_F)$$

$$= \frac{3}{2}\frac{N\mu^2}{k_B T_F}$$

where $\chi$ is the *Pauli Spin Susceptibility*.

Note that

(i) $\chi$ is temperature independent
(ii) $\chi$ is very small since $T_F \approx 10^4$.

## 13.4   Spin ordered systems

### 13.4.1   Spin ordering

Atoms in spin ordered systems have permanent magnetic dipoles with spins arranged with "order", for example, as Ferromagnet, Antiferromagnet and Ferrimagnet, illustrated in Figure 13.8, 13.9 and 13.10, respectively. The magnetic susceptibility is large and positive ($\chi_m >> 0$).

### 13.4.2   Spin-spin Interaction mechanisms

Each atom in a spin-ordered system carries a spin. The question is how do spins interact?Spin interactions are complex. However, let us consider a simple model of spins in $1-d$ with neighbouring spins $\mathbf{S}_i$ and $\mathbf{S}_j$. Two possible interaction mechanisms: the exchange interaction and dipole-dipole interaction are described below.

(i) Exchange interaction

$$H_{ex} = -J\mathbf{S}_i \cdot \mathbf{S}_j \qquad (13.35)$$

$$\text{where} \quad J = \int \psi^*\psi^*\frac{e^2}{r}\psi\psi dr \qquad (13.36)$$

is the exchange integral, and $J$ satisfies the following

$$
J = \begin{cases} > 0 & \text{for  parallel spins (ferromagnets)} \\[2ex] < 0 & \text{for  antiparallel spins (antiferromagnets)} \end{cases}
\tag{13.37}
$$

(ii) Dipole-dipole interaction

For dipole-dipole interaction, the interaction energy between neighbouring spins $\mathbf{S}_i$ and $\mathbf{S}_j$ is described by a Hamiltonian of the form

$$
H_{dip-dip} = \beta \left\{ \frac{\mathbf{S_i} \cdot \mathbf{S_j}}{r^3} - \frac{3(\mathbf{S_i} \cdot \mathbf{r})(\mathbf{S_j} \cdot \mathbf{r})}{r^5} \right\}
\tag{13.38}
$$

where      $r$ is the distance between the $i^{th}$ atom and $j^{th}$ atom

$\beta$ is a constant, characteristic of the magnetic material

## 13.4.3   Heisenberg model

The essence of the *Heisenberg model* is that in comparing the exchange intercation and the dipole-dipole interaction, the exchange interaction is the stronger of the two.  If the total Hamiltonian is

$$
\begin{aligned}
H &= H_{ex} + H_{dip-dip} \\
\text{and}\quad H_{ex} &\gg H_{dip-dip} \\
\text{Therefore}\quad H &= H_{ex} = -J\mathbf{S}_i \cdot \mathbf{S}_j
\end{aligned}
\tag{13.39}
$$

is the Hamiltonian in the Heisenberg model. For the whole crystal, the Heisenberg model becomes

$$
H = -J\sum_{i,j} \mathbf{S}_i \cdot \mathbf{S}_j \quad \text{in the \textit{absence} of an external magnetic field}
\tag{13.40}
$$

$$
H = -J\sum_{i,j} \mathbf{S}_i \cdot \mathbf{S}_j - g\mu_B H_0 \sum_i S_i^z \quad \text{in the \textit{presence} of an external magnetic field,}
\tag{13.41}
$$

where $J$ is the exchange integral, defined earlier in equation 13.36, $\mathbf{S}_i$ and $\mathbf{S}_j$ are the spins of the $i^{th}$ and $j^{th}$ atoms respectively, $g$ is Lande's $g$ factor, $\mu_B$ is the Bohr magneton, and $S_i^z$ is the $z$-component of the spin of the $i^{th}$ atom.

## 13.4.4   Molecular field theory

In this section, molecular field theory is applied to find the Magnetization, $M$, and the Susceptibility, $\chi$.

## Magnetization

According to *molecular field theory*, certain approximations to the Hamiltonian can be done as shown below.

$$
\begin{aligned}
H &= -J\sum_{i,j}\mathbf{S}_i\cdot\mathbf{S}_j - g\mu_B H_0\sum_i S_i^z \\
&= \{Jz<S^z> -g\mu_B H_0\}\sum_i S_i^z \\
&= -g\mu_B\left\{H_0 + \frac{Jz<S^z>}{g\mu_B}\right\}\sum_i S_i^z \\
&= -g\mu_B H_{eff}\sum_i S_i^z
\end{aligned}
$$

where $H_{eff}$ can be identified as

$$
\begin{aligned}
H_{eff} &= H_0 + \frac{Jz<S^z>}{g\mu_B} \\
&= H_0 + \frac{JzNg\mu_B<S^z>}{Ng^2\mu_B^2} \\
&= H_0 + \frac{JzM}{Ng^2\mu_B^2} \quad\quad\quad\quad (13.42)\\
&= H_0 + \lambda M \quad\quad\quad\quad\quad\quad (13.43)
\end{aligned}
$$

where $M = Ng\mu_B<S^z>$ has been used, and $\lambda = Jz/(Ng^2\mu_B^2)$ has been introduced. In equation (13.43), the first term is the external magnetic field and the second term $\lambda M$ is the *molecular field*.

Let us now calculate the *magnetization*, $M = Ng\mu_B<S^z>$, where the average spin $<S^z>$ is calculated using the Gibb's distribution from statistical mechanics.

$$
<S^z> = \frac{\sum_i S_i^z e^{-E/k_BT}}{\sum_i e^{-E/k_BT}} \quad\quad\quad\quad (13.44)
$$

where the energy $E = -g\mu_B H_{eff}S_i^z$. Hence the magnetization is given by

$$
\begin{aligned}
M &= Ng\mu_B\frac{\sum_i S_i^z e^{-E/k_BT}}{\sum_i e^{-E/k_BT}} \\
&= Ng\mu_B\frac{\sum_{i=-\frac{1}{2}}^{+\frac{1}{2}} S_i^z e^{g\mu_B H_{eff}S_i^z/k_BT}}{\sum_{i=-\frac{1}{2}}^{+\frac{1}{2}} e^{g\mu_B H_{eff}S_i^z/k_BT}} \\
&= Ng\mu_B\frac{\left\{\frac{1}{2}e^{a/2} - \frac{1}{2}e^{-a/2}\right\}}{e^{a/2} + e^{-a/2}} \\
&= \frac{Ng\mu_B}{2}\tanh\left[\frac{a}{2}\right] \quad \text{where } g = 2 \\
M &= N\mu_B\tanh\left[\frac{\mu_B H_{eff}}{k_BT}\right] \quad\quad\quad\quad (13.45)
\end{aligned}
$$

Let us study the expression for magnetization when $H_0 = 0$, and hence $H_{eff} = \lambda M$. For $T = T_C$,

$$M = N\mu_B \tanh \left[ \frac{\mu_B \lambda M}{k_B T_C} \right] \tag{13.46}$$

Using the approximation $\tanh x \approx x$ in equation (13. ) gives the critical temperature as

$$T_C = \frac{N\mu_B^2 \lambda}{k_B} \tag{13.47}$$

which is also known as the *Curie temperature*. Using this, the expression for the magnetization can also be written in the form

$$
\begin{aligned}
M &= N\mu_B \tanh \left( \frac{\mu_B \lambda M}{k_B T} \right) \\
&= N\mu_B \tanh \left( \frac{M N \mu_B^2 \lambda}{T N \mu_B k_B} \right) \\
&= M_s \tanh \left( \frac{M T_C}{T M_s} \right) \quad \text{where} \quad M_s = N\mu_B \\
\frac{M}{M_s} &= \tanh \left( \frac{T_C M}{T M_s} \right) \tag{13.48}
\end{aligned}
$$

There are three limits of physical interest to equation (13.48).

*(i) Low temperature limit, $T \ll T_C$*

Using the definition of $\tanh x$

$$
\begin{aligned}
\tanh x &= \frac{e^x - e^{-x}}{e^x - e^{-x}} \\
\frac{M}{M_s} &= \tanh \left( \frac{T_C M}{T M_s} \right) \\
&= \frac{e^{\frac{T_C M}{T M_s}} - e^{-\frac{T_C M}{T M_s}}}{e^{\frac{T_C M}{T M_s}} + e^{-\frac{T_C M}{T M_s}}} \\
&\approx 1 - 2 e^{-2\frac{T_C M}{T M_s}} \quad \text{as } T \to 0 \tag{13.49}
\end{aligned}
$$

which shows an exponential deviation.

*(ii) Close to $T_C$, $T \le T_C$*

Using the definition of $\tanh x$

$$
\begin{aligned}
\tanh x &= x - \frac{1}{3}x^3 + \frac{2}{5}x^5 \cdots \\
\frac{M}{M_s} &\approx \frac{T_C M}{T M_s} - \frac{1}{3}\left( \frac{T_C M}{T M_s} \right)^3
\end{aligned}
$$

$$1 = \frac{T_C}{T} - -\frac{1}{3}\left(\frac{T_C}{T}\right)^3 \left(\frac{M}{M_s}\right)^2$$

$$\left(\frac{M}{M_s}\right)^2 = 3\left(\frac{T_C}{T} - 1\right)\left(\frac{T}{T_C}\right)^3$$

$$\frac{M}{M_s} = 3^{1/2}\left(\frac{T}{T_C}\right)\left(\frac{T_C - T}{T_C}\right)^{1/2} \qquad (13.50)$$

which when compared with equation (13. )

$$\frac{M(T)}{M(0)} = \mathcal{B}'\left(-\epsilon\right)^\beta$$

implies that

$$\beta = \frac{1}{2} \qquad (13.51)$$

*(iii) High temperature limit, $T \gg T_C$*

$$M = 0 \text{ ie Paramagnetic}$$

## Magnetic Susceptibility

To obtain the magnetic susceptibility, recall equation (13.45), and consider the case when there is a finite external magnetic field, $H_0 \neq 0$, and hence $H_{eff} = H_0 + \lambda M$. Hence the magnetization is given by

$$M = N\mu_B \tanh\left[\frac{\mu_B H_{eff}}{k_B T}\right]$$

$$= N\mu_B \tanh\left[\frac{\mu_B (H_0 + \lambda M)}{k_B T}\right]$$

$$= \frac{N\mu_B^2 (H_0 + \lambda M)}{k_B T} \quad \text{where} \quad \tanh x \approx x \text{ has been used}$$

$$M = \frac{N\mu_B^2 H_0}{k_B T} + \frac{N\mu_B^2 \lambda M}{k_B T}$$

$$M\left[1 - \frac{N\mu_B^2 \lambda}{k_B T}\right] = \frac{N\mu_B^2 H_0}{k_B T}$$

$$M = \frac{N\mu_B^2 H_0}{k_B T - N\mu_B^2 \lambda} \qquad (13.52)$$

from which the magnetic susceptibility, $\chi_m$ is given by

$$\chi_m = \frac{M}{H_0}$$

$$= \frac{N\mu_B^2}{k_B T - N\mu_B^2 \lambda}$$

$$= \frac{N\mu_B^2}{k_B (T - T_C)} \qquad (13.53)$$

where $T_C$ is the critical temperature (or Curie temperature) given by equation (13.47). Equation (13.53) is known as *Curie-Weiss law* which shows that as $T \to T_C$, the susceptibility becomes very large.

### 13.4.5  Spin wave theory

Spins in spin ordered systems generate spin waves. Quanta of spin waves are known as *magnons*. We shall study spin waves using the Heisenberg model, where the Hamiltonian is given by

$$
\begin{aligned}
H &= -J \sum_{i,j} \mathbf{S}_i \cdot \mathbf{S}_j \\
&= -J \{\mathbf{S}_n \cdot \mathbf{S}_{n-1} + \mathbf{S}_n \cdot \mathbf{S}_{n+1}\} \quad \text{considering the } n^{th} \text{ site} \\
&= -J \mathbf{S}_n \cdot \{\mathbf{S}_{n-1} + \mathbf{S}_{n+1}\} \\
&= -(-g\mu_B \mathbf{S}_n) \cdot \left(\frac{-J}{g\mu_B}\right)(\mathbf{S}_{n-1} + \mathbf{S}_{n+1})
\end{aligned}
$$

$$
H = -\mu_n \cdot \mathbf{H}^n_{eff} \tag{13.54}
$$

where  $\mu_n = -g\mu_B \mathbf{S}_n$  is the magnetic moment at the $n^{th}$ spin

$$
\mathbf{H}^n_{eff} = -\frac{J}{g\mu_B}(\mathbf{S}_{n-1} + \mathbf{S}_{n+1}) \tag{13.55}
$$

and $\mathbf{H}^n_{eff}$ can be identified as the effective magnetic field acting on the $n^{th}$ spin.

### Equation of motion

The equation of motion is essentially the *Torque eqation*, where the rate of change of the spin momentum $\mathbf{S}_n$ is equal to the torque acting on the system, and is given by

$$
\hbar \frac{d\mathbf{S}_n}{dt} = \mu_n \wedge \mathbf{H}^n_{eff} \tag{13.56}
$$

$$
\frac{d\mathbf{S}_n}{dt} = \left(\frac{-g\mu_B}{\hbar}\right)\mathbf{S}_n \wedge \left(-\frac{J}{g\mu_B}\right)(\mathbf{S}_{n-1} + \mathbf{S}_{n+1})
$$

$$
= \frac{J}{\hbar} \{\mathbf{S}_n \wedge \mathbf{S}_{n-1} + \mathbf{S}_n \wedge \mathbf{S}_{n+1}\}
$$

Expanding

$$
\frac{dS^x_n}{dt} = \frac{J}{\hbar} \{S^y_n(S^z_{n-1} + S^z_{n+1}) - S^z_n(S^y_{n-1} + S^y_{n+1})\}
$$

$$
\frac{dS^y_n}{dt} = \frac{J}{\hbar} \{S^z_n(S^x_{n-1} + S^x_{n+1}) - S^x_n(S^z_{n-1} + S^z_{n+1})\}
$$

$$
\frac{dS^z_n}{dt} = \frac{J}{\hbar} \{S^x_n(S^y_{n-1} + S^y_{n+1}) - S^y_n(S^x_{n-1} + S^x_{n+1})\}
$$

### Approximations

The amplitude of the spin wave excitations is small. The spins are mostly pointing in one direction most of the time, and let this be the $z$-direction.

$$
\mathbf{S} = S^x \mathbf{i} + S^y \mathbf{j} + S^z \mathbf{k} \approx S^z_n \mathbf{k} \approx S^z_{n+1} \mathbf{k} \tag{13.57}
$$

and products of $S^x$ and $S^y$ are negligible. Hence, we have *linearised equations*

$$\frac{dS_n^x}{dt} = \frac{JS}{\hbar}\{2S_n^y - S_{n-1}^y - S_{n+1}^y)\}$$

$$\frac{dS_n^y}{dt} = -\frac{JS}{\hbar}\{2S_n^x - S_{n-1}^x - S_{n+1}^x)\}$$

$$\frac{dS_n^z}{dt} = 0$$

Try solutions of the form

$$S_n^x = U_0 e^{i(kna-\omega t)}$$

$$S_{n\pm1}^x = U_0 e^{i[k(n\pm1)a-\omega t]}$$

$$S_n^y = V_0 e^{i(kna-\omega t)}$$

$$S_{n\pm1}^y = V_0 e^{i[k(n\pm1)a-\omega t]}$$

Hence, we obtain

$$-i\omega U_0 e^{i(kna-\omega t)} - \frac{JS}{\hbar}\left\{2e^{ikna} - e^{i[k(n-1)a} - e^{i[k(n+1)a}\right\}V_0 e^{-i\omega t} = 0$$

$$-i\omega V_0 e^{i(kna-\omega t)} + \frac{JS}{\hbar}\left\{2e^{ikna} - e^{i[k(n-1)a} - e^{i[k(n+1)a}\right\}U_0 e^{-i\omega t} = 0$$

Rearranging, we obtain

$$-i\omega U_0 - \frac{JS}{\hbar}\left\{2 - e^{i[ka} - e^{-ika}\right\}V_0 = 0$$

$$-i\omega V_0 + \frac{JS}{\hbar}\left\{2 - e^{i[ka} - e^{-ika}\right\}U_0 = 0$$

Hence, we obtain

$$-i\omega U_0 - \frac{JS}{\hbar}\left\{2 - 2C\cos ka\right\}V_0 = 0$$

$$-i\omega V_0 + \frac{JS}{\hbar}\left\{2 - 2C\cos ka\right\}U_0 = 0$$

Putting in a matrix form, we obtain

$$\begin{bmatrix} -i\omega & -\frac{2JS}{\hbar}(1-\cos ka) \\ \frac{2JS}{\hbar}(1-\cos ka) & -i\omega \end{bmatrix}\begin{bmatrix} U_0 \\ V_0 \end{bmatrix} = \begin{bmatrix} 0 \\ 0 \end{bmatrix} \qquad (13.58)$$

The equations have a solution if and only if the determinant of the $2 \times 2$ matrix vanishes.

$$-\omega^2 + 4\left(\frac{JS}{\hbar}\right)^2(1-\cos ka)^2 = 0$$

$$\hbar\omega = 2JS(1-\cos ka) \qquad (13.59)$$

which is the dispersion relation for spin waves in $1d$.

In the long wavelength limit, we have

$$
\begin{aligned}
1 - \cos ka &= 1 - \left[ 1 - \frac{1}{2}(ka)^2 + \cdots \right] \\
&\approx \frac{1}{2}ka^2
\end{aligned}
$$

Hence, the dispersion equation becomes

$$
\begin{aligned}
\hbar\omega &= 2JS\frac{1}{2}k^2a^2 \\
&= (JSa^2)k^2 \\
&= Ak^2
\end{aligned}
$$

which is a parabolic function of $k$.

**Density of states for magnons, $D(\omega)$**

Density of states, $D(\omega)$, is the number of allowed modes between the frequency interval $d\omega$ and $\omega + d\omega$.

In a volume $\left(\frac{2\pi}{L}\right)^3$ of $k$-space there is one mode

In unit volume, there are $\frac{L^3}{8\pi^3}$ modes,

A sphere of radius $k$ has volume, $V_k = \frac{4}{3}\pi k^3$, has $\frac{L^3}{8\pi^3} \times 4\pi k^2 dk$ modes, or equivalently

$$
\begin{aligned}
\frac{L^3}{8\pi^3} \times 4\pi k^2 dk \quad \text{modes} &= \frac{L^3}{2\pi^2}k^2 dk \quad \text{modes} \\
&= D(\omega)d\omega \\
D(\omega)d\omega &= \frac{L^3}{2\pi^2}k^2 dk \\
D(\omega) &= \frac{V}{2\pi^2}\frac{k^2}{(d\omega/dk)}
\end{aligned}
$$

Hence, we have

$$
\begin{aligned}
\hbar\omega &= (JSa^2)k^2 \\
\frac{d\omega}{dk} &= 2\frac{JSa^2}{\hbar}k \\
&= \frac{2JSa^2}{\hbar}\left(\frac{\hbar\omega}{JSa^2}\right)^{1/2} \\
\frac{d\omega}{dk} &= 2\left(\frac{JSa^2}{\hbar}\right)^{1/2}\omega^{1/2} \\
\text{Hence } D(\omega) &= \frac{V}{2\pi^2}\left(\frac{\hbar\omega}{JSa^2}\right)\frac{\hbar^{1/2}}{2(JSa^2)^{1/2}\omega^{1/2}} \\
D(\omega) &= \frac{V}{4\pi^2}\left(\frac{\hbar}{JSa^2}\right)^{3/2}\omega^{1/2}
\end{aligned}
$$

is the *magnon Density of states.*

**Magnetization, M**

Magnetization, $M$ is given by

$$M = g\mu_B \sum_i S_i^z \tag{13.60}$$

$$\text{But} \quad \sum_i S_i^z = NS - \sum n_k \Rightarrow M = g\mu_B[NS - \sum n_k]$$

$$\text{where} \quad \sum_k n_k = \int D_1(\omega) <n(\omega)> d\omega$$

and $\quad D_1(\omega)$ is the magnon density of states per unit volume

$\quad n(\omega)$ is the Bose-Einstein factor

$$\sum_k n_k = \int_0^\infty \frac{1}{4\pi^2}\left(\frac{\hbar}{JSa^2}\right)^{3/2} \omega^{1/2}\frac{1}{e^{\hbar\omega/k_BT}-1}d\omega$$

$$= \frac{1}{4\pi^2}\left(\frac{\hbar}{JSa^2}\right)^{3/2}\int_0^\infty \frac{\omega^{1/2}}{e^{\hbar\omega/k_BT}-1}d\omega$$

$$\text{Let} \quad \frac{\hbar\omega}{k_BT}=x \Rightarrow d\omega = \frac{k_BT}{\hbar}dx$$

$$\text{Hence} \quad \sum_k n_k = \frac{1}{4\pi^2}\left(\frac{\hbar}{JSa^2}\right)^{3/2}\left(\frac{k_BT}{\hbar}\right)^{3/2}\int_0^\infty \frac{x^{1/2}}{e^x-1}dx$$

where the integral can be recognized to be a standard integral of the form

$$\int_0^\infty \frac{x^{n-1}}{e^x-1}dx = \Gamma(n)\zeta(n) \tag{13.61}$$

$$\text{where} \quad \Gamma(n) \quad \text{is} \quad \text{the Gamma function}$$

$$\zeta(n) \quad \text{is} \quad \text{the Riemann-Zeta function}$$

$$\text{But} \quad n-1 = \frac{1}{2} \Rightarrow n = \frac{3}{2}$$

$$\int_0^\infty \frac{x^{1/2}}{e^x-1}dx = \Gamma(3/2)\zeta(3/2)$$

$$\text{where} \quad \Gamma(3/2) = \frac{\sqrt{\pi}}{2}$$

$$\zeta(3/2) = 2.612$$

$$\int_0^\infty \frac{x^{1/2}}{e^x-1}dx = \frac{\sqrt{\pi}}{2}\times 2.612$$

$$= 2.3148$$

$$= \frac{2,3148}{(4\pi^2)}\times(4\pi^2)$$

$$= 0.0586\times(4\pi^2)$$

$$\text{Hence} \quad \sum_k n_k = \frac{1}{4\pi^2}\left(\frac{k_B T}{JSa^2}\right)^{3/2} 0.0586(4\pi^2)$$

$$= 0.0586\left(\frac{k_B T}{JSa^2}\right)^{3/2}$$

$$\text{Magnetisation,} \quad M = g\mu_B[NS - \sum n_k]$$

$$= NSg\mu_B[1 - \frac{\sum n_k}{NS}]$$

$$M = M_s[1 - \frac{0.0586}{NS}\left(\frac{k_B T}{JSa^2}\right)^{3/2}] \qquad (13.62)$$

$$\text{where} \quad M_s = NSg\mu_B$$

where it can be noted that the magnetization shows a $T^{3/2}$ deviation.

**Total thermal energy due to magnons, $U$**

The total thermal energy, $U$, due to magnons is given by

$$U = \int_0^\infty D_1(\omega) < E(\omega) > d\omega \qquad (13.63)$$

where $\quad D_1(\omega) \quad$ is the magnon density of states per unit volume

and $\quad < E(\omega) > = n(\omega)\hbar\omega \quad$ is the mean energy, ignoring the zero point energy

and $\quad n(\omega) = \dfrac{1}{e^{\hbar\omega/k_B T} - 1}$

$$U = \int_0^\infty \frac{1}{4\pi^2}\left(\frac{\hbar}{JSa^2}\right)^{3/2} \frac{\omega^{1/2}\hbar\omega}{e^{\hbar\omega/k_B T} - 1}d\omega$$

Let $\quad \dfrac{\hbar\omega}{k_B T} = x \Rightarrow d\omega = \dfrac{k_B T}{\hbar}dx$

$$U = \frac{1}{4\pi^2}\left(\frac{\hbar}{JSa^2}\right)^{3/2}\int_0^\infty \frac{x(k_B T)}{e^x - 1}x^{1/2}\left(\frac{k_B T}{\hbar}\right)^{1/2}\left(\frac{k_B T}{\hbar}\right)dx$$

$$= \frac{1}{4\pi^2}\left(\frac{1}{JSa^2}\right)^{3/2}(k_B T)^{5/2}\int_0^\infty \frac{x^{3/2}}{e^x - 1}dx$$

where the integral can be recognized to be a standard integral of the form

$$\int_0^\infty \frac{x^{n-1}}{e^x - 1}dx = \Gamma(n)\zeta(n)$$

where $\quad \Gamma(n) \quad$ is $\quad$ the Gamma function

$\quad\quad\quad \zeta(n) \quad$ is $\quad$ the Riemann-Zeta function

But $\quad n - 1 = \dfrac{3}{2} \Rightarrow n = \dfrac{5}{2}$

$$\int_0^\infty \frac{x^{3/2}}{e^x - 1}dx = \Gamma(5/2)\zeta(5/2)$$

$$
\begin{aligned}
\text{where} \quad \Gamma(5/2) &= \frac{3}{4}\sqrt{\pi} \\
\zeta(5/2) &= 1.341 \\
\int_0^\infty \frac{x^{3/2}}{e^x - 1} dx &= \frac{3}{4}\sqrt{\pi} \times 1.341 \\
&= 1.01\sqrt{\pi} \\
U &= \frac{1.01\sqrt{\pi}}{4\pi^2} \frac{(k_B T)^{5/2}}{(JSa^2)^{3/2}} = 0.25 \frac{(k_B T)^{5/2}}{(JSa^2)^{3/2}}
\end{aligned}
\tag{13.64}
$$

**Specific heat due to magnons, $C_V$**

The Specific Heat, $C_V$, is defined as

$$
\begin{aligned}
C_V &= \frac{\partial U}{\partial T} \\
&= \frac{5}{2} \times 0.25 k_B \frac{(k_B T)^{3/2}}{(\pi JSa^2)^{3/2}} \\
&= AT^{3/2}
\end{aligned}
\tag{13.65}
$$

which in view of $C_V \propto T^{3/2}$ is known as BLOCH $T^{3/2}$ LAW.

Hence, the total specific heat at low temperatures is due to the contributions from magnons and phonons, in the form

$$
C_V = AT^{3/2} + BT^3
\tag{13.66}
$$

or

$$
\frac{C_V}{T^{3/2}} = A + BT^{3/2}
\tag{13.67}
$$

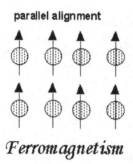

Figure 13.8: Spin ordered system for a Ferromagnet

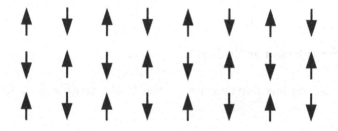

Figure 13.9: Spin ordered system for an Antiferromagnet.

Figure 13.10: Spin ordered system for a Ferrimagnetic

## 13.5  Exercises

**13.1.** Explain diamagnetic, paramagnetic, ferromagnetic and antiferromagnetic materials, illustrating your answer with suitable diagrams, and give two examles of each.

**13.2.** Magnetic dipole-dipole interaction can be modelld as follows. A magnetc field of a magnetic dipole moment $\mathbf{m_1}$ produces a magnetic field of components $\mathbf{H_r} = H_r \hat{\mathbf{u}}_r$ and $H_\theta = H_\theta \hat{\mathbf{u}}_\theta$ at a distance $R$ away.
(a) Show that the magnetic field at $R$ due to the electric dipole $\mathbf{p_1}$ is given by

$$\mathbf{H} = \frac{1}{4\pi R^3} \left\{ 3\mathbf{u}_r (\mathbf{m_1} \cdot \mathbf{u}_r) - \mathbf{m_1} \right\}$$

(b) Show that the interaction energy between two magnetic dipoles of dipole moments $\mathbf{p_1}$ and $\mathbf{p_2}$ at a distance $R$ apart is given by

$$U(R) = \frac{1}{4\pi} \left\{ \frac{\mathbf{m_1} \cdot \mathbf{m_2}}{R^3} - \frac{3(\mathbf{m_1} \cdot \mathbf{R})(\mathbf{m_2} \cdot \mathbf{R})}{R^5} \right\}$$

**13.3.** (a) Find an expression for magnetization, $M = N\overline{\mu}$, using classical theory of paramagnetism, where $\overline{\mu}$ is the average magnetic dipole moment, and show that

$$M = N\mu L(x)$$
$$\text{where} \quad L(x) = \coth(x) - \frac{1}{x}$$

is known as the Langevin function.

(b) (i) Illustrated graphically, $L(x)$ vs $x$
(ii)Hence show that the paramagnetic susceptibility, $\chi_m$ is given byCuries's law

$$\chi_m = \frac{C}{T}$$

where $C = \frac{N\mu^2}{3k_B}$ is he Curie's constant. (b) Illustrate graphically $\chi$ vs $T$.

**13.4.** (a) Find an expression for Magnetization, $M = N\overline{\mu_z}$, using quantum theory of paramagnetism, where $\overline{\mu}_z$ is the average quantised magnetic dipole moment, and show that

$$M = Ng\mu_B J B_J(x) \tag{13.68}$$
$$\text{where} \quad B_J(x) = \left\{ \left( \frac{2J+1}{2J} \right) \coth \left( \frac{2J+1}{2J} \right) x - \frac{1}{2J} \coth \frac{x}{2J} \right\} \tag{13.69}$$

is the Brillouin function.

(b) (i) Illustrated graphically, $B_J(x)$ vs $x$
(ii) Show that in the limit $J \to \infty$, $B_\infty(x) = L(x)$.

**13.5.** (a) Using molecular field theory, show that the magnetization, $M = Ng\mu_B < S^z >$, where the average spin $< S^z >$ is given by

$$< S^z >= \frac{\sum_i S_i^z e^{-E/k_B T}}{\sum_i e^{-E/k_B T}}$$

is given by

$$M = N\mu_B \tanh\left[\frac{\mu_B H_{eff}}{k_B T}\right]$$

(b) Plot a schematic graph for magnetization with temperature and discuss the spin orientations for $T < T_c$ and $T > T_c$.

**13.6.** (a) Show that the magnetization, $M$, due to conduction electrons in the presence of a magnetic field $H$ is given by

$$M = \frac{3}{2}\frac{N\mu^2}{k_B T_F}H$$

(b) 9i) Show that the *Pauli Spin Susceptibility* is given by

$$\chi = \frac{3}{2}\frac{N\mu^2}{k_B T_F}$$

where (ii) Comment on the order of magnitude of $\chi$.

**13.7.** (a) Give expressions for the interaction energy between spins $\vec{S}_i$ and $\vec{S}_j$ in a spin ordered system for
(i) the exchange interaction
(ii) the dipole-dipole interaction
(b) Using the Heisenberg model and molecular field theory, show that the effective field, $H_{eff}$, is given by

$$H_{eff} = H_0 + \lambda M$$

where $H_0$ is the external magnetic field and $\lambda M$ is the molecular field.
(c) (i) Show that the spontaneous magnetization occurs below a critical temperature $T_c$ given by

$$T_c = \frac{N\mu_B^2 \lambda}{k_B}$$

where you may assume that the magnetization M is given by

$$M = N\mu_B \tanh\left[\frac{\mu_B H_{eff}}{k_B T}\right]$$

**13.8.** (a) Explain what is meant by the Heisenberg model, and show that it leads to the following Hamiltonian

$$H = -\mu_n \cdot \mathbf{H}_{eff}^n$$

$$\text{where} \quad \mu_n = -g\mu_B \mathbf{S}_n \quad \text{is the magnetic moment at the } n^{th} \text{ spin}$$

$$\mathbf{H}_{eff}^n = -\frac{J}{g\mu_B}(\mathbf{S}_{n-1} + \mathbf{S}_{n+1})$$

(b) (i) Hence, show that spin waves for magnons is given by

$$\hbar\omega = 2JS(1 - \cos ka)$$

(ii) Illustrate a graph of dispersion curves for magnons.

**13.9.** (a) Using spin wave theory, show that the magnetization, $M = g\mu_B \sum_i S_i^z$, where

$$\sum_i S_i^z = NS - \sum n_k$$

shows a $T^{3/2}$ deviation in the form

$$M = M_s[1 - \frac{0.0586}{NS}\left(\frac{k_BT}{JSa^2}\right)^{3/2}]$$

$$\text{where} \quad M_s = NSg\mu_B$$

*Hints:*
(i) Density of states per unit volume for magnons in 1-d, $D(\omega) = \frac{V}{4\pi^2}\left(\frac{\hbar}{JSa^2}\right)^{3/2}\omega^{1/2}$
(ii) The following standard integral may be useful.

$$\int_0^\infty \frac{x^{n-1}}{e^x - 1}dx = \Gamma(n)\zeta(n)$$

$$\Gamma(3/2) = \frac{\sqrt{\pi}}{2}$$

$$\zeta(3/2) = 2.612$$

**13.10.** Consider a 1-d chain of a spin ordered system for magnons. Assuming that the zero point energy is negligible, show that

(a) The total thermal energy due to magnons is given by

$$U = \frac{1}{4\pi^2}\left(\frac{1}{JSa^2}\right)^{3/2}(k_BT)^{5/2}\int_0^\infty \frac{x^{3/2}}{e^x - 1}dx$$

where $x = \hbar\omega/k_BT$ and all other symbols are in the usual notation.

(b) The specific heat, $C_V$, is given by the Bloch $T^{3/2}$ law

$$C_V = \frac{5}{2} \times 0.25k_B\frac{(k_BT)^{3/2}}{(\pi JSa^2)^{3/2}} = AT^{3/2}$$

*Hints:*
(i) Density of states per unit volume for magnons in 1-d, $D(\omega) = \frac{V}{4\pi^2}\left(\frac{\hbar}{JSa^2}\right)^{3/2}\omega^{1/2}$
(ii) The following standard integral may be useful.

$$\int_0^\infty \frac{x^{n-1}}{e^x - 1}dx = \Gamma(n)\zeta(n)$$

$$\Gamma(5/2) = \frac{3}{4}\sqrt{\pi}$$

$$\zeta(5/2) = 1.341$$

# Chapter 14

# Semiconductor Devices

## 14.1   Introduction

When an intrinsic semiconductor is doped, it can become an n-type or p-type semiconductor. A p-type and n-type semiconductor are used to fabricate a pn junction. The pn junction is the basis of many of semiconductor devices. The energy bands for an intrinsic, n-type and p-type semiconductor are illustrated in Figure 14.1.

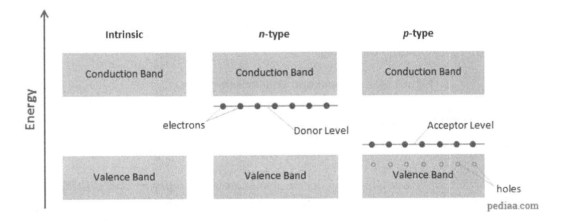

Figure 14.1: Energy bands for intrinsic, n-type and p-type semiconductor.

## 14.2   Two terminal devices

### 14.2.1   pn junction diode

Let us analyse what happens in the pn junction. In the p-region, there are fixed acceptors and holes, while in n-side there are donors and electrons. Holes *diffuse* from the p-side to the n-side. The hole concentrations varies from $p = p_{p0}$ in the p-side to $p = p_{no}$ in the n-side. Similarly, electrons *diffuse* from the n-side to the p-side. The electron concentrations varies from $n = n_{n0}$ in the n-side to

$n = n_{p0}$ in the p-side. The relationship between $p = p_{p0}$ and $p = p_{no}$ or $n = n_{n0}$ and $n = n_{0p}$ can be obtained as follows.

There are *three* cases of physical interest.

## Case 1: Open circuit

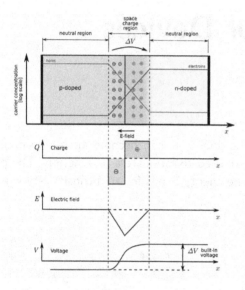

Figure 14.2: pn-junction equilibrium graphs

Let the width of the depletion region be

$$W_0 = W_{p0} + W_{no} \qquad (14.1)$$

There are two important processes happening in the depletion layer, *diffusion* and *drift* due to an electric field $\vec{E}$. Applying Gauss law for electric fields in the various regions of the depletion layer, as illustrated in Figure 14.2, we have

$$\nabla \cdot \vec{E} \;=\; \frac{\rho}{\epsilon_0 \epsilon_r} = \frac{\rho}{\epsilon} \qquad (14.2)$$

In the region $-W_{p0} < x < 0$, $\quad \dfrac{dE}{dx} \;=\; -\dfrac{eN_a}{\epsilon}$

Integrating and applying boundary conditions (bcs)

$$E(x) \;=\; -\frac{eN_a}{\epsilon}(x + W_{p0}) \qquad (14.3)$$

which is a linear variation with a negative slope.

In the region $0 < x < W_{n0}$, $\quad \dfrac{dE}{dx} \;=\; -\dfrac{eN_a}{\epsilon}$

Integrating and applying boundary conditions (bcs)

$$E(x) = \frac{eN_d}{\epsilon}(x - W_{n0})$$

which is a linear variation with a positive slope.

Note that at $x = 0$, the magnitude of the electric field is maximum with a value

$$E(x = 0) = -\frac{eN_aW_{p0}}{\epsilon} = -\frac{eN_dW_{n0}}{\epsilon}$$

$$N_aW_{p0} = N_dW_{n0}$$

$$
\begin{aligned}
W_0 &= W_{p0} + W_{no} \\
&= W_{p0}\left(1 + \frac{W_{n0}}{W_{p0}}\right) \\
&= W_{p0}\left(1 + \frac{N_a}{N_d}\right) \\
W_0 &= W_{p0}\left(\frac{N_a + N_d}{N_d}\right) \text{ or } W_{p0} = \frac{N_dW_0}{N_a + N_d} \\
\text{and } W_{n0} &= \frac{N_aW_0}{N_a + N_d}
\end{aligned}
$$

But

$$\nabla \cdot \vec{E} = \frac{\rho}{\epsilon_0 \epsilon_r} = \frac{\rho}{\epsilon}$$

$$\vec{E} = -\nabla V$$

$$\nabla^2 V = -\frac{\rho}{\epsilon} \text{ which is known as Poisson's equation.}$$

Let us solve Poisson's equation for the potential $V(x)$ in the various regions of the depletion layer of the pn junction,

$$
\begin{aligned}
\text{In the region } -\infty < x < -W_{p0}, \quad \frac{d^2V(x)}{dx^2} &= 0 \\
-W_{p0} < x < 0, \quad \frac{d^2V(x)}{dx^2} &= \frac{eN_a}{\epsilon} \\
0 < x < W_{n0}, \quad \frac{d^2V(x)}{dx^2} &= -\frac{eN_d}{\epsilon} \\
W_{n0} < x < \infty, \quad \frac{d^2V(x)}{dx^2} &= 0
\end{aligned}
$$

The electric field in 1d is related to the potential in all regions,

$$E(x) = -\frac{dV(x)}{dx}$$

In the region $-W_{p0} < x < 0$ (p-side of the depletion layer), the solutions are obtained as follows

$$\frac{d^2V(x)}{dx^2} = \frac{eN_a}{\epsilon}$$

Integrating and applying boundary conditions (bcs)

$$\frac{dV(x)}{dx} = \frac{eN_ax}{\epsilon} + \frac{eN_aW_{p0}}{\epsilon}$$

$$E(x) = -\frac{dV(x)}{dx} = -\frac{eN_ax}{\epsilon} - \frac{eN_aW_{p0}}{\epsilon}$$

Integrating, again, and applying bcs

$$V(x) = \frac{eN_ax^2}{2\epsilon} + \frac{eN_aW_{p0}x}{\epsilon} - \frac{eN_aW_{p0}^2}{\epsilon} + V_p$$

Hence

$$V(0) - V(-W_{p0}) = \frac{eN_aW_{p0}^2}{2\epsilon}$$

In the region $0 < x < W_{n0}$ (n-side of the depletion layer), the solutions are obtained as follows

$$\frac{d^2V(x)}{dx^2} = -\frac{eN_d}{\epsilon}$$

Integrating and applying boundary conditions (bcs)

$$\frac{dV(x)}{dx} = -\frac{eN_dx}{\epsilon} + \frac{eN_dW_{n0}}{\epsilon}$$

$$E(x) = -\frac{dV(x)}{dx} = \frac{eN_dx}{\epsilon} - \frac{eN_dW_{n0}}{\epsilon}$$

Integrating, again, and applying bcs

$$V(x) = -\frac{eN_dx^2}{2\epsilon} + \frac{eN_dW_{n0}x}{\epsilon} + \frac{eN_dW_{n0}^2}{\epsilon} + V_n$$

Hence

$$V(W_{n0}) - V(0) = \frac{eN_dW_{n0}^2}{2\epsilon}$$

Thus, the in built potential, $V_0$, is given by

$$V(W_{n0}) - V(-W_{p0}) = V_0 = \frac{eN_dW_{n0}^2}{2\epsilon} + \frac{eN_aW_{p0}^2}{2\epsilon}$$

Recall

$$N_dW_{n0} - N_aW_{p0} = 0 \Rightarrow W_{p0}) = \frac{N_d}{N_a}W_{n0}$$

Solving for $W_{n0}$ and $W_{p0}$

$$W_{n0}^2 = \frac{2\epsilon V_0}{e}\frac{N_a}{N_d[N_a + N_d]}$$

$$W_{p0}^2 = \frac{2\epsilon V_0}{e}\frac{N_d}{N_a[N_a + N_d]}$$

or

$$W_{n0} = \left\{\frac{2\epsilon V_0}{e}\frac{N_a}{N_d[N_a + N_d]}\right\}^{1/2}$$

$$W_{p0} = \left\{\frac{2\epsilon V_0}{e}\frac{N_d}{N_a[N_a + N_d]}\right\}^{1/2}$$

$$W_0 = \left\{ \frac{2\epsilon V_0}{e} \frac{[N_a + N_d]}{N_a N_D} \right\}^{1/2} \tag{14.4}$$

which gives the depletion capacitance as

$$C_{dep} = \frac{\epsilon A}{W_0} = \frac{A}{V_0^{1/2}} \left[ \frac{e\epsilon N_a N_d}{2(N_a + N_d)} \right]^{1/2} \tag{14.5}$$

Assuming Maxwell-Boltzmann statistics, the number of particles satify

$$n(E) = e^{-E/k_B T}$$

$$\frac{n_2}{n_1} = \frac{e^{-E_2/k_B T}}{e^{-E_1/k_B T}} = e^{-(E_2 - E_1)/k_B T}$$

Hence

$$\frac{n_{p0}}{n_{n0}} = e^{-eV_0/k_B T}$$

$$\frac{p_{n0}}{p_{p0}} = e^{-eV_0/k_B T}$$

$$n_{p0} = n_{n0} e^{-eV_0/k_B T}$$

$$p_{n0} = p_{p0} e^{-eV_0/k_B T}$$

Rearranging

$$V_0 = \left( \frac{k_B T}{e} \right) \ln \left( \frac{n_{n0}}{n_{p0}} \right)$$

$$V_0 = \left( \frac{k_B T}{e} \right) \ln \left( \frac{p_{p0}}{p_{n0}} \right) \tag{14.6}$$

## Case 2: Forward bias

Consider what happens in the *forward bias*, that is, when a battery with a voltage $V$ is connected across the $pn$ junction such that the positive terminal of the battery is connected to the $p$-side and the negative terminal of the battery is connected to the $n$-side. Essentially, the analysis is as before, but with the following modification.

$$V_o \rightarrow (V_0 - V)$$

Hence for holes,

$$p_n(0) = p_{p0} e^{-e(V_o - V)/k_B T}$$

$$= p_{n0} e^{eV/k_B T} \text{ known as the law of the junction for holes}$$

Similarly for electrons,

$$n_p(0) = n_{p0} e^{-e(V_o - V)/k_B T}$$

$$= n_{n0} e^{eV/k_B T} \text{ known as the law of the junction for electrons}$$

Note that the depletion capacitance is now given by

$$C_{dep} = \frac{\epsilon A}{W_0} = \frac{A}{(V_0 - V)^{1/2}} \left[ \frac{e\epsilon N_a N_d}{2(N_a + N_d)} \right]^{1/2} \tag{14.7}$$

The total current density due to holes and electrons is given by

$$
\begin{aligned}
J &= J_{D,hole} + J_{D,elec} \\
&= \left( \frac{eD_h}{L_h N_d} + \frac{eD_e}{L_e N_e} \right) n_i^2 \left[ e^{eV/k_B T} - 1 \right] \\
J &= J_{so} \left[ e^{eV/k_B T} - 1 \right] \quad \text{known as the Shockley equation}
\end{aligned}
$$

where

$$
J_{so} = \left( \frac{eD_h}{L_h N_d} + \frac{eD_e}{L_e N_e} \right) n_i^2
$$

Generally, the diode current is given by

$$
I = I_0 \left[ e^{eV/\eta k_B T} - 1 \right] \tag{14.8}
$$

where $\eta$ is the diode ideality factor, $1 < \eta < 2$. The Shockley equation and IV characteritic are illustrated (theory and experiment) in Figure 14.3.

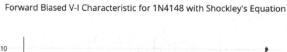

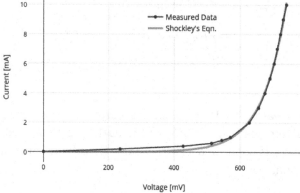

Figure 14.3: Shockley equation for a pn junction Forward bias (Source:https://ploy.ly/javario).

## Case 3: Reverse bias

Consider what happens in the *reverse bias*, that is, when a battery with a voltage $V$ is connected across the $pn$ junction such that the negative terminal of the battery is connected to the $p$-side and the positive terminal of the battery is connected to the $n$-side. In the reverse bias, the current is very small, as illustrated in Figure 14.4. The large current in the forward bias and the small current in the reverse bias can also be understood in terms of the energy band diagram of the pn junction as illustrated in Figure 14.5.

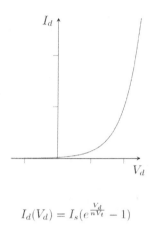

$$I_d(V_d) = I_s(e^{\frac{V_d}{nV_t}} - 1)$$

Figure 14.4: pn junction diode IV characteristic.

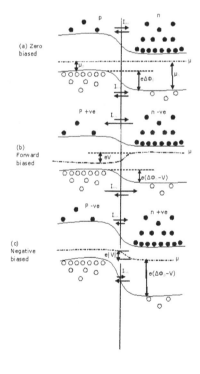

Figure 14.5: Energy bands for a pn junction.

## 14.2.2   pn junction Light Emitting Diode (LED)

A light emitting diode (LED) is essentially a pn junction, usually made from a direct gap semiconductor of energy gap, $E_g$, illustrated in figure 14.6. When there is electron-hole recombination, the

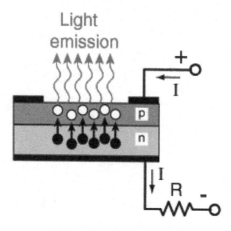

Figure 14.6: LED

electron makes a transition accompanied by an emission of a photon of frequency $\nu$ such that

$$Eg = h\nu \qquad (14.9)$$

The emission of light by a LED can also be illustrated by the energy band diagram as in Figure 14.7. An illustration of wavelengths emitted by a LED and respective band gaps are shown in Table 14.1. The relationship between band gaps, wavelength and lattice constants of a crystal, can however be complex, as discussed in https://www.ecse.rpc.edu/Schubert, EF and illustrated in Figure 14.8.

**Table 14.1:** Colour, Wavelengths, $\lambda$ (nm) and Band gap energy, $E$ (eV).

| Colour | Wavelength, $\lambda$ (nm) | Band gap energy, $E$ (eV) |
|---|---|---|
| Infrared | $\lambda > 760$ | $E < 1.63$ |
| Red | $610 < \lambda < 760$ | $1.63 < E < 2.03$ |
| Orange | $590 < \lambda < 610$ | $2.03 < E < 2.10$ |
| Yellow | $570 < \lambda < 590$ | $2.10 < E < 2.18$ |
| Green | $500 < \lambda < 570$ | $2.18 < E < 2.48$ |
| Blue | $450 < \lambda < 500$ | $2.48 < E < 2.76$ |
| Violet | $400 < \lambda < 450$ | $2.76 < E < 3.10$ |
| Ultraviolet | $\lambda < 400$ | $3.10 < E$ |

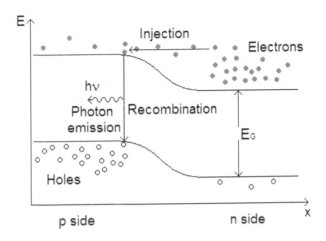

Figure 14.7: Energy band diagram for a LED

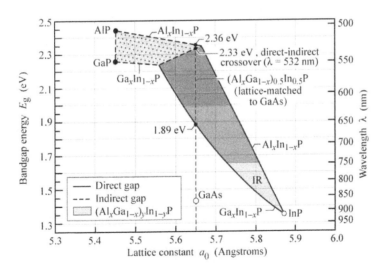

Figure    14.8:    Band    gap,    wavelength    and    lattice    constant    (Source: https://www.ecse.rpc.edu/Schubert, EF)

### 14.2.3    pn junction Laser diode(LD)

The most important laser diodes are based on the III-V semiconductors, such as GaAs or $Ga_{1-x}Al_xAs$, which emits at $0.75\mu m$ to $0.88\mu m$ depending on the molar fraction $x$. In the design of a laser diode, the pn junction (See Figure 14.9) must be heavily doped in such a way that the Fermi level on the n-side lies in the conduction band whereas the Fermi level in the p-side lies in the valence band. Lasing occurs when the supply of free electrons exceeds the losses in the cavity. An energy band diagram to illustrate how a laser diode works is illustrated in Figure 14.10. The forward bias is large enough to cause poulation inversion so that stimulated emission occurs.

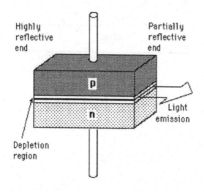

Figure 14.9: pn junction Laser diode

**Semiconductor Laser Diode**

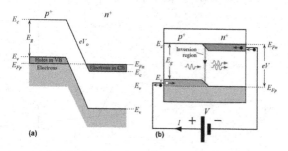

(a) The energy band diagram of a degenerately doped $pn$ with no bias. (b) Band diagram with a
sufficiently large forward bias to cause population inversion and hence stimulated emission.

Figure 14.10: An energy band diagram for a laser diode.

### 14.2.4   pn junction photodiode

A pn photodiode converts incident light energy into electrical energy. When light shines on a photodiode as illustrated in Figure 14.11, an electric current flows in the circuit. This can also be understood in terms of the energy band diagram in Figure 14.12.

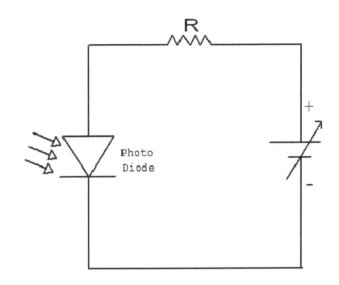

Figure 14.11: pn junction Photodiode

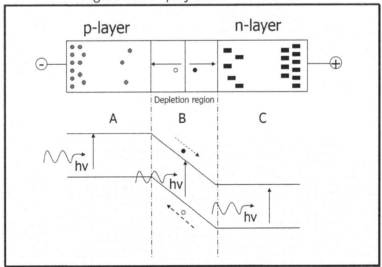

Figure 14.12: Photodiode and energy bands

## 14.2.5   pn junction Solar cell

A solar cell (also known as a *photovoltaic cell*) converts incident light energy into electrical energy, is illustrated in Figure 14.13. The solar energy band diagram is illustrated in Figure 14.14. The IV characteristics of a pn junction solar cell are illustrated in Figures 14.15 and 14.16.

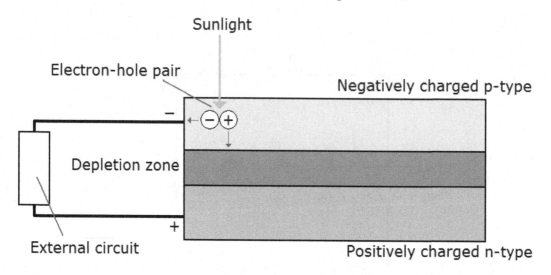

Figure 14.13: pn junction Solar Cell

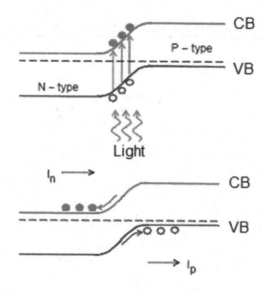

Figure 14.14: Solar cell energy band diagram.

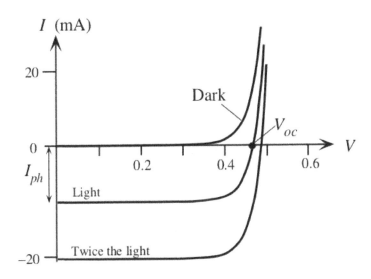

Figure 14.15: IV characteristics of a pn junction solar cell

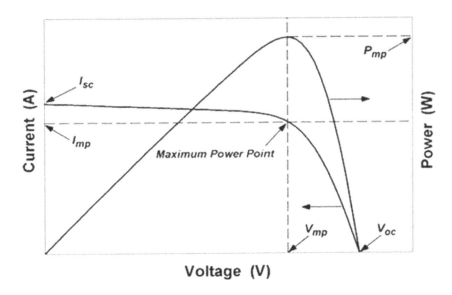

Figure 14.16: IV characteristic (inverted) of a pn junction solar cell.

## 14.2.6   PIN diode

The PIN diode is illustrated n Figure 14.17 consists of p-type, a wide undoped intrinsic semiconductor and an n-type semiconductor, and is a photodiode. the p-type and n-type regions are heavily doped. The high electric field present in the depletion layer causes photo-generated carriers to move and give rise to a photocurrent. The energy band diagram for a PIN diode is illustrated in Figure 14.18.

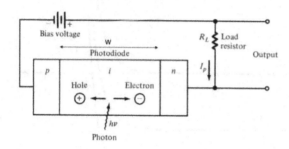

Figure 14.17:  PIN Diode

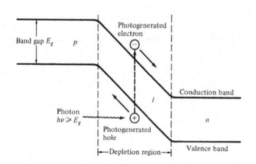

Figure 14.18: Energy band diagram for PIN diode

### 14.2.7   pn junction Avalanche photodiode

The Avalanche photodiode (APD) is a highly sensitive semiconductor sensor of light. A simplified diagram of an APD is shown in Figure 14.19. The $n^+$-layer is a thin and has a window through which an incident photon enters. Next to this, there are three p-type layers, with the first being thin, the second is thick and lightly doped called the $\pi$-layer, while the third is a heavily doped $p^+$-layer. The diode is reverse biased so as to increase the fields in the depletion regions. The electric field is maximum at the $n^+p$ junction, then decreases slowly through the $p$-layer, $\pi$-layer and vanishes at the end of the $p^+$-layer.

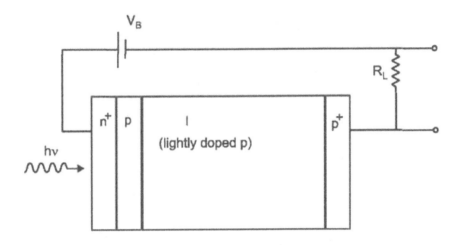

Figure 14.19: Structue of an Avalanche Photodiode (APD).

The avalanche occurs in the $\pi$-layer. Typical applications of the APD include fiber optic telecommunication, laser range finders, positron emission tomography (PET).

## 14.2.8   pn junction Tunnel diode (or Esaki diode)

The tunnel diode (or Esaki diode) is a pn junction which is heavily doped and this enables quantum tunnelling to occur in the forward bias. The IV characterisic of a tunnel diode is illustrated in Figure 14.20. The energy bands to illustrate the working of the tunnel diode is illustrated in Figure 14.21.

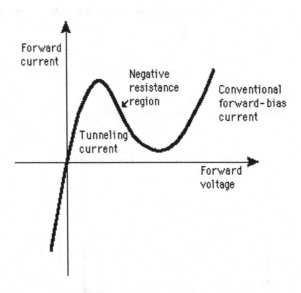

Figure 14.20: IV characteristic of the Tunnel (or Esaki) Diode.

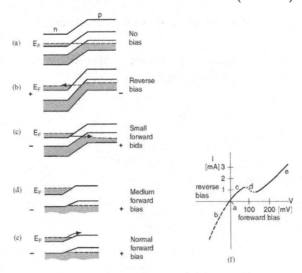

Figure 14.21: Energy bands for the tunnel diode.

### 14.2.9   pn junction Zener diode

The Zener diode is fabricated in such a way that the pn junction is heavily doped, so that it allows a current to flow like the normal pn junction, but it also permits a current to flow in the reverse bias when its Zener voltage has reached the Zener breakdown. A Zener diode in a circuit is shownin Figure 14.22, and the IV characteristic is illustrated in Figure 14.23. The Zener reverse breakdown is due to quantum mechanical tunneling of electrons. Zener diodes are widely used to stabilize power supply.

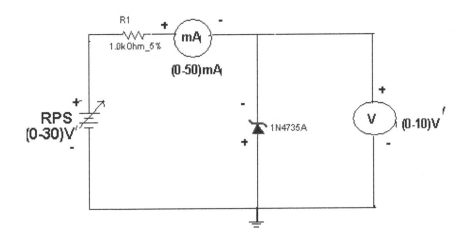

Figure 14.22:  Zener Diode

**Zener Diode I-V Characteristics Curve**

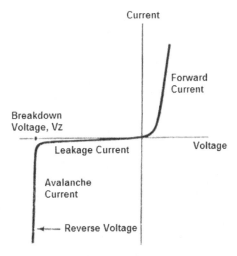

Figure 14.23: IV characteristics of the Zener Diode

### 14.2.10   Semiconductor Quantum Well Lasers

A quantum well laser has typically an ultra-thin narrow band gap semiconductor of band gap $E_{g1}$ sandwiched between two wider gap semiconductors with a band gap $E_{g2}$ as illustrated in Figure 14.24. For example, this could be a thin layer of $GaAS$ with band gap $E_{g1}$ sandwiched between two layers of $Ga_{1-x}Al_xAs$ with bandgap $E_{g2}$. Emission of light can be undestood in terms of a quantum transition from the conduction band (CB) to the valence band (VB) accompanied by emission of a photon, as illustrated in Figure 14.25.

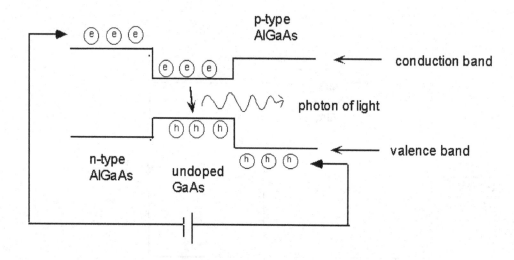

Figure 14.24: Quantum Well Laser

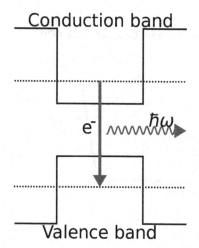

Figure 14.25: Electronic transition from the CB to the VB leading to the emission of a photon for a Quantum Well Laser.

## 14.3   Three terminal devices

### 14.3.1   Introduction

A transistor is a three terminal device that acts essentially as an electrically controlled switch and as an amplilfier. There are two types of transistors: the Bipolar Junction Transistors (BJT) and the Field Effect Transistors (FET), as illustrated in Figure 14.26. The BJT are in turn divided into npn transistors and pnp transistors (See Figure 14.27), while the FET transistors consist of JFET and MOSFET (See Figure 14.28).

There have been a lot of growth in the electronics industry over the last fifty years or so. It has been observed that over the years, the dimensions of the transistor have become smaller and smaller because of improved techniques of fabrication leading to miniaturization, thus enabling the packing of several transistors in integrated circuits. This has led to Moore's law of 1965 which says that the number of transistors would double approximately every two years, as illustrated in Figure 14.29. the reduction in size is illustrated by the reduction in gate length illustrated in Figure 14.30.

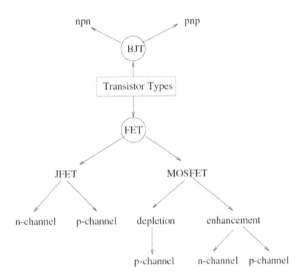

Figure 14.26: Classification of Transistors: BJT (npn and pnp) and FET (JFET and MOSFET)

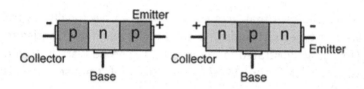

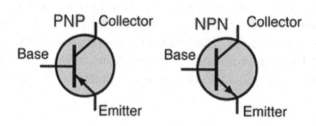

Figure 14.27: BJT (pnp and npn transistors)

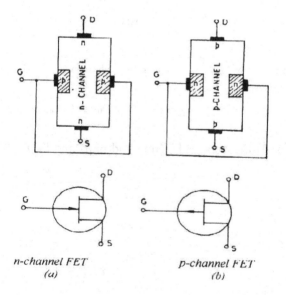

Figure 14.28: FET (n channel and p channel)

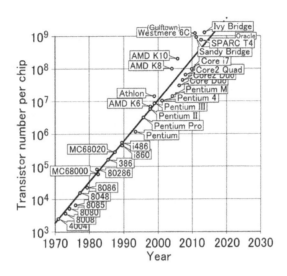

Figure 14.29: Moore's Law (Source: https://commons.wikimedia.org)

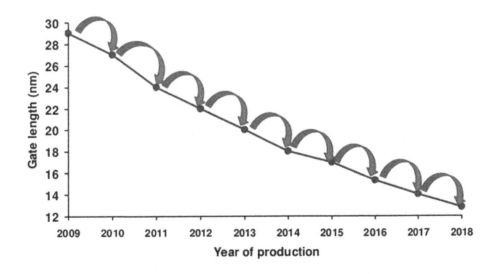

Figure 14.30: Transistor Gate length and Moore's law.

## 14.3.2   Bipolar Junction Transistor (BJT)

**npn Transistor**

A npn transistor consists of a np junction and an pn junction as illustrated in Figure 14.31, where the np junction is forward biased and the pn junction is reverse biased as shown in Figure 14.31. The $I_C$ (collector current) vs $V_{CE}$ (collector-emitter voltage) is shown in Figure 14.32. The energy band diagram of an npn transistor is shown in Figure 14.33.

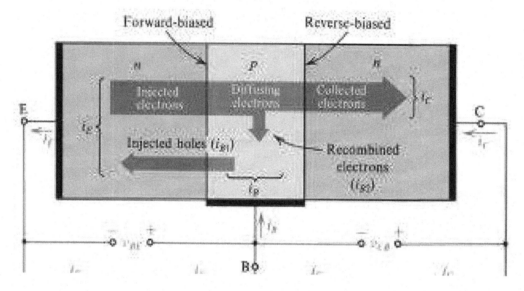

Figure 14.31: npn transistor.

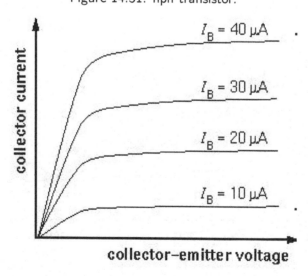

Figure 14.32: npn transistor characteristics

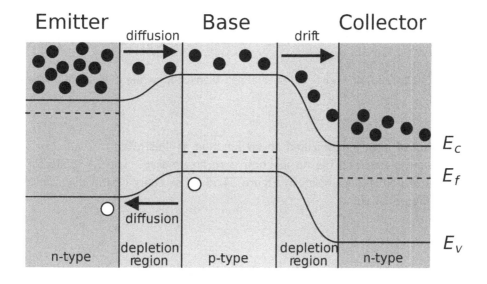

Figure 14.33: Energy bands for an npn transistor.

When the base-emitter junction is forward biased, a collector current $I_C$ is produced dependent on $V_{BE}$, given by

$$I_C = I_S(e^{qV_{BE}/k_BT} - 1) \tag{14.10}$$

The collector current, $I_C$, the base current, $I_B$ and the emitter current, $I_E$, are related by

$$I_B + I_C = I_E \tag{14.11}$$

**Current Gain Parameters:** $\beta$ and $\alpha$ are defined by

$$\beta = \frac{I_C}{I_B}$$

$$\alpha = \frac{I_C}{I_E}$$

Note that

$$\alpha = \frac{I_C}{I_B + I_C}$$

$$= \frac{I_C/I_B}{1 + I_C/I_B}$$

$$\alpha = \frac{\beta}{1 + \beta} \tag{14.12}$$

which can be rearranged to give

$$\alpha(1 + \beta) = \beta$$
$$\alpha + \alpha\beta = \beta$$
$$\alpha = \beta(1 - \alpha)$$

or

$$\beta = \frac{\alpha}{1-\alpha} \tag{14.13}$$

Practically, $\beta$ is large, 50 to 200, or higher for specialised transistor, while $\alpha$ is small, close to unit.

## pnp Transistor

A pnp transistor consists of a pn junction and an np junction as illustrated in Figure 14.34, where the pn junction is forward biased and the np junction is reverse biased. The $I_C$ (collector current) vs $V_{CE}$ (collector-emitter voltage) is shown in Figure 14.35. The energy band diagram of an pnp transistor is shown in Figure 14.36.

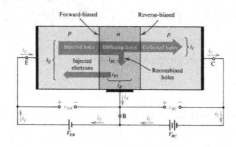

Figure 14.34: pnp transistor.

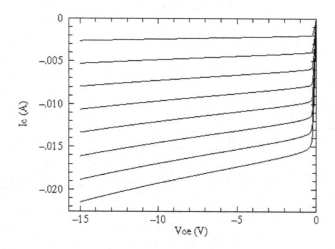

Figure 14.35: IV characterics for a pnp transistor.

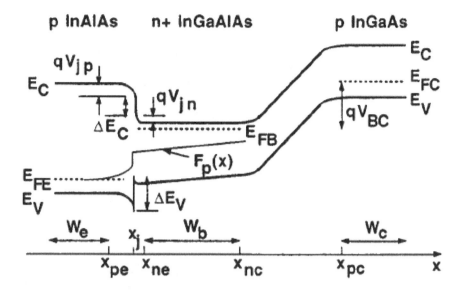

Figure 14.36: Energy bands for a pnp transistor.

**Phototransistor**

A phototransistor is a light sensitive transistor. An incident photon is absorbed in the space charge layer between the base and the collector and an electron-hole pair is generated. The photoransistor characteristics are shown in Figure 14.37.

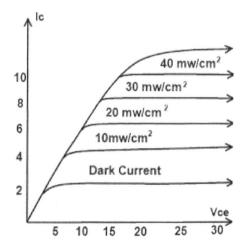

Figure 14.37: $I_C V_{CE}$ Characteristic of a phototransistor

### 14.3.3   Field Effect Transistor (FET)

The FET are of two types:  The Junction FET (JFET) and the Metal-Oxide-Semiconductor FET (MOSFET). The JFET operate in two modes:  N-channel JFET and P-channel (JFET) (See Figure 14.38).  The MOSFET is illustrated in Figure 14.39.  the energy bands for a MOS is illustrated in Figure 14.40.

**JFET**

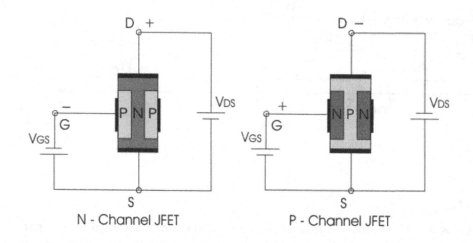

Figure 14.38: n-channel JFET and p-channel JFET.

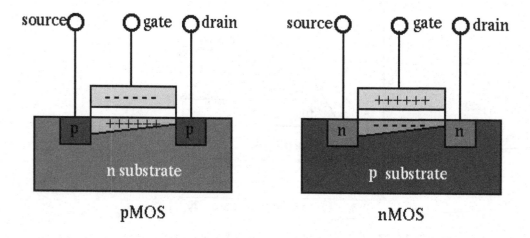

Figure 14.39: MOSFET: pMOS and nMOS.

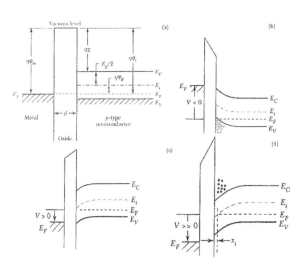

Figure 14.40: Energy bands for MOS

## 14.4   Exercises

**14.1**. (a) The development of the electronics industry has experienced miniaturization leading to the dense parking of electronic components in integrated circuits. What does Moore's law say about this observation?

(b) Calculate the Compton wavelength of the electron, $\frac{h}{m_e c}$, and comment whether this has impact on the process of miniaturization in the electronics industry.

**14.2**. The energy band gap of Si depends on temperature as

$$E_g = 1.17 eV - 4.73 \times 10^{-4} \frac{T^2}{T + 636}$$

Calculate the concentration of electrons in the conduction band of an intrinsic Si at $T = 77$K if at 300K, $n_i = 1.05 \times 10^{10}$ cm$^{-3}$

**14.3**. (a) State, without proof, the expression for the electrical conductivity of an intrinsic semiconductor.

(b) Estimate (in eV) the energy gap of an intrinsic semiconductor if the specimen is observed to have an electrical conductivity at 300 K to be 1000 times its electrical conductivity at liquid nitrogen temperature (77 K).

**14.4**. (a) Show that the Fermi energy, $E_F$, of an intrinsic semiconductor, is given by

$$E_F = \frac{1}{2}(E_c + E_v) + \frac{3}{4}k_B T ln(\frac{m_h^*}{m_e^*})$$

and you may assume that the electron and hole concentrations in the conduction band and valence band of an intrinsic semiconductor respectively are given by

$$n_e = N_e e^{[(E_F - E_c)/k_B T]}$$

$$n_h = N_h e^{[-(E_F - E_v)/k_B T]}$$

where

$$N_e = 2(\frac{m_e^* k_B T}{2\pi \hbar^2})^{3/2}$$

$$N_h = 2(\frac{m_h^* k_B T}{2\pi \hbar^2})^{3/2}$$

where all symbols are in the usual notation.

**14.5**. (a) Explain what is meant by *doping* in semiconductor technology, and give two examples.

(b) According to the hydrogenic model of impurity levels in a semiconductor, the energy levels, $E_n$, and the radius, $a_n$ of a donor, are given by

$$E_n = -\frac{m^* e^{*4}}{2(4\pi \epsilon_0)^2 \hbar^2} \frac{1}{n^2}$$

$$a_n = \frac{(4\pi \epsilon_0) \hbar^2 n^2}{m^* e^{*2}}$$

where $m^*$ is the effective mass, $e^* = e/\sqrt{\epsilon}$ is effective charge, and $n = 1, 2, \cdots$ denotes the levels.

Given that in InAs, $m^* = 0.02m_e$ and $\epsilon = 14.5$, calculate:
(i) the first energy level (in eV) of the donor impurity.
(ii) the first level radius (in $\overset{\circ}{A}$) of the donor impurity.

**14.6.** (a) Calculate the donor ionisation energy (in eV) for silicon, given that the effective mass of electrons, $m_e^* = 0.2m_e$ and the dielectric constant is 17.
(b) Calculate the radius of the ground state orbit from the data mentioned above.

**14.7.** Show that the effective mass $m^*$ of an electron in some allowed energy band in a semiconductor satisfies the relation

$$\frac{1}{m^*} = \frac{1}{\hbar^2}\frac{d^2 E}{dk^2}$$

where all symbols are in the usual notation.

**14.8.** (a) Describe briefly the *depletion layer* of a $pn$ junction, including a well labelled diagram.

(b) A semiconductor of dielectric constant $\epsilon$ has a $pn$ junction with a cross-sectional area $A$, $N_a$ acceptors in the $p$-side and $N_d$ donors in the $n$-side, and a depletion layer of width $W$ given by

$$W = \left[\frac{2\epsilon(V_0 - V)}{e}\frac{(N_a + N_d)}{N_a N_d}\right]^{1/2}$$

A symmetrical GaAs $pn$ junction has the following parameters: $N_a = 10^{23}\text{m}^{-3}$, $N_d = 10^{23}\text{m}^{-3}$, $\epsilon = 1.15 \times 10^{-10}\text{Fm}^{-1}$. Suppose the built-in voltage, $V_0 = 1.27$ V and the forward voltage across the $pn$ junction diode, $V = 0.8$ V, calculate the depletion layer width.

**14.9.** A very thin GaAs quantum well is sandwiched between two wider gap semiconductor layers of AlGaAs to form a quantum well laser diode. The energy levels for an electron in the one-dimensional infinite potential well of thickness $L$ are given by

$$E_n = \frac{n^2 \hbar^2 \pi^2}{2m^* L^2}$$

where the effective mass $m^* = 0.07m_e$, $L = 10$ nm and $n = 1, 2, \ldots$.

(i) Calculate the energies (in eV) for $n = 1$ and $n = 2$
(ii) Calculate the wavelength of the emitted photon for an energy gap of $E_g = 1.42$ eV.

**14.10.** (a) What is the Hall effect?
(b) In a Hall effect experiment on a rectangualr block of silver (dimensions: 1.5 cm wide X 0.05 mm thick X 5.0 cm long) a p.d of $59\mu$V is developed across the width, when a magnetic field of 1.25 T is applied in the direction of the thickness, and a current of 28 A along the length. Calculate the Hall coefficient of silver.

# Chapter 15

# Quantum Optics

## 15.1    Introduction

Quantum optics is a field of study that uses quantum mechanics to investigate phenomena involving light and its interaction with matter. As such its results are important in optoelectronics and quantum electronics. As an example, lasers are an important application.

The word LASER is an acronym for Light Amplification by Stimulated Emision of Radiation. A laser is a light source, but very different from a conventional light source. Light from a laser source is:
(i) Monochromatic
(ii) Coherent
(iii) narrow and highly collimated
(iv) Parallel
(v) very intense

Although laser technology has several applications, it is important to emphasize a cautionary note that lasers must be used with important safety protocols to avoid injury.

## 15.2    Emission and absorption

Light interacts with an atom in three ways: absorption, spontaneous emission and stimulated emission, as illustrated in Figure 15.1.
In *absorption*, an atom in an energy level $E_1$ absorbs a photon of energy $h\nu$, and its energy is raised to a higher quantised energy level $E_2$, such that

$$E_2 - E_1 = h\nu \qquad (15.1)$$

In *spontaneous emission*, an atom at a higher energy level $E_2$ may on its own accord emit a photon of energy $h\nu$ and leave the atom in a lower energy level $E_1$.

In *stimulated emission*, an atom at a higher energy level $E_2$ may be stimulated by a photon of energy $h\nu$ to emit another photon of the same energy, and finally leave the atom in a lower energy level $E_1$.

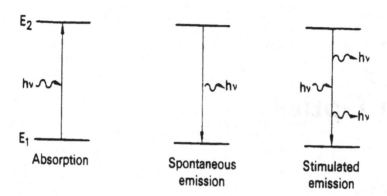

**Figure 15.1:** Light interacts with an atom in three ways: absorption, spontaneous emission and stimulated emission.

## 15.3   Einstein's $A$ and $B$ coefficients

Einstein's theory of absorption, spontaneous emission and stimulated emision is based on a physically reasonable model, and was developed in 1917. Consider a cavity with $N$ atoms, such that $N_1$ atoms are at a lower energy level with energy $E_1$ and $N_2$ atoms are at a higher energy level with energy $E_2$. Let the total energy density in a closed cavity be $U(\omega)$, consisting of a thermal part $U_T(\omega)$ and external radiation $U_E(\omega)$, *ie,*

$$N_1 + N_2 = N \tag{15.2}$$

and

$$U(\omega) = U_T(\omega) + U_E(\omega) \tag{15.3}$$

Let $A_{21}$ be the transition rate that the atom will spontaneously fall from state 2 to the lower state 1 and emit a photon of energy $\hbar\omega$, and

$B_{12}U(\omega)$ be the transition rate of an upward transition from state 1 to 2 in presence of radiation $U(\omega)$ due to absorption of a photon of energy $\hbar\omega$, and

$B_{21}U(\omega)$ be the transition rate of a stimulated emission from state 2 to 1 in presence of radiation $U(\omega)$ and a photon of energy $\hbar\omega$ is emitted

These processes are illustrated in Figure 15.2.

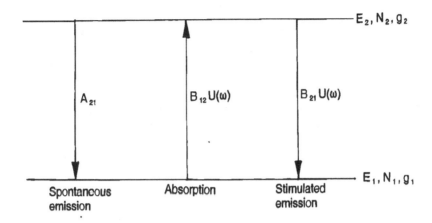

**Figure 15.2:** Absorption, spontaneous emission and stimulated emission as transitions between two levels of an atom.

The rate equation is

$$\frac{dN_1}{dt} = -\frac{dN_2}{dt} = N_2 A_{21} + N_2 B_{21} U(\omega) - N_1 B_{12} U(\omega) \tag{15.4}$$

For thermal equilibrium

$$\frac{dN_1}{dt} = -\frac{dN_2}{dt} = 0 \tag{15.5}$$

and hence

$$N_2 A_{21} + N_2 B_{21} U(\omega) - N_1 B_{12} U(\omega) = 0 \tag{15.6}$$

which can be rearranged to give

$$U(\omega) = \frac{A_{21}}{(N_1/N_2)B_{12} - B_{21}} \tag{15.7}$$

The ratio $N_1/N_2$ can be obtained using statistical mechanics, where according to Boltzmann's law, the population levels $N_1$ and $N_2$ with degeneracies $g_1$ and $g_2$ respectively are given by

$$N_1 \propto g_1 e^{-E_1/K_B T} \tag{15.8}$$
$$N_2 \propto g_2 e^{-E_2/K_B T} \tag{15.9}$$

which reduces to

$$\frac{N_1}{N_2} = \frac{g_1}{g_2} e^{(E_2 - E_1)/k_B T} \tag{15.10}$$

$$\frac{N_1}{N_2} = \frac{g_1}{g_2} e^{\hbar\omega/k_B T} \tag{15.11}$$

Using equation (15.11) in (15.7), the energy density can be written as

$$U(\omega) = \frac{A_{21}/B_{21}}{(g_1/g_2)(B_{12}/B_{21})e^{\hbar\omega/k_BT} - 1} \qquad (15.12)$$

However, according to Planck's radiation law, the energy density which was given in equation (7.6) can also be written in the form

$$U(\omega) = \frac{\hbar\omega^3}{\pi^2 c^3} \frac{1}{[e^{\hbar\omega/k_BT} - 1]} \qquad (15.13)$$

The result of the energy density from Einstein's theory given in equation (15.12) and that from Planck's radiation law given in equation (15.13) agree if and only if

$$\frac{A_{21}}{B_{21}} = \frac{\hbar\omega^3}{\pi^2 c^3} \qquad (15.14)$$

and

$$\frac{g_1}{g_2}\frac{B_{12}}{B_{21}} = 1 \qquad (15.15)$$

and hence equations (15.12) and (15.13) give

$$U(\omega) = \frac{\hbar\omega^3}{\pi^2 c^3}\bar{n} = \frac{A_{21}}{B_{21}}\bar{n} \qquad (15.16)$$

where

$$\bar{n} = \frac{1}{[e^{\hbar\omega/k_BT} - 1]} \qquad (15.17)$$

From the above discussion we can obtain the following results

$$A_{21} + B_{21}U(\omega) = A_{21}(\bar{n} + 1) \qquad (15.18)$$

$$\frac{A_{21}}{B_{21}U(\omega)} = \frac{1}{\bar{n}} \qquad (15.19)$$

where equation (15.18) gives the *sum* of the two emission rates and equation (15.19) gives the *ratio* of the two emission rates, namely the spontaneous and the stimulated emissions.

## 15.4   Lasers

In a laser, atoms or molecules of a gas (such as He-Ne, Argon, $CO_2$, $N_2$), liquid or solid (such as ruby, neodymium/yttrium aluminium garnet (Nd:YAG)) are excited in a *laser cavity*, of say length $L$. The laser cavity has reflecting surfaces at its ends so that photons can be reflected back and forth. One of the reflecting surfaces must be partially transparent so that light can escape the cavity and be generated as a laser beam which can be used for several applications. A standing electromagnetic wave fills the space between these two mirrors. This wave must satisfy the boundary condition that its electric field is zero at the position of the mirrors. The condition for the standing waves to be formed is that the distance $L$ between the mirrors and the wavelength $\lambda$ of the wave are related by

$$L = \left(\frac{n+1}{2}\right)\lambda \qquad n = 0,1,2,\cdots \qquad (15.20)$$

Within the laser cavity, conditions are created such that there is *population inversion*. If $N_1$ and $N_2$ are the numbers of atoms at energy levels $E_1$ and $E_2$, then according to statistical mechanics, at a temperature $T$ the ratio is given by

$$N_2 = N_1 e^{-(E_2 - E_1)/K_B T} \tag{15.21}$$

where population inversion occurs if $N_2 > N_1$, that is more atoms are at the higher energy level than there are at the lower energy level. Thus, population inversion occurs when a system exists such that there are more atoms in the excited state than in the lower states, as illustratd in Figure 15.3 and 15.4. This concept is of fundamental importance in laser science and technology because population inversion is a necessary step in the operation of a laser.

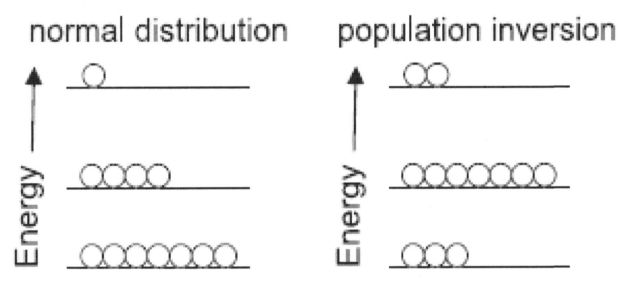

**Figure 15.3:** Population inversion in asystem with $N_2 > N_1$.

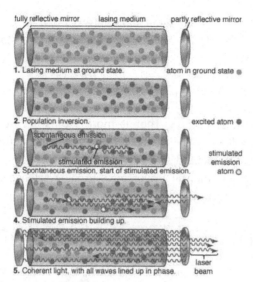

**Figure 15.4:** Population Inversion, Stimulated emission and Laser beam (Source: Encyclopedia Britannica).

Apart from the semiconductor diode laser discussed here, there are several types of lasers: Gas lasers, solid state lasers, dyw lasers etc. Wavelengths of some typical lasers are shown in Table 15.1.

**Table 15.1:** Wavelengths of some typical lasers.

| | |
|---|---|
| *Gas lasers* | |
| He-Ne | $6328\text{Å}$ |
| Argon ion | $4880\text{Å}$, $5145\text{Å}$ |
| $N_2$ | $3370\text{Å}$ |
| $CO_2$ | 9.6 to $10.6\mu$m |
| | |
| *Solid state lasers* | |
| Ruby($Cr^{3+}$) | $6943\text{Å}$ |
| Nd/YAG | $1.06\mu$m plus other near infrared lines |
| GaAs | $9000\text{Å}$ |
| GaAsP | $6500\text{Å}$ |
| | |
| *Dye lasers* | |
| Rhodamine G | $5710\text{Å}$ |

Applications of lasers include

(i) Industrial applications such as cutting and boring metals

(ii) Medical applications such as surgery

(iii) Scientific research in biology, chemistry, physics and other branches

(iv) Communications

(v) Holography

(vi) Technologial devices such as laser printers, CD players, scanners, bar-code scanners used in supermarkets and many others.

## 15.5  Exercises

**15.1**. (a) What is a laser? What are the main characteristics of a laser beam?

(b) Explain briefly the Einstein's $A$ and $B$ coefficients, illustrating your answer with a two level laser system.

**15.2**. A laser beam can be so well collimated that it spreads out only as a result of diffraction. Suppose such a laser beam of wavelength, $\lambda = 632.8$ nm has a diameter of 2 mm, what will be the beam diameter at a distance of 1 km from the laser?

**15.3**. Suppose in a *laser* cavity there are $N_1$ paricles, with degeneracy $g_1$, and $N_2$ particles with degeneracy $g_2$, with energy levels $E_1$ and $E_2$ respectively, where $E_2 > E_1$, the total energy density $U(\omega)$, according to Planck's radiation law, is given by

$$U(\omega) = \frac{\hbar\omega^3}{\pi^2 c^3} \frac{1}{\left[e^{\hbar\omega//k_B T} - 1\right]}$$

(a) Assuming Boltzmann's statistics, write down expressions for $N_1$ and $N_2$, and hence give an expression for $\frac{N_1}{N_2}$.

(b) Write down the rate equation $\frac{dN_1}{dt}$, explaining all symbols used.

(c) Hence, write an expression for the energy density $U(\omega)$ in terms of the Einstein's $A$ and $B$ coefficients.

**15.4**. For the laser cavity describe in exercise 15.2 above, show that

$$A_{21} + B_{21}U(\omega) = A_{21}(\bar{n} + 1)$$
$$\frac{A_{21}}{B_{21}U(\omega)} = \frac{1}{\bar{n}}$$
$$\text{where} \quad \bar{n} = \frac{1}{\left[e^{\hbar\omega//k_B T} - 1\right]}$$

**15.5**. The tube of a laser has a mirror at each end. A standing electromagnetic wave fills the space between these two mirrors. This wave must satisfy the boundary condition that its electric field is zero at the position of the mirrors.

(a) Show that the distance $L$ between the mirrors and the wavelength $\lambda$ of the wave are related by

$$L = \left(\frac{n+1}{2}\right)\lambda \qquad n = 0,1,2,\cdots$$

(b) Calculate the value of $n$ for a He-Ne laser with $L = 30.0$ cm and $\lambda = 6328 \overset{\circ}{A}$.

**15.6**. The relative importance of stimulated emission $A$ two-level atom is placed within a cavity and is allowed to come into equilibrium with blackbody radiation of temperature $T$.

(a) Show that the condition for the rate of stimulated emission from the upper level being equal to the rate of spontaneous emission is

$$k_B T = \frac{\hbar\omega_{21}}{\ln 2}$$

where $\hbar\omega 21$ is the energy spacing of the levels.

(b) Find the temperature in electron volts (eV) and kelvin when this condition is met for transition in the following regions of the electromagnetic spectrum:

(i) radio frequencies at 50 MHz,

(ii) microwaves at 1 GHz,

(iii) visible light at 500nm,

(iv) X-rays of energy 1 keV

(c) Under this condition, what is the ratio of the population per state in the upper and lower level? Explain your answer in terms of the spontaneous and stimulated transition rates between the two levels.

**15.7.** The characteristic red light of a He-Ne laser is due to stimulated emission between neon levels at 20.66 eV and 18.70 eV. Calculate the wavelength and frequency of the emitted radiation.

**15.8.** A laser emits radiation at a wavelength of 555 nm. Photons are emitted at the rate of $8.5 \times 10^{18}$ s$^{-1}$. What is the power of the laser?

**15.9.** A laser opearting at 640 nm produces pulses at a rate of 85 MHz. Calculate the radiant power of each pulse llas 15fs each and the average radiant power of the laseris 2.6 W. How mant photons are produced by the laser per second?

**15.10.** Consider a three laser system illustated in the figure below, where there are three energy levels $E_1, E_2, E_3$ with $N_1, N_2, N_3$ atoms respectively and $N = N_1 + N_2 + N_3$

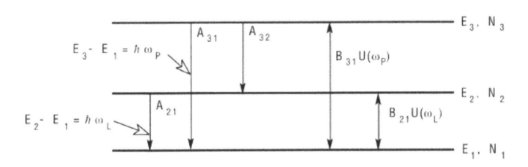

Figure for Question 15.10: Three level laser system.

(a) Verify the following rate equations

$$\frac{dN_1}{dt} = N_2 A_{21} + N_3 A_{31} + (N_2 - N_1)B_{21}U(\omega_L) + (N_3 - N_1)B_{31}U(\omega_P)$$

$$\frac{dN_2}{dt} = N_3 A_{31} - N_2 A_{21} - (N_2 - N_1)B_{21}U(\omega_L)$$

$$\frac{dN_3}{dt} = -N_3 A_{31} - N_3 A_{32} - (N_3 - N_1)B_{31}U(\omega_P)$$

(b) Show that in the steady state,

$$N_1 = \frac{N\left[R(A_{32} + A_{21} + B_{21}U(\omega_L)) - (A_{31} + A_{32})(A_{21} + B_{21}U(\omega_L))\right]}{[A_{31} + A_{32}]\,[A_{21} + B_{21}U(\omega_L)]}$$

$$N_2 = \frac{N}{[A_{31} + A_{32}]\,[A_{21} + B_{21}U(\omega_L)]}$$

$$\text{where}\quad R = \frac{U(\omega_P)B_{31}(N_1 - N_3)}{N}$$

# Chapter 16

# Introduction to Statistical Mechanics

## 16.1   Need for statistical laws in many particle systems

When one deals with particles in nuclei, atoms, molecules, solids, liquids, gases, one is dealing with *many particle systems*, that is, a system with very large number of particles. It may be recalled that the Avogadro's number, $N = 6.022 \times 10^{23}$. Such a system has $N = 6.022 \times 10^{23}$ degrees of freedom, and if we were to solve the dynamics of such a system by classical mechanics we would need $N = 6.022 \times 10^{23}$ equations of motion.

## 16.2   Phase space and its states

Consider a $2s$-dimensional space with $s$ generalised coordinates and $s$ generalised momenta, that is

$$\text{generalised coordinates: } q_1, q_2, \cdots q_s$$

and

$$\text{generalised momenta: } p_1, p_2, \cdots p_s$$

A *state* is a point in phase space. Since the dynamics of a state involves its evolution with time, there is need for an equation of motion. The equation of motion involves a distribution function of the system, $\rho(p, q)$, given by

$$\rho(p, q) = \rho(p_1, p_2, \cdots p_s, q_1, q_2, \cdots q_s) \tag{16.1}$$

The probability, $\omega$, that $(q_i, p_i)$ has values lying between $q_i$ to $q_i + dq_i$ and $p_i$ to $p_i + dp_i$ is

$$
\begin{aligned}
d\omega &= \rho(p_1, p_2, \cdots p_s, q_1, q_2, \cdots q_s) dp_1 dp_2 \cdots dp_s dq_1 dq_2 \cdots dq_s \\
&= \rho(p_1, p_2, \cdots p_s, q_1, q_2, \cdots q_s) dp dq \\
&= \rho(p, q) dp dq
\end{aligned} \tag{16.2}
$$

where an element of phase space $dpdq$ is given by

$$dpdq = dp_1 dp_2 \cdots dp_s dq_1 dq_2 \cdots dq_s$$

374 CHAPTER 16. INTRODUCTION TO STATISTICAL MECHANICS

Using the normalisation condition that the total probability must be unit, we obtain

$$\int d\omega = \int \rho(p,q)dpdq = 1 \tag{16.3}$$

The *mean value* of an observable $f(p,q)$ can either be a *time average* or *an ensemble average*.

The *time average* is defined as

$$<f> = \lim_{T\to\infty} \frac{1}{T} \int_0^T f(t)dt \tag{16.4}$$

The *ensemble average* is defined as

$$<f> = \frac{\int f(p,q)\rho(p,q)dpdq}{\int \rho(p,q)dpdq} \tag{16.5}$$

for a *a continuous system*, and

$$<f> = \frac{\sum_n f_n(p,q)\rho_n(p,q)}{\sum_n \rho_n(p,q)} \tag{16.6}$$

for a *a discrete system*.

A system is said to be in *statistical equilibrium* if all the macroscpic physical quantities are approximately equal to their mean values. If some external force is applied to the system, the system will be perturbed and displaced from equilibrium. A *relaxation time*, $\tau$, is the time in which a system that has been displaced will make a transition to statistical equilibrium.

The distribution function $\rho(p,q)$ satisfies an equation analogous to the continuity equation

$$\nabla.\vec{j} + \frac{\partial \rho}{\partial t} = 0 \tag{16.7}$$

in $3d$. However, equation (16.7) has to be extended to the phase space ($2s$-dimensional space), and becomes

$$\frac{\partial \rho}{\partial t} + \sum_{i=1}^s \left[ \frac{\partial(\rho\dot{q}_i)}{\partial q_i} + \frac{\partial(\rho\dot{p}_i)}{\partial p_i} \right] = 0$$

$$\frac{\partial \rho}{\partial t} + \sum_{i=1}^s \left\{ \left[ \dot{q}_i \frac{\partial \rho}{\partial q_i} + \dot{p}_i \frac{\partial \rho}{\partial p_i} \right] + \rho \left[ \frac{\partial \dot{q}_i}{\partial q_i} + \frac{\partial \dot{p}_i}{\partial p_i} \right] \right\} = 0 \tag{16.8}$$

The derivatives of the generalised coordinates, $\dot{q}_i$ and generalised momenta, $\dot{p}_i$, are related to derivatives of the Hamiltonian $H$, through Hamilton's equations, given as

$$\dot{q}_i = \frac{\partial H}{\partial p_i} \tag{16.9}$$

$$\dot{p}_i = -\frac{\partial H}{\partial q_i} \tag{16.10}$$

and hence, one obtains

$$\frac{\partial \dot{q}_i}{\partial q_i} = \frac{\partial^2 H}{\partial q_i \partial p_i} = \frac{\partial}{\partial p_i}\left(\frac{\partial H}{\partial q_i}\right) = -\frac{\partial \dot{p}_i}{\partial p_i} \tag{16.11}$$

which leads to

$$\left[\frac{\partial \dot{q}_i}{\partial q_i} + \frac{\partial \dot{p}_i}{\partial p_i}\right] = 0 \tag{16.12}$$

and hence the term in the second square brackets in equation (16.8) vanishes, and hence

$$\frac{\partial \rho}{\partial t} + \sum_{i=1}^{s}\left[\frac{\partial \rho}{\partial q_i}\dot{q}_i + \frac{\partial \rho}{\partial p_i}\dot{p}_i\right] = 0 \tag{16.13}$$

From equation (16.13), since the differential of the distribution function with time is zero, it is well known from calculus that this can only happen if the distribution function is a constant. This is known as *Liouville's theorem*, stating that the distribution function is a constant along the phase trajectory of the subsystem.

## 16.3 Gibb's distribution

Recall the macrocanonical distribution $\rho$ given by

$$\rho = \text{constant } \delta(E - E_0)$$

If the energy $E$ is represented as

$$E = E_n + E' \tag{16.14}$$

then

$$\begin{aligned}
\rho &= \text{constant } \delta(E_n + E' - E_0) \\
d\omega_n &= \rho dpdq \\
&= \text{constant } \delta(E_n + E' - E_0)dpdq
\end{aligned} \tag{16.15}$$

Let $dpdq \propto d\Gamma'$ Hence

$$\begin{aligned}
\omega_n &= \int \text{constant } \delta(E_n + E' - E_0)dpdq \\
&= \int \text{constant } \delta(E_n + E' - E_0)\frac{d\Gamma'}{dE'}dE' \\
&= \rho_n
\end{aligned}$$

and $\rho_n$ is known as the *Gibb's distribution for probability*.

But

$$\frac{d\Gamma'}{dE'} = \frac{\Delta\Gamma'}{\Delta E'} \tag{16.16}$$

Using the probability interpretation of entropy, $S'$, given by

$$S' = k_B \ln \Delta\Gamma'$$
$$\Delta\Gamma' = e^{S'/k_B} \tag{16.17}$$

Using equations (16.16) and (16.17), we obtain

$$\rho_n = \text{constant} \times \int \delta(E_n + E' - E_0)\frac{e^{S'/k_B}}{\Delta E'}dE' \tag{16.18}$$

Using the property of the delta-function

$$\int f(x)\delta(x-a)dx = f(a)$$

in equation (16.18), we obtain

$$\rho_n = \text{constant} \times \left(\frac{e^{S'/k_B}}{\Delta E'}\right)_{E'=E_0-E_n}$$

Expanding $S'(E')$ using Taylor's expansion,

$$S'(E') = S'(E_0 - E_n)$$
$$= S'(E_0) - \frac{\partial s'}{\partial E}.E_n - \cdots \tag{16.19}$$

and using this in equation (16.18), we obtain

$$\rho_n = \text{constant} \times \frac{e^{S'/k_B}}{\Delta E'}e^{-E_n.\frac{\partial S'}{\partial E}/k_B}$$

But using the second law of thermodynamics

$$\partial S = \frac{\partial E}{T} \quad \text{or} \quad \frac{\partial S}{\partial E} = \frac{1}{T}$$

we obtain

$$\rho_n = Ae^{-E_n/k_BT} \tag{16.20}$$

where $A$ is a constant that can be determined as follows.

Note that using the normalisation condition, we have

$$\sum_n \rho_n = 1 = A\sum_n e^{-E_n/k_BT}$$

from which $A$ is obtained as

$$A = \frac{1}{\sum_n e^{-E_n/k_BT}} \tag{16.21}$$

and hence using equation (16.21) in (16.20), we obtain

$$\rho_n = \frac{e^{-E_n/k_BT}}{\sum_n e^{-E_n/k_BT}} \tag{16.22}$$

$$= \frac{e^{-E_n/k_BT}}{Z} \tag{16.23}$$

where $Z$ is known as the *partition function* given by

$$Z = \sum_n e^{-E_n/k_BT} \tag{16.24}$$

If the energy distribution is continuous, the partition function is given by

$$Z = \frac{1}{\tau}\int e^{-E_n/k_BT}dpdq \tag{16.25}$$

where $\tau$ is some normalisation factor. The partition function is a very important function in statistical mechanics, since from it, one can obtain several functions of physical interest.

Related to partition function, $Z$, is another function known as the *Grand partition function*, $Q$, defined as

$$Q = \sum_n e^{-(E_n-\mu N)/k_BT} \tag{16.26}$$

where $\mu$ is the chemical potential.

## 16.3.1 Applications of the Gibb's Distribution

### Thermodynamic Expectation Value

The *Thermodynamic Expectation Value* of some observable $O$ is denoted by $<O>$ or $\bar{O}$ is given by

$$<O> = \sum_n O_n\rho_n \tag{16.27}$$

$$= \frac{\sum_n O_n e^{-E_n/k_BT}}{\sum_n e^{-E_n/k_BT}} \tag{16.28}$$

$$= \frac{\sum_n O_n e^{-E_n/k_BT}}{Z} \tag{16.29}$$

where $\rho_n$ is given in equation (16.23) and $Z$ is the partition function given in equation (16.24)

### Number of Particles, $n_i$

Suppose $n_i$ is the number of particles in a given energy level of energy $E_i$, and that the total number of particles in the system is $N$. From this, we obtain

$$\sum_i n_i = N \tag{16.30}$$

But

$$\sum_i \rho_i = 1 \tag{16.31}$$

and hence

$$\begin{aligned}
\sum_i n_i &= N \sum_i \rho_i \\
&= \sum_i N \rho_i
\end{aligned}$$

$$\sum_i n_i = \frac{\sum_i N e^{-E_i/k_B T}}{\sum_i e^{-E_i/k_B T}} \tag{16.32}$$

$$n_i = \frac{N}{Z} e^{-E_i/k_B T} \tag{16.33}$$

**Mean Energy, $< E >$**

$$\begin{aligned}
< E > &= \sum_i E_i \rho_i \\
&= \frac{\sum_i E_i e^{-E_i/k_B T}}{\sum_i e^{-E_i/k_B T}} \tag{16.34} \\
&= \frac{\sum_i E_i e^{-E_i/k_B T}}{Z} \tag{16.35}
\end{aligned}$$

which can only also be derived by noting that the mean energy is an average energy in the form

$$\begin{aligned}
< E > &= \frac{1}{N} \sum_i n_i E_i \\
&= \frac{1}{N} \sum_i \frac{N}{Z} e^{-E_i/k_B T} E_i \\
&= \frac{N}{N} \frac{\sum_i E_i e^{-E_i/k_B T}}{Z} \\
&= \frac{\sum_i E_i e^{-E_i/k_B T}}{Z}
\end{aligned}$$

which is the same as equation (16.29).

## 16.4   The partition function and some thermodynamic functions

It has been stated that the partition function $Z$ is a very important function in statistical mechanics. This is demonstrated in this section by showing how $Z$ is related to the thermodynamic functions: Internal energy $U$, Free energy $F$ and entropy $S$.

$$U = N k_B T^2 \frac{\partial}{\partial T} ln Z \tag{16.36}$$

$$F = Nk_BTlnZ \tag{16.37}$$

$$S = Nk_BlnZ + \frac{U}{T} \tag{16.38}$$

## 16.5 Condition equations

Consider a system with $N$ particles distributed such that there are $n_i$ particles at an energy level $E_i$. Condition equations are a set of three equations: the population condition, the energy condition and the stability condition, described below.

**Population condition**

In an isolated system, the number of particles $N$ is a constant given by

$$N = \sum_i n_i \tag{16.39}$$

Hence

$$\begin{aligned} \delta N &= \sum_i \delta n_i \\ \delta N &= 0 \\ \sum_i \delta n_i &= 0 \end{aligned} \tag{16.40}$$

that is, the change $\delta N$ is zero, which is known as the *population condition*. What equation 16.40 is saying is that the sum of the change of energy levels is zero, that is, if there is an increase in the population of an energy level, there must be an equal decease in the population of some other energy level or levels. This is the conservation of particles in the system.

**Energy condition**

The internal energy $U$ of an isolated system is a constant, and is given by

$$U = \sum_i n_i E_i \tag{16.41}$$

Hence

$$\begin{aligned} \delta U &= \sum_i (E_i \delta n_i + n_i \delta E_i) \\ \delta U &= 0 \end{aligned} \tag{16.42}$$

that is, the change $\delta U$ is zero, which is known as the *energy condition*. The energy condition is essentially the conservation of energy of an isolated system.

**Stability condition**

For a system to be in equilibrium with its surroundings, there must be a maximum probability $W$. Hence

$$\text{if } W \text{ is maximum}, \Rightarrow \delta W = 0 \qquad (16.43)$$

Hence

$$lnW \text{ is maximum}, \Rightarrow \delta(lnW) = 0 \qquad (16.44)$$

that is, the change $\delta(lnW)$ is zero, which is known as the *stability condition*. What the stability condition is saying is that a system in equilibrium must have the maximum number of microstates which in turn translate to a maximum probability for the system to exist in that state.

## 16.6   Stirling's approximation

There are several times one is required to evaluate $\ln n!$, and Stirling's approximation becomes useful in the form

$$\ln n! = n \ln n - n \qquad (16.45)$$

This can easily be proved as below.

Consider the graph of $\ln x$ vs $x$ as shown Figure 16.1.

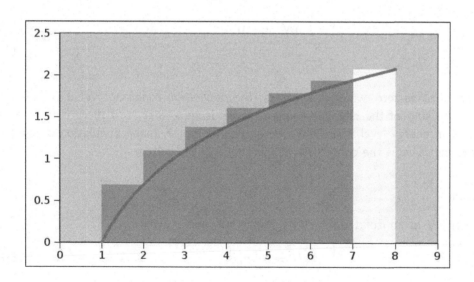

Figure 16.1: Graph of $\ln x$ vs $x$.

$$
\begin{aligned}
\text{Area under the curve} \ &= \ \int \ln x \, dx \\
&\approx \ \text{Area of the rectangles} \\
&= \ \ln 1 + \ln 2 + \ln 3 + \cdots + \ln n \\
&= \ \ln n!
\end{aligned}
$$

Let us integrate $\ln x$ by parts.

$$
\begin{aligned}
\int \ln x\, dx &= \int u\, dv \\
\text{Let } u &= \ln x \quad \text{and} \quad dv = du \\
\text{Hence } du &= \frac{1}{x} dx \\
v &= x \\
\int \ln x\, dx &= uv - \int v\, du \\
&= x \ln(x) - \int x\left(\frac{1}{x}\right) dx \\
&= x \ln x - x + C \qquad\qquad (16.46) \\
\text{Area under the curve} &= \int_1^n \ln x\, dx \\
&= [x \ln x - x]_1^n \\
\ln n! &= n \ln n - n \qquad\qquad (16.47)
\end{aligned}
$$

which completes the proof of Stirling's approximation.

## 16.7  Normal Distribution Curve

The normal distrbution, also known as the Gaussian distribution or Bell curve is described by a probability density function given by

$$
f(x) = \frac{1}{\sqrt{2\sigma^2\pi}} e^{-\frac{(x-\mu)^2}{2\sigma^2}} \qquad\qquad (16.48)
$$

$$
\begin{aligned}
\text{where } \mu \quad &\text{is} \quad \text{the mean} \\
\sigma \quad &\text{is} \quad \text{the standard deviation} \\
\sigma^2 \quad &\text{is} \quad \text{the variance}
\end{aligned}
$$

The normal distribution curve is symmetrical, as illustrated in Figure 16.2. About 68% of the are is within one standard deviation, about 95% is within two standard deviations, and about 99.7% is within three standard deviations, as illustrated in Figure 16.3.

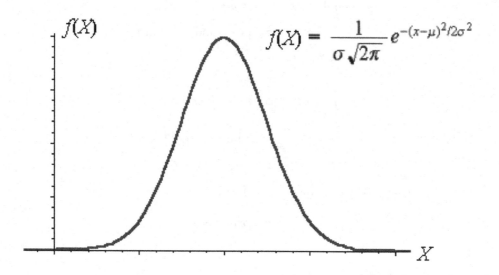

Figure 16.2: Normal Distribution Curve.

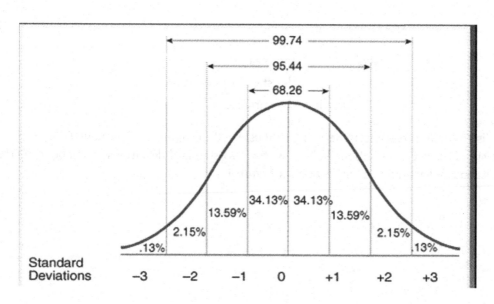

Figure 16.3: Normal Distribution Curve.

## 16.8  Exercises

**16.1**. (a) Explain briefly what is meant by an *ensemble average*, defining all the symbols used.
(b) A certain distribution function, known as the normal distribution is given by

$$f(x) = A \exp^{-\frac{1}{2}\beta x^2}$$

in 1-dimension, and $A$ is a constant.
(i) Express the constant $A$ in terms of $\beta$ by using the fact $f(x)$ satisfies the normalisation condition, given by

$$\int_{\infty}^{\infty} f(x) dx = 1$$

(ii) Calculate the mean value of the observable $x$.

**16.2**. (a) State the expression for the *Gibb's distribution for probability*, which is also known by several other names such as the *canonical distribution* or *Boltzmann's law*, defining all the symbols used.
(b) Consider a system with two states having energy eigenvalues $\epsilon_0$ and $\epsilon_1$, with $\epsilon_1 > \epsilon_0$.
(i) Show that the probability that the system is the lower and upper levels is given by

$$\rho_0 = \frac{1}{1 + \exp^{-\theta/T}}$$

and

$$\rho_1 = \frac{\exp^{-\theta/T}}{1 + \exp^{-\theta/T}}$$

respectively, where $\theta = (\epsilon_1 - \epsilon_0)/k_B$.
(ii) Plot accurately the graph of $\rho_1$ vs $T/\theta$.
(iii) A two level system has an energy gap of 0.1 eV. Calculate the probability that the system is in the higher energy level when the system is in equilibrium with a heat reservoir at 300 K.

**16.3**. (a) (i) Give an expression for the partition function $Z$, defining all the symbols used.
(ii) Show that the mean energy $< E >$ is related to $Z$ by

$$< E >= k_B T^2 \frac{\partial}{\partial T} (\ln Z)$$

(b) A certain system has particles with energy $E_n$ given by $E_n = n\epsilon$ where $n = 0, 1, 2, \ldots\ldots\ldots$
(i) Show that the mean energy is given by

$$< E >= \frac{\epsilon}{e^{\epsilon/k_B T} - 1}$$

(ii) hence find the limiting value of the mean energy when $\epsilon \ll k_B T$.

**16.4**. (a) Give an expression for the mean value of an observable $O, < O >$, using the *Gibb's distribution*, defining all the symbols used.

(b) Consider a system satisfying the *Gibb's distribution* such that there are only two energy levels, $E_n = -\alpha O_n$ with $n = 0, 1$ and that the observables $O_n$ satisfy

$$O_1 + O_0 = 0$$
$$O_1 - O_0 = 1$$

(i) Find the values $O_0$ and $O_1$.
(ii) Show that

$$< O >= \frac{1}{2} \tanh(\frac{\alpha}{2k_B T})$$

(iii) Illustrate schematically a graph of $< O >$ vs $T$.

**16.5**. According to the Gibb's distribution, the thermodynamic expectation value of an observable $O$, is given by the

$$< O > \ = \ \frac{\sum_0^\infty O_n e^{-E_n/k_B T}}{Z}$$

$$\text{where } Z \ = \ \sum_0^\infty e^{-E_n/k_B T} \text{ is the Partition Function.}$$

Use the expression for the energy of a quantum harmonic oscillator, $E_n = (n + \frac{1}{2})\hbar\omega$, to show that the mean energy is given by

$$< E(\omega) > \ = \ \left(n(\omega) + \frac{1}{2}\right)\hbar\omega$$

$$\text{where } n(\omega) \ = \ \frac{1}{e^{\hbar\omega/k_B T} - 1} \text{ is the Bose-Einstein Factor}$$

Note that this result is used in the Einstein's theory of specific heat in Chapter 11.

**16.6**. According to the Gibb's distribution, the average spin $< S^z >$, is given by

$$< S^z >= \frac{\sum_i S_i^z e^{-E/k_B T}}{\sum_i e^{-E/k_B T}}$$

Use the expression for the energy, $E = -g\mu_B H_{eff} S_i^z$, to show that the magnetization is given by

$$M \ = \ Ng\mu_B \frac{\sum_i S_i^z e^{-E/k_B T}}{\sum_i e^{-E/k_B T}}$$

$$M \ = \ N\mu_B \tanh\left[\frac{\mu_B H_{eff}}{k_B T}\right]$$

Note that this result is used in Molecular field theory of magnetization in Chapter 13.

**16.7**. (a) State Stirling's approximation and prove it.

(b) In the table below, calculate $\ln n!$ by a direct method and by Stirling's approximation and calculate the magnitude of the difference.

| n | $\ln n!$ | $n \ln n - n$ | Difference, $\|\ln n! - (n \ln n - n)\|$ |
|---|---|---|---|
| 0 | | | |
| 1 | | | |
| 2 | | | |
| 3 | | | |
| 4 | | | |
| 5 | | | |
| 6 | | | |
| 7 | | | |
| 8 | | | |
| 9 | | | |
| 10 | | | |

**16.8.** Show that the normal distribution function, $f(x) = \frac{1}{\sqrt{2\sigma^2 \pi}} e^{-\frac{(x-\mu)^2}{2\sigma^2}}$, satisfies

$$\int_\infty^\infty f(x)dx = 1$$

**16.9.** Show that the normal distribution function, $f(x) = \frac{1}{\sqrt{2\sigma^2 \pi}} e^{-\frac{(x-\mu)^2}{2\sigma^2}}$, satisfies the following differential equation

$$\sigma^2 \frac{df(x)}{dx} + f(x)(x - \mu) = 0$$

**16.10.** (a) Show that the fourier transform of the normal distribution function, $f(x) = \frac{1}{\sqrt{2\sigma^2 \pi}} e^{-\frac{(x-\mu)^2}{2\sigma^2}}$, is given by

$$f(k) = \int_{-\infty}^{\infty} f(x)e^{-ikx}dx$$
$$= e^{i\mu k} e^{-\frac{1}{2}\sigma^2 k^2}$$

(b) Comment on the case when in (a) above, $\mu = 0$.

# Chapter 17

# Statistical Mechanics Distribution Functions

## 17.1 Maxwell-Boltzmann Statistics

### 17.1.1 Introduction

Consider a large number of *identical* and *distinguishable* particles. By identical, we mean that the particles have the same structure. By distingushable, we mean that we can distinguish between one identical and another.

### 17.1.2 Thermodynamic probability for a perfect gas

Consider $N$ particles arranged such that there are $n_1, n_2, n_3 \cdots$ particles in different energy levels of energies $E_1, E_2, E_3 \cdots$ respectively. The number of particles satisfies

$$\sum_i n_i = N \tag{17.1}$$

What are the number of ways of placing particles in the different energy levels? This is found as follows.

The number of ways of placing $n_1$ particles out of $N$ in the state with energy $E_1$ is given by

$$^N C_{n_1} = \frac{N!}{n_1!(N - n_1)!} \tag{17.2}$$

and $(N - n_1)$ particles remain.

The number of ways of placing $n_2$ particles out of $(N - n_1)$ in the state with energy $E_2$ is given by

$$^{(N-n_1)} C_{n_2} = \frac{(N - n_1)!}{n_2!(N - n_1 - n_2)!} \tag{17.3}$$

and $(N - n_1 - n_2)$ particles remain.

The number of ways of placing $n_3$ particles out of $(N - n_1 - n_2)$ in the state with energy $E_3$ is given by

$$^{(N-n_1-n_2)}C_{n_3} = \frac{(N - n_1 - n_2)!}{n_3!(N - n_1 - n_2 - n_3)!} \tag{17.4}$$

and $(N - n_1 - n_2 - n3)$ particles remain.

and so on. This process is continued until all the particles have occupied the various energy levels.

The *thermodynamic probability* for a perfect gas, $W$ is defined as the number of distinguishable different ways of placing $n_1, n_2, n_3 \cdots$ particles in different energy levels of energies $E_1, E_2, E_3 \cdots$ respectively. Hence, $W$ is obtained by multiplying the different possibilities given in equations (17.2), (17.3), (17.4) and so on, given by

$$\begin{aligned}
W &= \frac{N!}{n_1!(N - n_1)!} \times \frac{(N - n_1)!}{n_2!(N - n_1 - n_2)!} \times \frac{(N - n_1 - n_2)!}{n_3!(N - n_1 - n_2 - n_3)!} \times (N - n_1 - n_2 - n_3)! \cdots \\
&= \frac{N!}{n_1!n_2!n_3!\cdots} \\
W &= \frac{N!}{\prod_i n_i!} \tag{17.5}
\end{aligned}$$

Let us recall the condition equations.

The population condition is given by

$$\sum_i \delta n_i = 0 \tag{17.6}$$

The energy condition is given by

$$\sum_i E_i \delta n_i = 0 \tag{17.7}$$

and the stability condition is given by

$$\delta(\ln W) = 0 \tag{17.8}$$

Using the expression for the thermodynamic probability, the stability condition can be recast in another form as shown below.

$$\begin{aligned}
\ln W &= \ln \frac{N!}{\prod_i n_i!} \\
&= \ln N! - \sum_i \ln n_i! \\
&= N \ln N - N - \sum_i n_i \ln n_i + \sum_i n_i \quad \text{using Stirling's approximation,} \\
&= N \ln N - \sum_i n_i \ln n_i \tag{17.9}
\end{aligned}$$

from which we obtain

$$
\begin{aligned}
\delta(\ln W) &= -\sum_i \delta n_i - \sum_i \ln n_i \delta n_i \\
&= -\sum_i \ln n_i \delta n_i
\end{aligned}
\tag{17.10}
$$

and hence the stability equation can also be written as

$$
\sum_i \ln n_i \delta n_i = 0
\tag{17.11}
$$

### 17.1.3 Lagrange's Method of Undetermined Multipliers and Maxwell-Boltzmann Distribution

The Lagranges'Method of Undetermined Multipliers consists of using the condition equations and introducing two undetermined multipliers, $\alpha$ and $\beta$, as shown below.

Multiply equation (17.6) by $-\ln\alpha$
Multiply equation (17.7) by $\beta$
Multiply equation (17.11) by $1$

and *add* the resulting equations, to obtain

$$
\sum_i (-\ln\alpha + \beta E_i + \ln n_i)\delta n_i = 0
\tag{17.12}
$$

and hence

$$
\begin{aligned}
(-\ln\alpha + \beta E_i + \ln n_i) &= 0 \\
\ln\frac{n_i}{\alpha} &= -\beta E_i \\
n_i &= \alpha e^{-\beta E_i}
\end{aligned}
\tag{17.13}
$$

What is $\alpha$? This can be found as follows.

$$
\begin{aligned}
\sum_i n_i &= N \\
&= \sum_i \alpha e^{-\beta E_i} \quad \text{using equation (17.13)} \\
&= \alpha \sum_i e^{-\beta E_i}
\end{aligned}
\tag{17.14}
$$

and hence $\alpha$ is found as

$$
\alpha = \frac{N}{\sum_i e^{-\beta E_i}}
$$

using equations (17.13) and (17.14), we obtain

$$n_i = \frac{Ne^{-\beta E_i}}{\sum_i e^{-\beta E_i}} \qquad (17.15)$$

What is $\beta$? This can be found as follows. The mean energy, $\bar{E}$ or $<E>$ can be written in the form

$$\begin{aligned}
\bar{E} &= \frac{1}{N}\sum_i n_i E_i \\
&= \frac{1}{N}\frac{\sum_i Ne^{-\beta E_i}E_i}{\sum_i e^{-\beta E_i}} \\
&= \frac{\sum_i E_i e^{-\beta E_i}}{\sum_i e^{-\beta E_i}} \\
&= \sum_i E_i \rho_i \qquad (17.16)
\end{aligned}$$

where

$$\rho_i = \frac{e^{-\beta E_i}}{\sum_i e^{-\beta E_i}} \qquad (17.17)$$

Comparing the above equation with the Gibb's distribution given in equation (16.22) in chapter 16, we have

$$\rho_i = \frac{e^{-E_i/k_B T}}{\sum_i e^{-E_i/k_B T}} \qquad (17.18)$$

and by comparing equations (17.17) and (17.18), we can identify $\beta$ as

$$\beta = \frac{1}{k_B T} \qquad (17.19)$$

Therefore $\alpha$ can be written in the form

$$\alpha = \frac{N}{\sum_i e^{-\beta E_i}} = \frac{N}{\sum_i e^{-E_i/k_B T}} = \frac{N}{Z} \qquad (17.20)$$

where $Z$ is the partition function.

The number of particles in each energy level, $n_i$, is therefore given by

$$n_i = \frac{N}{Z}e^{-E_i/k_B T} \qquad (17.21)$$

which is known as the *Maxwell-Boltzmann distribution*

## 17.1.4   Applications of Maxwell-Boltzmann Statistics

### Ideal gas

The *Total Energy*, $E$, of molecules of an *ideal gas* is given by

$$E = K.E. + P.E. \qquad (17.22)$$

where $KE$ is the kinetic energy and $PE$ is the potential energy. The potential energy can be considered to be negligible if the intermolecular interaction is considered to be very weak. The total energy is therefore dominated by the kinetic energy, and hence

$$E = \frac{1}{2}mv_i^2 \tag{17.23}$$

where $m$ is the mass of each molecule, and $v_i$ is the velocity of the $i^{th}$ molecule.

The *Partition function*, $Z$, for an ideal gas is given by

$$\begin{aligned} Z &= \sum_i e^{-E_i/k_BT} \\ &= \sum_i e^{-mv_i^2/2k_BT} \end{aligned}$$

Using the following prescription which relates a summation and an integration, we have

$$\sum_i \cdots = \frac{1}{\tau}\int \cdots d\Gamma \tag{17.24}$$

where $\tau$ is some normalisation factor, and $d\Gamma$ is an element of phase space.

Using the integral fornulation, the partition function is given by

$$\begin{aligned} Z &= \frac{1}{\tau}\int e^{-\frac{1}{2}mv^2/k_BT}d\Gamma \\ &= \frac{1}{\tau}\int\int\int\int\int\int e^{-\frac{1}{2}mv^2/k_BT}dxdydzdv_xdv_ydv_z \\ &= \frac{1}{\tau}\int_{-\infty}^{\infty}\int_{-\infty}^{\infty}\int_{-\infty}^{\infty}dxdydz\int_{-\infty}^{\infty}e^{-\frac{1}{2}mv_x^2/k_BT}dv_x\int_{-\infty}^{\infty}e^{-\frac{1}{2}mv_y^2/k_BT}dv_y\int_{-\infty}^{\infty}e^{-\frac{1}{2}mv_z^2/k_BT}dv_z \end{aligned} \tag{17.25}$$

where the phase space has been taken in $6 - dimensions$ so that $d\Gamma = dxdydzdv_xdv_ydv_z$. The integral over $dxdydz$ gives the volume $V$, that is

$$\int_{-\infty}^{\infty}\int_{-\infty}^{\infty}\int_{-\infty}^{\infty}dxdydz = V \tag{17.26}$$

The integrals over $dv_i$ for $i = x, y, z$ are standard integrals of the form

$$\int_0^{\infty}e^{-\alpha y^2}y^n dy = \frac{1}{2}\Gamma\left(\frac{n+1}{2}\right)\alpha^{-(n+1)/2} \tag{17.27}$$

which for $n = 0$, $(n+1)/2 = 1/2$, and noting that $\Gamma(1/2) = \sqrt{\pi}$, we obtain

$$\begin{aligned} \int_0^{\infty}e^{-\alpha y^2}dy &= \frac{1}{2}\sqrt{\pi}\alpha^{-1/2} \\ &= \frac{1}{2}\sqrt{\frac{\pi}{\alpha}} \quad \text{and hence} \\ \int_{-\infty}^{\infty}e^{-\alpha y^2}dy &= \sqrt{\frac{\pi}{\alpha}} \end{aligned}$$

and hence

$$\int_{-\infty}^{\infty} e^{-\frac{1}{2}mv_i^2/k_B T} dv_i = \left(\frac{2\pi k_B T}{m}\right)^{1/2} \quad \text{where} \quad v_i = v_x, v_y, v_z \tag{17.28}$$

Using equations (17.26) and (17.28) in (17.25), we obtain

$$Z = \frac{V}{\tau}\left(\frac{2\pi k_B T}{m}\right)^{3/2} \tag{17.29}$$

which is the partition function for an ideal gas.

From the Maxwell-Boltzmann distribution given in equation (17.21), we can insert the expression for $Z$ to obtain

$$n_i = \frac{N}{Z}e^{-E_i/k_B T}$$

$$n_i = \frac{N\tau}{V}\left(\frac{m}{2\pi k_B T}\right)^{3/2} e^{-\frac{1}{2}mv_i^2/k_B T}$$

which leads to

$$dn = 4\pi N\left(\frac{m}{2\pi k_B T}\right)^{3/2} v^2 e^{-\frac{1}{2}mv^2/k_B T} dv$$

or

$$\frac{dn}{dv} = 4\pi N\left(\frac{m}{2\pi k_B T}\right)^{3/2} v^2 e^{-\frac{1}{2}mv^2/k_B T} dv \tag{17.30}$$

A graph of $dn/dv$ against $v$ is illustrated in Figure (17.1), and three velocities of interest are shown.

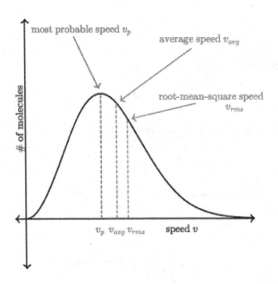

Figure 17.1: $\frac{dn}{dv}$ vs $v$, showing $v_{mp}, \bar{v}, v_{rms}$.

(i) Most probable velocity, $v_{mp}$

(ii) Mean velocity, $\bar{v}$

(iii) Root mean square velocity, $v_{rms}$

The *Most probable velocity, $v_{mp}$,* can be obtained by finding the velocity for which $dn/dv$ is a maximum. Let $dn/dv = y$, where $y$ can be identified from equation (17.30). Hence,

$$
\begin{aligned}
\frac{dy}{dv} &= 4\pi N \left(\frac{m}{2\pi k_B T}\right)^{3/2} \left\{ 2v e^{-\frac{1}{2}mv^2/k_B T} - \frac{2mv}{2k_B T} v^2 e^{-\frac{1}{2}mv^2/k_B T} \right\} \\
&= 4\pi N \left(\frac{m}{2\pi k_B T}\right)^{3/2} e^{-\frac{1}{2}mv^2/k_B T} v \left\{ 2 - \frac{mv^2}{k_B T} \right\} \\
&= 0 \ \text{for a maximum when } v = v_{mp}
\end{aligned}
$$

from which we obtain

$$
v_{mp} = \sqrt{\frac{2k_B T}{m}} \tag{17.31}
$$

as an expresion for the *most probable velocity.*

The *mean velocity, $\bar{v}$,* can be obtained as follows

$$
\begin{aligned}
\bar{v} &= \frac{\int_0^\infty v\,dn}{\int_0^\infty dn} \\
&= \frac{1}{N}\int_0^\infty v\,dn \\
&= \sqrt{\frac{8k_B T}{\pi m}} \tag{17.32}
\end{aligned}
$$

as an expresion for the *mean velocity.*

The *root mean quare velocity, $v_{rms} = \sqrt{\bar{v^2}}$,* can be obtained as follows

$$
\begin{aligned}
\bar{v^2} &= \frac{\int_0^\infty v^2\,dn}{\int_0^\infty dn} \\
&= \frac{1}{N}\int_0^\infty v^2\,dn \\
&= \frac{3k_B T}{m} \tag{17.33}
\end{aligned}
$$

from which an expresion for the *root mean square velocity* is obtained as

$$
v_{rms} = \sqrt{\bar{v^2}} = \sqrt{\frac{3k_B T}{m}} \tag{17.34}
$$

**Molecules in the gravitational field: The Barometric Equation**

The *Total Energy*, $E$, of molecules in the gravitational field is given by

$$E = \frac{1}{2}mv^2 + mgz \tag{17.35}$$

where the first term is the kinetic energy due to a molecule of mass $m$ moving with avelocity $v$. The second term is the potential energy, which is due to the *gravitational potential energy* due to a molecule of mass $m$ at a height $z$ and $g$ is the acceleration due to gravity.

The *Partition function*, $Z$, for molecules in the gravitational field is given by

$$Z = \sum_i e^{-E_i/k_B T}$$

$$= \sum_i e^{-(\frac{1}{2}mv_i^2 + mgz)/k_B T}$$

Using the following prescription which relates a summation and an integration, we have

$$\sum_i \cdots = \frac{1}{\tau}\int \cdots d\Gamma \tag{17.36}$$

where $\tau$ is some normalisation factor, and $d\Gamma$ is an element of phase space.

Using the integral fornulation, the partition function is given by

$$Z = \frac{1}{\tau}\int e^{-(\frac{1}{2}mv^2 + mgz)/k_B T}d\Gamma$$

$$= \frac{1}{\tau}\int\int\int\int\int\int e^{-\frac{1}{2}mv^2/k_B T}dxdydzdv_xdv_ydv_z$$

$$= \frac{1}{\tau}\int_{-\infty}^{\infty}\int_{-\infty}^{\infty}dxdy$$

$$\times \int_{-\infty}^{\infty}e^{-mgz/k_B T}dz\int_{-\infty}^{\infty}e^{-\frac{1}{2}mv_x^2/k_B T}dv_x\int_{-\infty}^{\infty}e^{-\frac{1}{2}mv_y^2/k_B T}dv_y\int_{-\infty}^{\infty}e^{-\frac{1}{2}mv_z^2/k_B T}dv_z \tag{17.37}$$

where the phase space has been taken in $6 - dimensions$ so that $d\Gamma = dxdydzdv_xdv_ydv_z$. The integral over $dxdy$ gives the area $A$, that is

$$\int_{-\infty}^{\infty}\int_{-\infty}^{\infty}dxdy = A \tag{17.38}$$

The integral over $z$ gives

$$\int_{-\infty}^{\infty}e^{-mgz/k_B T}dz = \left[\frac{e^{-mgz/k_B T}k_B T}{-mg}\right]_0^{\infty}$$

$$= \frac{k_B T}{mg} \tag{17.39}$$

The integrals over $dv_i$ for $i = x, y, z$ are standard integrals, and as was shown in section (17. ), the result is

$$\int_{-\infty}^{\infty} e^{-\frac{1}{2}mv_i^2/k_BT} dv_i = \left(\frac{2\pi k_BT}{m}\right)^{1/2} \quad \text{where} \quad v_i = v_x, v_y, v_z \tag{17.40}$$

Using equations (17.38), (17.39) and (17.40) in (17.37), we obtain

$$Z = \frac{Ak_BT}{\tau mg}\left(\frac{2\pi k_BT}{m}\right)^{3/2} \tag{17.41}$$

which is the partition function for molecules in the gravitational field.

From the Maxwell-Boltzmann distribution given in equation (17.21),

$$n_i = \frac{N}{Z}e^{-E_i/k_BT}$$

we can insert the expression for $Z$ from (17. ) to obtain

$$dn = \frac{Nmg}{Ak_BT}e^{-mgz/k_BT}dxdydz \quad \text{and} \quad \text{hence}$$

$$n = \frac{dn}{dxdydz} = \frac{Nmg}{Ak_BT}e^{-mgz/k_BT} \tag{17.42}$$

Using the initial conditions that when $z = 0$,

$$n(z = 0) = n_0 = \frac{Nmg}{Ak_BT}$$

and hence equation (17.42) becomes

$$n = n_0 e^{-mgz/k_BT}$$

Using $p = nk_BT$ and equation (17.42), we obtain

$$p = \frac{Nmg}{A}e^{-mgz/k_BT}$$

Using the initial conditions that when $z = 0$,

$$p(z = 0) = p_0 = \frac{Nmg}{A}$$

and hence we obtain

$$p = p_0 e^{-mgz/k_BT}$$

which is known as the *Law of atmospheres* or the *Barometric equation* which is well known by the observation that presure decreases as height increases.

**Principle of Equipartition of Energy**

The principle of equipartition of energy (also known as *equipartition theorem*) states that the mean value of energy for each degree of freedom is equal to $\frac{1}{2}k_B T$. Let us prove this theorem using the results of Maxwell-Boltzmann statistics.

The total energy $E$ is a function of the generalised coordinates $q$ and momenta $p$, that is,

$$E = E((q_1, q_2, \cdots q_s, p_1, p_2, \cdots p_s) \tag{17.43}$$

Two assumptions are made:

(i) The total energy is a sum two parts

$$E = \epsilon_i(p_i) + f(q_1, q_2, \cdots q_s, p_1, p_2, \cdots p_s) \tag{17.44}$$

(ii) where $f$ does not depend on $P_i$.

The function $\epsilon_i(p_i)$ is quadratic in $p_i$, that is

$$\epsilon_i(p_i) = bp_i^2 \tag{17.45}$$

where $b$ is some constant.

According to Maxwell-Boltzmann statistics, the *mean energy*, using equation (17.16) is given by

$$\begin{aligned}
\bar{\epsilon}_i &= \frac{\sum_i \epsilon_i e^{-E_i/k_B T}}{Z} \\
&= \frac{\sum_i \epsilon_i e^{-E_i/k_B T}}{\sum_i e^{-E_i/k_B T}}
\end{aligned}$$

Using the following prescription which relates a summation and an integration, we have

$$\sum_i \cdots = \frac{1}{\tau} \int \cdots dp dq$$

where $\tau$ is some normalisation factor, and $dp dq$ is an element of phase space.

$$\begin{aligned}
\bar{\epsilon}_i &= \frac{\int \epsilon_i e^{-E/k_B T} dp dq}{\int e^{-E/k_B T} dp dq} \\
&= \frac{\int \epsilon_i e^{-\beta E} dp dq}{\int e^{-\beta E} dp dq}
\end{aligned} \tag{17.46}$$

where $\beta = 1/(k_B T)$. Using the expression for energy from equation (17.44) in (17.46), we obtain

$$
\begin{aligned}
\bar{\epsilon}_i &= \frac{\int \epsilon_i e^{-(\epsilon_i + f)/k_B T} dp dq}{\int e^{-(\epsilon_i + f)/k_B T} dp dq} \\
&= \frac{\int \epsilon_i e^{-\beta \epsilon_i} dp_i \int \epsilon_i e^{-\beta f} dq_1 \cdots dq_s dp_1 \cdots dp_s}{\int e^{-\beta \epsilon_i} dp_i \int \epsilon_i e^{-\beta f} dq_1 \cdots dq_s dp_1 \cdots dp_s} \\
&= \frac{\int \epsilon_i e^{-\beta \epsilon_i} dp_i}{\int e^{-\beta \epsilon_i} dp_i} \\
&= \frac{b \int p_i^2 e^{-\beta b p_i^2} dp_i}{\int e^{-\beta b p_i^2} dp_i}
\end{aligned}
\tag{17.47}
$$

Using the following standard integrals in equation (17.47)

$$
\begin{aligned}
\int_{\infty}^{\infty} e^{-\alpha x^2} dx &= \sqrt{\frac{\pi}{\alpha}} \\
\int_{\infty}^{\infty} e^{-\alpha x^2} x^2 dx &= \frac{1}{2\alpha}\sqrt{\frac{\pi}{\alpha}}
\end{aligned}
$$

we obtain

$$
\begin{aligned}
\bar{\epsilon}_i &= \frac{b \frac{1}{2\beta b}\sqrt{\frac{\pi}{\beta b}}}{\sqrt{\frac{\pi}{\beta b}}} \\
&= \frac{1}{2\beta} \\
\bar{\epsilon}_i &= \frac{1}{2}k_B T
\end{aligned}
\tag{17.48}
$$

which is the principle of equipartition of energy.

## 17.2 Fermi-Dirac Statistics

### 17.2.1 Introduction

In this section we consider particles which satisfy *Fermi-Dirac statistics*, known as *Fermions*. Fermions have the following properties.

(i) Fermions are *indistinguishable* particles.

(ii) Fermions satisfy *Pauli exclusion principle* which states that no two particles can uccupy the same state.

(iii) Fermions have a symmetric wavefunction.

(iv) Fermions have a spin of half integer.

(v) Fermions obey Fermi-Dirac statistics.

## 17.2.2   Thermodynamic probability for fermions

Consider $n_i$ particles obeying Fermi-Dirac statistics in $g_i$. What is the thermodynamic probability of such indistinguisahble particles satisfyng the Pauli exclusion principle? This can be found as follows.

The number of ways of placing $n_1$ particles in $g_i$ states without repetition (since each state can not contain more than one particle) is the same as the number of choosing $n_i$ from $g_i$, given by

$$
\begin{aligned}
w_i &= {}^{g_i}C_{n_i} \\
&= \frac{g_i!}{n_i!(g_i - n_i)!}
\end{aligned}
\tag{17.49}
$$

The *thermodynamic probability* for fermions, $W$ is obtained by multiplying the different possibilities $W_i$, which is

$$
\begin{aligned}
W &= \prod_i W_i \\
&= \prod_i \frac{g_i!}{n_i!(g_i - n_i)!}
\end{aligned}
\tag{17.50}
$$

Let us recall the condition equations.

The population condition is given by

$$
\sum_i \delta n_i = 0
\tag{17.51}
$$

The energy condition is given by

$$
\sum_i E_i \delta n_i = 0
\tag{17.52}
$$

and the stability condition is given by

$$
\delta(lnW) = 0
\tag{17.53}
$$

Using the expression for the thermodynamic probabilty, the stability condition can be recast in another form as shown below.

$$
\begin{aligned}
\ln W &= \ln \prod_i \frac{g_i!}{n_i!(g_i - n_i)!} \\
&= \sum_i \{\ln g_i! - \ln n_i! - \ln(g_i - n_i)!\} \\
&= \sum_i \{g_i \ln g_i - g_i - n_i \ln n_i + n_i - (g_i - n_i)\ln(g_i - n_i) + g_i - n_i\} \quad \text{using Stirling's approximation,} \\
&= \sum_i \{g_i \ln g_i - n_i \ln n_i - (g_i - n_i)\ln(g_i - n_i)\}
\end{aligned}
\tag{17.54}
$$

from which we obtain

$$
\begin{aligned}
\delta(\ln W) &= \sum_i \left\{ \ln(g_i - n_i) - \ln n_i \right\} \delta n_i \\
&= \sum_i \ln \left( \frac{g_i - n_i}{n_i} \right) \delta n_i
\end{aligned}
\tag{17.55}
$$

and hence the stability equation can also be written as

$$
\sum_i \ln \left( \frac{g_i - n_i}{n_i} \right) \delta n_i = 0
\tag{17.56}
$$

### 17.2.3  Lagrange's Method of Undetermined Multipliers and Fermi-Dirac Distribution

The Lagranges'Method of Undetermined Multipliers consists of using the condition equations and introducing two undetermined multipliers, $B$ and $\beta$, as shown below.

Multiply equation (17.51) by $-\ln B$
Multiply equation (17.52) by $-\beta$
Multiply equation (17.53) by $1$

and *add* the resulting equations, to obtain

$$
\sum_i \left\{ -\ln B - \beta E_i + \ln \left( \frac{g_i - n_i}{n_i} \right) \right\} \delta n_i = 0
\tag{17.57}
$$

and hence

$$
\begin{aligned}
\left\{ \ln \left( \frac{g_i - n_i}{n_i} \right) - \beta E_i - \ln B \right\} &= 0 \\
\ln \left( \frac{g_i - n_i}{n_i} \right) - \ln B &= \beta E_i \\
\ln \left( \frac{g_i - n_i}{n_i B} \right) &= \beta E_i \\
(g_i - n_i) &= B n_i e^{\beta E_i} \\
n_i \left\{ B e^{\beta E_i} + 1 \right\} &= g_i \\
f(E_i) = \frac{n_i}{g_i} &= \frac{1}{B e^{\beta E_i} + 1}
\end{aligned}
\tag{17.58}
$$

What is $\beta$? This can be found as follows. Comparing equations (17.57) and results from thermodynamics, we can identify $\beta$ as

$$
\beta = \frac{1}{k_B T}
\tag{17.59}
$$

and hence

$$\ln B = -\frac{E_F}{k_B T}$$
$$B = e^{-E_F/k_B T} \tag{17.60}$$

and hence equation (17. ) becomes

$$f(E_i) = \frac{1}{e^{(E_i - E_F)/k_B T} + 1}$$

or

$$f(E) = \frac{1}{e^{(E - E_F)/k_B T} + 1} \tag{17.61}$$

which is known as the *Fermi-Dirac distribution*

**Electron gas**

## 17.3   Bose-Einstein Statistics

### 17.3.1   Introduction

In this chapter we consider particles which satisfy *Bose-Einstein statistics*, known as *Bosons*. Bosons have the following properties.

(i) Bosons are *indistinguishable* particles.

(ii) any number of bosons can occupy the same state. This means that bosons *do not* satisfy *Pauli exclusion principle*.

(iii) Bosons have a symmetric wavefunction.

(iv) Bosons have an integer spin.

(v) Bosons obey Bose-Einstein statistics.

### 17.3.2   Thermodynamic probability for bosons

Consider $n_i$ particles obeying Bose-Einstein statistics in $g_i$ states. What is the thermodynamic probability of such indistinguishable particles whose distribution is such that any number of states can occupy the same state? This can be found as follows.

The number of ways of placing $n_1$ particles in $g_i$ states such that any number of states can occupy the same state. If there re $g_i$ states, there are $(g_i - 1)$ compartments. Thus the total number of particles and compartments is $(n_i + g_i - 1)$. The question is in how many ways can one choose $(g_i - 1)$ compartments out of a total of $(n_i + g_i - 1)$? Let this be $W_i$ ways given by

$$
\begin{aligned}
W_i &= {}^{n_i+g_i-1}C_{g_i-1} \\
&= \frac{(n_i + g_i - 1)!}{(g_i - 1)!(g_i + n_i - 1 - g_i + 1)!} \\
&= \frac{(n_i + g_i - 1)!}{(g_i - 1)!n_i!}
\end{aligned} \tag{17.62}
$$

The *thermodynamic probability*, $W$ is obtained by multiplying the different possibilities $W_i$, which is

$$
\begin{aligned}
W &= \prod_i W_i \\
&= \prod_i \frac{(n_i + g_i - 1)!}{(g_i - 1)!n_i!}
\end{aligned} \tag{17.63}
$$

Let us recall the condition equations.

The population condition is given by

$$
\sum_i \delta n_i = 0 \tag{17.64}
$$

The energy condition is given by

$$
\sum_i E_i \delta n_i = 0 \tag{17.65}
$$

and the stability condition is given by

$$
\delta(lnW) = 0 \tag{17.66}
$$

Using the expression for the thermodynamic probabilty, the stability condition can be recast in another form as shown below.

$$
\begin{aligned}
\ln W &= \ln \prod_i \frac{(n_i + g_i - 1)!}{(g_i - 1)!n_i!} \\
&= \sum_i \{\ln(n_i + g_i - 1)! - \ln(g_i - 1)! - \ln n_i!\} \\
&= \sum_i \{(n_i + g_i - 1)\ln(n_i + g_i - 1) - (n_i + g_i - 1) - (g_i - 1)\ln(g_i - 1) + g_i - 1 - n_i \ln n_i) + n_i\}
\end{aligned}
$$

where we have used the Stirling's approximation,

$$
= \sum_i \{(n_i + g_i)\ln(n_i + g_i) - g_i \ln g_i - n_i \ln n_i\} \tag{17.67}
$$

where we have used $g_i \gg 1, n_i \gg 1, (n_i + g_i) \gg 1$

from which we obtain

$$
\begin{aligned}
\delta(\ln W) &= \sum_i \{\ln(g_i - n_i) - \ln n_i\} \delta n_i \\
&= \sum_i \ln\left(\frac{g_i - n_i}{n_i}\right) \delta n_i
\end{aligned} \tag{17.68}
$$

and hence the stability equation can also be written as

$$\sum_i \ln\left(\frac{g_i + n_i}{n_i}\right)\delta n_i = 0 \tag{17.69}$$

### 17.3.3 Lagrange's Method of Undetermined Multipliers and Bose-Einstein Distribution

The Lagranges'Method of Undetermined Multipliers consists of using the condition equations and introducing two undetermined multipliers, $B$ and $\beta$, as shown below.

Multiply equation (17.64) by $-\ln B$
Multiply equation (17.65) by $-\beta$
Multiply equation (17.66) by $1$

and *add* the resulting equations, to obtain

$$\sum_i \left\{-\ln B - \beta E_i + \ln\left(\frac{g_i + n_i}{n_i}\right)\right\}\delta n_i = 0 \tag{17.70}$$

and hence

$$\left\{\ln\left(\frac{g_i + n_i}{n_i}\right) - \beta E_i - \ln B\right\} = 0$$

$$\ln\left(\frac{g_i + n_i}{n_i}\right) - \ln B = \beta E_i$$

$$\ln\left(\frac{g_i + n_i}{n_i B}\right) = \beta E_i$$

$$(g_i + n_i) = B n_i e^{\beta E_i}$$

$$n_i\left\{B e^{\beta E_i} - 1\right\} = g_i$$

$$\bar{n} = \frac{n_i}{g_i} = \frac{1}{B e^{\beta E_i} - 1} \tag{17.71}$$

What is $\beta$? This can be found as follows. Comparing equations (17.70) and resilts from thermodynamics, we can identify $\beta$ as

$$\beta = \frac{1}{k_B T} \tag{17.72}$$

and hence

$$\ln B = -\frac{\mu}{k_B T}$$

$$B = e^{-\mu/k_B T} \tag{17.73}$$

and hence equation (17. ) becomes

$$\bar{n}_i = \frac{1}{e^{(E_i - \mu)/k_B T} - 1}$$

or

$$n(\bar{E}) = \frac{1}{e^{(E-\mu)/k_B T} - 1} \qquad (17.74)$$

which is known as the *Bose-Einstein distribution*

### 17.3.4 Bosons at very low temperatures

In this section, the theory of Bose-Einstein statistics is applied to study the following quantities:

- Number of particles

- Total energy

- Specific heat

Let us consider each of these below.

*(i) Number of particles*

The number of particles $N$ for bosons is given by

$$N = \int_0^\infty g(E) n(E) dE \qquad (17.75)$$

where $g(E)$ is the density of states and $n(E)$ is the Bose-Einstein distribution function given by equation (17.74). $g(E)$ is given by

$$g(E) = \frac{gV}{2^{1/2}\pi^2} \frac{m^{3/2} E^{1/2}}{\hbar^3} \qquad (17.76)$$

and hence

$$N = \int_0^\infty \frac{gV}{2^{1/2}\pi^2} \frac{m^{3/2} E^{1/2}}{\hbar^3} \frac{1}{e^{(E-\mu)/k_B T} - 1} dE \qquad (17.77)$$

Consider a temperature $T = T_C$ for which $\mu = 0$, and let $z = E/k_B T_C$. Hence

$$N = \frac{gV (mk_B T_C)^{3/2}}{2^{1/2}\pi^2\hbar^3} \int_0^\infty \frac{z^{1/2}}{e^z - 1} dz \qquad (17.78)$$

The integral is of a standard form

$$\int_0^\infty \frac{z^{n-1}}{e^z - 1} dz = \Gamma(n)\zeta(n) \qquad (17.79)$$

where $\Gamma(n)$ is the Gamma function and $\zeta(n)$ is the Riemann-Zeta function. With $n - 1 = \frac{1}{2}$, we obtain

$$\begin{aligned}
\int_0^\infty \frac{z^{1/2}}{e^z - 1} dz &= \Gamma(3/2)\zeta(3/2) \\
&= \frac{\sqrt{\pi}}{2} 2.612 \\
&= 2.31 \qquad (17.80)
\end{aligned}$$

which when used in equation (17.78 ) gives

$$N = \frac{gV\,(mk_BT_C)^{3/2}}{2^{1/2}\pi^2\hbar^3} \times 2.31 \tag{17.81}$$

The critical temperature $T_C$ is obtained by rearranging equation (17.81) to obtain

$$
\begin{aligned}
T_C &= \frac{\left(0.612\pi^2\right)^{3/2}}{g^{3/2}}\,\frac{\hbar^2}{mk_B}\left(\frac{N}{V}\right)^{2/3} \\
&= \frac{3.31}{g^{3/2}}\frac{\hbar^2}{mk_B}\left(\frac{N}{V}\right)^{2/3}
\end{aligned}
\tag{17.82}
$$

So far we have been considering what happens at a temperature $T_C$. It is of interest to consider what happens at *any temperature, $T \neq 0$, which is still very low*, and let $y = E/k_BT$. The number of particles is then given by

$$
\begin{aligned}
N_{\epsilon>0} &= \frac{gV\,(mk_BT)^{3/2}}{2^{1/2}\pi^2\hbar^3}\int_0^\infty \frac{y^{1/2}}{e^y-1}dy \\
&= T^{3/2}\frac{gV\,(mk_B)^{3/2}}{2^{1/2}\pi^2\hbar^3} \times 2.31 \\
&= T^{3/2}\frac{N}{T_C^{3/2}} \\
&= N\left(\frac{T}{T_C}\right)^{3/2}
\end{aligned}
\tag{17.83}
$$

The number of particles $N$, the number of particles in the ground state $N_0$ and $N_{\epsilon>0}$ are related by

$$
\begin{aligned}
N &= N_0 + N_{\epsilon>0} \\
N_0 &= N - N_{\epsilon>0} \\
N_0 &= N\left[1 - \left(\frac{T}{T_C}\right)^{3/2}\right]
\end{aligned}
\tag{17.84}
$$

where equation (17.83) has been used to arrive to the above equation.

Note that as temperature *decreases*, the number of particles in the ground state *increases*. Not that as $T \to 0$, $N_0 \to N$, and at $T = 0$, $N_0 = N$, implying that all the particles in the system will have gone to the ground state. This is what is referred to as the *Bose-Einstein condensation*. This condensation occurs in momentum space, not real space.

## 17.3.5   Photon gas

A photon gas consists of *photons* which are quanta of electromagnetic radiation, and they have the following properties.

*(i) Photons satisfy Maxwell's equations.*

According to electromagnetic theory, electromagnetic radiation is composed of the electric field $\mathbf{E}$ and the magnetic field $\mathbf{B}$ which satisfy Maxwell's equations which are given below.

$$\nabla \cdot \mathbf{E} = \frac{\rho}{\epsilon_0} \tag{17.85}$$

$$\nabla \cdot \mathbf{B} = 0 \tag{17.86}$$

$$\nabla \wedge \mathbf{E} = -\frac{\partial \mathbf{B}}{\partial t} \tag{17.87}$$

$$\nabla \wedge \mathbf{B} = \mu_0 \mathbf{j} + \mu_0 \frac{\partial}{\partial t}[\epsilon_0 \epsilon(\omega) \mathbf{E}] \tag{17.88}$$

where

$$\rho \quad \text{is} \quad \text{the charge density,} \tag{17.89}$$

$$\epsilon_0 = 8.854 \times 10^{-9} Fm^{-1} \text{is the permittivity of free space,} \tag{17.90}$$

$$\mu_0 = 4\pi \times 10^{-7} Hm^{-1} \text{is the permeability of free space,} \tag{17.91}$$

$$\vec{j} \quad \text{is} \quad \text{the current density vector, and} \tag{17.92}$$

$$\epsilon(\omega) \quad \text{is} \quad \text{the dielectric function} \tag{17.93}$$

*(ii) Photon statistics*

Photons are *bosons* with spin 1, and they satisfy Bose-Einstein statistics such that the chemical potential, $\mu = 0$ and energy $E = \hbar\omega$. Hence the Bose-Einstein distribution given in equation (17.74) becomes

$$n(\omega) = \frac{1}{e^{\hbar\omega/k_B T} - 1} \tag{17.94}$$

which is also known as the *Planck's distribution*.

*(iii) Photon density of states*

The density of states for photons is given by

$$g(\omega) = \frac{V\omega^2}{\pi^2 c^3} \tag{17.95}$$

*(iv) Number of particles $N$ for photons*

The number of particles $N$ for photons is given by

$$N = \int_0^\infty g(\omega)n(\omega)d\omega \tag{17.96}$$

where $g(\omega)$ is the density of states given by equation (17.95) and $n(\omega)$ is the Bose-Einstein distribution function given by equation (17.94), and hence

$$N = \int_0^\infty \frac{V\omega^2}{\pi^2 c^3} \frac{1}{e^{\hbar\omega/k_B T} - 1} d\omega \tag{17.97}$$

Let $x = \hbar\omega/k_B T \Longrightarrow d\omega = \frac{k_B T}{\hbar}dx$, and hence equation (17.97) becomes

$$N = \frac{V}{\pi^2}\left(\frac{k_B T}{\hbar c}\right)^3 \int_0^\infty \frac{x^2}{e^x - 1}dx \qquad (17.98)$$

The integral is of a standard form

$$\int_0^\infty \frac{x^{n-1}}{e^x - 1}dz = \Gamma(n)\zeta(n)$$

where $\Gamma(n)$ is the Gamma function and $\zeta(n)$ is the Riemann-Zeta function. With $n - 1 = 2$, we obtain

$$\begin{aligned}
\int_0^\infty \frac{x^2}{e^x - 1}dx &= \Gamma(3)\zeta(3) \\
&= 2!1.202 \\
&= 2.404 \qquad (17.99)
\end{aligned}$$

which when used in equation (17.98) gives

$$\begin{aligned}
N &= \frac{V}{\pi^2}\left(\frac{k_B T}{\hbar c}\right)^3 2.404 \\
&= 0.244\left(\frac{k_B T}{\hbar c}\right)^3 \qquad (17.100)
\end{aligned}$$

*(v) Total radiation energy $U$ for photons*

The total radiation energy $U$ for photons is given by

$$U = \int_0^\infty E g(\omega)n(\omega)d\omega \qquad (17.101)$$

where $E$ is energy, $g(\omega)$ is the density of states given by equation (17.95) and $n(\omega)$ is the Bose-Einstein distribution function given by equation (17.94), and hence

$$\begin{aligned}
U &= \int_0^\infty \frac{V\omega^2}{\pi^2 c^3}\frac{1}{e^{\hbar\omega/k_B T} - 1}d\omega \\
&= \frac{V\hbar}{\pi^2 c^3}\int_0^\infty \frac{\omega^3}{e^{\hbar\omega/k_B T} - 1}d\omega \qquad (17.102)
\end{aligned}$$

Let $x = \hbar\omega/k_B T \Longrightarrow d\omega = \frac{k_B T}{\hbar}dx$, and hence equation (17.102) becomes

$$U = \frac{Vk_B^4 T^4}{\pi^2 c^3\hbar^3}\int_0^\infty \frac{x^3}{e^x - 1}dx \qquad (17.103)$$

The integral is of a standard form of equation (17.103) with $n - 1 = 3$, which gives

$$\begin{aligned}
\int_0^\infty \frac{x^3}{e^x - 1}dx &= \Gamma(4)\zeta(4) \\
&= 3!\frac{\pi^4}{90} \\
&= \frac{\pi^4}{15} \qquad (17.104)
\end{aligned}$$

which when used in equation (17.103) gives

$$\begin{aligned} U &= \frac{V k_B^4 T^4}{\pi^2 c^3 \hbar^3} \times \frac{\pi^4}{15} \\ &= \left( \frac{V \pi^2 k_B^4}{15 c^3 \hbar^3} \right) T^4 \end{aligned} \qquad (17.105)$$

which is *Stephan's radiation law*.

## 17.4   Exercises

**17.1**. (a) A system with $N$ particles at a temperature $T$ has energy levels given by $E_i$. Assuming that the system behaves as a perfect gas, give an expression for the number of particles $n_i$ in each energy level, defining all the symbols used.
(b) Consider a system of 4000 particles distributed into three equally spaced energy levels $0, \epsilon$ and $2\epsilon$, such that the total energy is $2300\epsilon$.
(i) If the system behaves as a perfect gas, determine the number of particles in each state for the most probable partition.
(ii) Hence calculate the temperature of the system when it is in thermal equilibrium, assuming $\epsilon = 0.02$ eV.

**17.2**. Consider an assembly of six particles in a system with seven energy levels $E_n = n\epsilon$, where $n = 0, 1, 2, 3, 4, 5, 6$ and total energy $6\epsilon$ and the degeneracy of each level is $g_i = 3$. Assuming that the particles satisfy *Maxwell-Boltzmann statistics*
(a) In how many ways can the particles be arranged? Each of these arrangements is known as a *macrostate*.
(b) (i) Calculate the thermodynamic probability for the particles for each macrostate. Each of these numbers is the number of *microstates* for each *macrostate*.
(ii) How many *microstates* are there in the most probable *macrostate*?
(c) Calculate the thermodynamic probability for the whole system.

**17.3**. (a) According to *Maxwell-Boltzmann statistics*, the number of particles with velocity components between $v_i$ and $v_i + dv_i$ $(i = x, y, z)$ is given by

$$dn = 4\pi N(\frac{m}{2\pi k_B T})^{3/2} v^2 e^{-\frac{1}{2}mv^2/k_B T} dv$$

where all symbols are in the usual notation.
(i) Show that the most probable velocity $v_{mp}$ is given by

$$v_{mp} = \sqrt{\frac{2k_B T}{m}}$$

(ii) Show that the mean velocity $\bar{v}$, is given by

$$\bar{v} = \sqrt{\frac{8k_B T}{\pi m}}$$

(iii) Show that the root mean square velocity $v_{r.m.s}$ is given by

$$v_{r.m.s} = \sqrt{\bar{v^2}} = \sqrt{\frac{3k_B T}{m}}$$

(b) Calculate the most probable velocity, the mean velocity and the root mean square velocity of oxygen molecules at 300 K in m s$^{-1}$ and km hr$^{-1}$.

(c) Sketch a graph of how $\frac{dn}{dv}$ varies with $v$, and on it label the mean velocity $\bar{v}$, most probable velocity $v_{mp}$ and the root mean square velocity $v_{rms}$.

**17.4.** Consider an assembly of six particles in a system with seven energy levels $E_n = n\epsilon$, where $n = 0, 1, 2, 3, 4, 5, 6$ and total energy $6\epsilon$ and the degeneracy of each level is $g_i = 3$. Assuming that the particles satisfy *Fermi-Dirac statistics*
(a) In how many ways can the particles be arranged? Each of these arrangements is known as a *macrostate*.
(b) (i) Calculate the thermodynamic probability for the particles for each macrostate. Each of these numbers is the number of *microstates* for each *macrostate*.
(ii) How many *microstates* are there in the most probable *macrostate*?
(c) Calculate the thermodynamic probability for the whole system.

**17.5.** Consider an assembly of six particles in a system with seven energy levels $E_n = n\epsilon$, where $n = 0, 1, 2, 3, 4, 5, 6$ and total energy $6\epsilon$ and the degeneracy of each level is $g_i = 3$. Assuming that the particles satisfy *Bose-Einstein statistics*
(a) In how many ways can the particles be arranged? Each of these arrangements is known as a *macrostate*.
(b) (i) Calculate the thermodynamic probability for the particles for each macrostate. Each of these numbers is the number of *microstates* for each *macrostate*.
(ii) How many *microstates* are there in the most probable *macrostate*?
(c) Calculate the thermodynamic probability for the whole system.

**17.6.** The total thermal energy due to photons is given by

$$U = \int_0^\infty E g(\omega) n(\omega) d\omega$$

where $E$ is energy, $g(\omega)$ is the density of states and $n(\omega)$ is the Bose-Einstein distribution function.
(a) Show that $U$ is given by

$$U = \frac{V\hbar}{\pi^2 c^3} \int_0^\infty \frac{\omega^3}{e^{\hbar\omega/k_B T} - 1} d\omega$$

(b) Show that if $x = \hbar\omega/k_B T$, then

$$U = \frac{V k_B^4 T^4}{\pi^2 c^3 \hbar^3} \int_0^\infty \frac{x^3}{e^x - 1} dx$$

(c) Hence prove *Stephan's radiation law*

$$U = \left( \frac{V \pi^2 k_B^4}{15 c^3 \hbar^3} \right) T^4$$

*Hints*:
(i) Density of states for photons, $g(\omega) = \frac{V\omega^2}{\pi^2 c^3}$
(ii) Bose-Einstein distribution for photons, $n(\omega) = \frac{1}{e^{\hbar\omega/k_B T} - 1}$

(iii) The following standard integral may be useful.

$$\int_0^\infty \frac{x^3}{e^x - 1} dx = \Gamma(4)\zeta(4) = 3!\frac{\pi^4}{90} = \frac{\pi^4}{15}$$

**17.7**. The Fermi-Dirac distribution function is given by

$$f(E) = \frac{1}{e^{(E-E_F)/k_BT} + 1}$$

(a) Plot $f(E)$ vs $E$ for $T = 0$ and $T \neq 0$, labelling the graph clearly.
(b) (i) Show that

$$-\frac{df(E)}{dE} = \frac{f(E)\,[1 - f(E)]}{k_BT}$$

(ii) Plot $-\frac{df(E)}{dE}$ vs $E$ for $T = 0$ and $T \neq 0$, labelling the graph clearly.

**17.8**. The thermodynamic potential $\Omega$ is related to the Grand partition function $Q$ by

$$\Omega = k_BT \ln Q$$

where

$$Q = \sum_n {}_i e^{(\mu - E_i)n_i/k_BT}$$

and all other symbols are in the usual notation.
If $n_i = -\frac{\partial \Omega}{\partial \mu}$, and assuming the above expression, derive expressions for $n_i$ for
(i) fermions
(ii) bosons

**17.9**. The entropy $S$ is related to the Grand partition function $Q$ by the relation

$$S = \frac{\partial}{\partial T}(k_BT \ln Q)_{\mu,V}$$

Hence show that the entropy of an *ideal fermion system* is

$$S = -k_B \sum_j \{n(E_j) \ln n(E_j) + [1 - n(E_j)] \ln[1 - n(E_j)]\}$$

where all other symbols are in the usual notation.

**17.10**. The entropy $S$ is related to the Grand partition function $Q$ by the relation

$$S = \frac{\partial}{\partial T}(k_BT \ln Q)_{\mu,V}$$

Hence show that the entropy of an *ideal boson system* is

$$S = -k_B \sum_j \{n(E_j) \ln n(E_j) - [1 + n(E_j)] \ln[1 + n(E_j)]\}$$

where all other symbols are in the usual notation.

# Chapter 18

# Transport Theory

## 18.1 Kinetic Theory of Transport Processes

### 18.1.1 Pressure of an Ideal gas

Macroscopic variables such as pressure, volume and temperature depend on microscopic properties. It is modelled that a gas consists of a very large number of molecules which move randomly. In this section we shall use the *kinetic theory of gases* to derive an expression for the pressure of an ideal gas.

The following assumptions are made:

- The forces of attraction between molecules are negligible except during collisions.

- The volume occupied by the molecules is negligible compared to the volume of the container.

- The molecules undergo elastic collisions with each other. during the collisions both kinetic energy and momentum are conserved.

- The time interval between collisions is much greater than the time of an individual collision.

- Only binary collisions are considered since it is assumed that the probability of three or more molecules colliding is negligibly small

Consider an ideal gas in a cubical container of side $d$, consisting of $N$ molecules in the volume $V = d^3$. If a molecule of mass $m$ moves with avelocity $\mathbf{v} = v_x \mathbf{i} + v_y \mathbf{j} + v_z \mathbf{k}$ to the right, and collides with the wall elastically, its $x$-component of velocity is reversed. The change in velocity of the molecule is given by

$$\Delta p_x = -mv_x - (mv_x) = -2mv_x \tag{18.1}$$

and the corresponding momentum delivered to the wall is $2mv_x$ system the momentum of the system is conserved. For a molecule to make two successive collisions with the same wall, it must travel a distance $2d$ along the $x$-axis in time $\Delta t$. But $\Delta t = 2d/v_x$. If F is the average force, the impulse is given by $F\Delta t = \Delta p = 2mv_x$. Hence

$$F = \frac{2mv_x}{\Delta t}$$

$$= \frac{2mv_x}{2d/v_x}$$

$$= \frac{mv_x^2}{d} \tag{18.2}$$

The total force is the sum of all such terms for all the molecules. The total pressure $P$ is the total force divided by the total area $d^2$, given by

$$
\begin{aligned}
P &= \frac{\sum F}{d^2} \\
&= \frac{1}{d^2} \sum_i \frac{mv_{xi}^2}{d} \\
&= \frac{m}{d^3}\left(v_{x1}^2 + v_{x2}^2 + \cdots\right) \\
P &= \frac{Nm}{V}\bar{v_x^2}
\end{aligned}
\tag{18.3}
$$

where we have introduced $V = d^3$ and an average value $\bar{v_x^2}$ given by

$$\bar{v_x^2} = \frac{(v_{x1}^2 + v_{x2}^2 + \cdots)}{N}$$

Since there is no preferred direction, average values of the square velocities in the $x, y, z$ directions are equal, that is

$$\bar{v_x^2} = \bar{v_y^2} = \bar{v_z^2} = \frac{1}{3}\bar{v^2}$$

and hence the expression for pressure from equation (18.3) becomes

$$P = \frac{1}{3}\frac{Nm}{V}\bar{v^2} \tag{18.4}$$

or

$$P = \frac{2}{3}\frac{N}{V}\left(\frac{1}{2}m\bar{v^2}\right) \tag{18.5}$$

and hence the pressure is proportional to the number of molecules per unit volume and to the translational kinetic energy per molecule.

### 18.1.2   Coefficient of Thermal Conductivity

Consider a container with $n$ molecules per unit volume. A reasonable estimate will give one third $(n/3)$ of the molecules in each of the $x, y,$ and $z$ directions. Half of these, $n/6$, have mean velocity $\bar{v}$ in the $+z$ direction and and the other half have a mean velocity $\bar{v}$ in the $-z$ direction. Consequently, $n\bar{v}/6$ molecules in unit time cross a unit area in the plane $z =$ constant from below, and from above. Temperature varies along the $z$ direction, that is, $T = T(z)$. The mean energy $\bar{\epsilon}$ is a function of temperature, and hence mean energy varies along $z$ also, that is, $\bar{\epsilon} = \bar{\epsilon}(z)$. On the average molecules which cross the plane *from below* experienced their last collision at a distance $l$ *below* the plane, with an average energy $\bar{\epsilon}(z - l)$, where $l$ is the mean free path. Hence the mean energy transported per unit time per unit area across the plane in the *upward* direction is given by

$$\frac{1}{6}n\bar{v}\bar{\epsilon}(z - l) \tag{18.6}$$

Similarly, the average molecules which cross the plane *from above* experienced their last collision at a distance $l$ *above* the plane, with average energy $\epsilon(z + l)$. Hence the mean energy transported per unit time per unit area across the plane in the *downward* direction is given by

$$\frac{1}{6}n\bar{v}\bar{\epsilon}(z + l) \tag{18.7}$$

The *net flux* of energy per unit time per unit area from below to above the plane is $Q_z$ which is given by subtracting Equation (18.7) from (18.6),

$$
\begin{aligned}
Q_z &= \frac{1}{6}n\bar{v}\bar{\epsilon}_x(z - l) - \frac{1}{6}n\bar{v}\bar{\epsilon}_x(z + l) \\
&= \frac{1}{6}n\bar{v}\left\{\bar{\epsilon}(z - l) - \bar{\epsilon}(z + l)\right\} \\
&= \frac{1}{6}n\bar{v}m\left\{\left[\bar{\epsilon}(z) - \frac{\partial\bar{\epsilon}}{\partial z}l\cdots\right] - \left[\bar{\epsilon}(z) + \frac{\partial\bar{\epsilon}}{\partial z}l\cdots\right]\right\} \\
&= \frac{1}{6}n\bar{v}\left\{-2\frac{\partial\bar{\epsilon}}{\partial z}l\right\} \\
&= -\frac{1}{3}n\bar{v}l\frac{\partial\bar{\epsilon}}{\partial z} \\
&= -\frac{1}{3}n\bar{v}l\frac{\partial\bar{\epsilon}}{\partial T}\frac{\partial T}{\partial z} \\
&= -\frac{1}{3}n\bar{v}lc\frac{\partial T}{\partial z}
\end{aligned} \tag{18.8}
$$

$$\tag{18.9}$$

where we have introduced

$$c = \frac{\partial\bar{\epsilon}}{\partial T}$$

and in equation (18.9) a Taylor's expansion of $\bar{\epsilon}(z \pm l)$ has been done, neglecting higher order terms. Equation (18.9) may be compared with the equation relating the energy flux and the temperature gradient

$$Q_z = -\kappa\frac{\partial T}{\partial z} \tag{18.10}$$

from which the coefficient of thermal conductivity $\kappa$ is identified as

$$\kappa = \frac{1}{3}n\bar{v}cl \tag{18.11}$$

### 18.1.3  Coefficient of Viscosity

Consider a container with $n$ molecules per unit volume. A reasonable estimate will give one third $(n/3)$ of the molecules in each of the $x, y,$ and $z$ directions. Half of these, $n/6$, have mean velocity $\bar{v}$

in the $+z$ direction and and the other half have a mean velocity $\bar{v}$ in the $-z$ direction. Consequently, $n\bar{v}/6$ molecules in unit time cross a unit area in the plane $z =$constant from below, and from above. However, on the average molecules which cross the plane *from below* experienced their last collision at a distance $l$ *below* the plane, with an average $x$-component of velocity $u_x(z-l)$, where $l$ is the mean free path. Hence the mean $x$-component of momentum transported per unit time per unit area across the plane in the *upward* direction is given by

$$\frac{1}{6}n\bar{v}mu_x(z-l) \tag{18.12}$$

Similarly, the average molecules which cross the plane *from above* experienced their last collision at a distance $l$ *above* the plane, with an average $x$-component of velocity $u_x(z+l)$. Hence the mean $x$-component of momentum transported per unit time per unit area across the plane in the *downward* direction is given by

$$\frac{1}{6}n\bar{v}mu_x(z+l) \tag{18.13}$$

The *net transfer* of the $x-$ component of momentum per unit time per unit area from below to above the plane is the stress $P_{zx}$ which is given by subtracting (18. ) from (18. ),

$$
\begin{aligned}
P_{zx} &= \frac{1}{6}n\bar{v}mu_x(z-l) - \frac{1}{6}n\bar{v}mu_x(z+l) \\
&= \frac{1}{6}n\bar{v}m\left\{u_x(z-l) - u_x(z+l)\right\} \\
&= \frac{1}{6}n\bar{v}m\left\{\left[u_x(z) - \frac{\partial u_x}{\partial z}l\cdots\right] - \left[u_x(z) + \frac{\partial u_x}{\partial z}l\cdots\right]\right\} \tag{18.14}\\
&= \frac{1}{6}n\bar{v}m\left\{-2\frac{\partial u_x}{\partial z}l\right\} \tag{18.15}\\
&= -\frac{1}{3}n\bar{v}ml\frac{\partial u_x}{\partial z} \tag{18.16}
\end{aligned}
$$

where in equation (18.14) a Taylor's expansion of $u_x(z \pm l)$ has been done, neglecting higher order terms. Equation (18.16) may be compared with the equation relating the stress and the velocity gradient

$$P_{zx} = -\eta\frac{\partial u_x}{\partial z} \tag{18.17}$$

from which the coefficient of viscosity $\eta$ is identified as

$$\eta = \frac{1}{3}n\bar{v}ml \tag{18.18}$$

## 18.1.4  Coefficient of Self-diffusion

Consider a container with $n$ molecules per unit volume. A reasonable estimate will give one third $(n/3$ of the molecules in each of the $x, y$, and $z$ directions. Half of these, $n/6$, have mean velocity $\bar{v}$ in the $+z$ direction and and the other half have a mean velocity $\bar{v}$ in the $-z$ direction. Consequently, $n\bar{v}/6$ molecules in unit time cross a unit area in the plane $z =$constant from below, and from above. However, on the average molecules which cross the plane *from below* experienced their last collision at a distance $l$ *below* the plane, where $l$ is the mean free path. Hence the mean number of molecules transported per unit time per unit area across the plane in the *upward* direction is given by

$$\frac{1}{6}\bar{v}n(z-l) \tag{18.19}$$

Similarly, the average molecules which cross the plane *from above* experienced their last collision at a distance $l$ *above* the plane. Hence the mean number of molecules transported per unit time per unit area across the plane in the *downward* direction is given by

$$\frac{1}{6}\bar{v}n(z+l) \tag{18.20}$$

The *net transfer* of molecules transferred per unit time per unit area from below to above the plane is $J_z$ which is given by subtracting (18. ) from (18. ),

$$
\begin{aligned}
J_z &= \frac{1}{6}\bar{v}n(z-l) - \frac{1}{6}\bar{v}n(z+l) \\
&= \frac{1}{6}\bar{v}\left\{n(z-l) - n(z+l)\right\} \\
&= \frac{1}{6}\bar{v}\left\{\left[n(z) - \frac{\partial n}{\partial z}l\cdots\right] - \left[n(z) + \frac{\partial n}{\partial z}l\cdots\right]\right\} \tag{18.21} \\
&= \frac{1}{6}\bar{v}\left\{-2\frac{\partial n}{\partial z}l\right\} \tag{18.22} \\
&= -\frac{1}{3}\bar{v}l\frac{\partial n}{\partial z} \tag{18.23}
\end{aligned}
$$

where in equation (18.21) a Taylor's expansion of $n(z\pm l)$ has been done, neglecting higher order terms. Equation (18.23) may be compared with the equation relating the flux and the concentration gradient

$$J_z = -D\frac{\partial n}{\partial z} \tag{18.24}$$

from which the coefficient of self-diffusion $D$ is identified as

$$D = \frac{1}{3}\bar{v}l \tag{18.25}$$

### 18.1.5 Coefficient of Electrical conductivity

Electrical conductivity was discussed in chapter 12. For ease of reference and the context of this chapter we re-discuss the problem here. An electric current is due to electrons in a conductor, usually a metal. Consider a metal of length $l$ and cross sectional $A$ as illustrated in Figure 18.1. Let there be $n$ electrons per unit volume.

Let the conduction move with a drift velocity $v$.
The number of electrons in a length $l$ is $nAl$. If each electron is of charge $e$, then the total charge, $q$ is given by

$$q = nAle$$

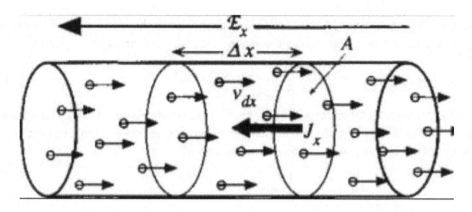

Figure 18.1: An electric current is due electrons in a metal

Crossing the length $l$ takes time, $t = \frac{l}{v}$. Then

$$
\begin{aligned}
\text{The electric current, } I \; &= \; \frac{q}{t} \\
&= \; \frac{nAle}{t} \\
&= \; nAev \\
&= \; jA \quad \text{where } j = nev \text{ is the current density, that is } j = \frac{I}{A}
\end{aligned}
$$

$$
\begin{aligned}
\text{Under an electric field } E \text{, there is a force, } F \; &= \; ma = eE \\
\text{But acceleration is given by, } a \; &= \; \frac{v}{\tau} \quad \text{where } \tau \text{ is relaxation time} \\
\text{Hence } \quad m\frac{v}{\tau} \; &= \; eE \\
v \; &= \; \frac{\tau eE}{m} \\
\text{Hence, the current density, } j \; &= \; nev = ne\left(\frac{\tau eE}{m}\right) \\
j \; &= \; \left(\frac{ne^2\tau}{m}\right)E \\
j \; &= \; \sigma E \quad \text{which is a form of Ohm's law} \quad (18.26) \\
\text{where } \quad \sigma \; &= \; \frac{ne^2\tau}{m} \quad \text{which is the electrical conductivity} \\
\rho \; &= \; \frac{1}{\sigma} = \frac{m}{ne^2\tau} \quad \text{where } \rho \text{ is the Resistivity} \, (18.27)
\end{aligned}
$$

## 18.2 Boltzmann Equation without collisions

Consider N particles in a volume $V$. Suppose that at a time $t$ a particle has coordinates $\mathbf{r}$ and velocity $\mathbf{v}$. The distribution function $f(\mathbf{r}, \mathbf{v}, t)$ satisfies

$$\int f(\mathbf{r}, \mathbf{v}, t) d^3\mathbf{v} = N(\mathbf{r}, t) \tag{18.28}$$

Let us find what kind of equation the distribution function $f(\mathbf{r}, \mathbf{v}, \mathbf{t})$ satisfies. This can be done as follows. Suppose there is a force $\mathbf{F}(\mathbf{r}, t)$ acting on the particles, each of which is of mass $m$. The effect of the force is to cause motion so that at a later time $t' = t + dt$, the particles will be at a new position $\mathbf{r}'$ and with a new velocity $\mathbf{v}'$, given by

$$\mathbf{r}' = \mathbf{r} + \dot{\mathbf{r}}dt = \mathbf{r} + \mathbf{v}dt \tag{18.29}$$

$$\mathbf{v}' = \mathbf{v} + \dot{\mathbf{v}}dt = \mathbf{v} + \frac{1}{m}\mathbf{F}dt \tag{18.30}$$

In the *absence* of collisions, all the particles in the volume element $d^3\mathbf{r}d^3\mathbf{v}$ will be found in the volume element $d^3\mathbf{r}'d^3\mathbf{v}'$ at a time $t'$ later. Hence

$$f(\mathbf{r}', \mathbf{v}', t)d^3\mathbf{r}'d^3\mathbf{v}' = f(\mathbf{r}, \mathbf{v}, t)d^3\mathbf{r}d^3\mathbf{v} \tag{18.31}$$

The two elements of volume $d^3\mathbf{r}d^3\mathbf{v}$ and $d^3\mathbf{r}'d^3\mathbf{v}'$ may become distorted because of the motion and are related by

$$d^3\mathbf{r}'d^3\mathbf{v}' = |J|d^3\mathbf{r}d^3\mathbf{v} \tag{18.32}$$

where $|J|$ is the *Jacobian* of the transformation from the variables $\mathbf{r}, \mathbf{v}$ to $\mathbf{r}', \mathbf{v}'$, given by

$$J = \frac{\partial(x', y', z', v'_x, v'_y, v'_z)}{\partial(x, y, z, v_x, v_y, v_z)} \tag{18.33}$$

$$= \begin{vmatrix} 1 & 0 & 0 & dt & 0 & 0 \\ 0 & 1 & 0 & 0 & dt & 0 \\ 0 & 0 & 1 & 0 & 0 & dt \\ \frac{1}{m}\frac{\partial F_x}{\partial x}dt & \cdots & \cdots & 1 & 0 & 0 \\ \cdots & \cdots & \cdots & 0 & 1 & 0 \\ \cdots & \cdots & \cdots & 0 & 0 & 1 \end{vmatrix}$$

$$= 1 + O(dt^2) \tag{18.34}$$

in which case $J$ can be taken to be unit, and hence

$$d^3\mathbf{r}'d^3\mathbf{v}' = d^3\mathbf{r}d^3\mathbf{v} \tag{18.35}$$

Using equations (18.32) and (18.35), we obtain

$$f(\mathbf{r}', \mathbf{v}', \mathbf{t}) = f(\mathbf{r}, \mathbf{v}, t) \tag{18.36}$$

$$f(\mathbf{r} + \dot{\mathbf{r}}dt, \mathbf{v} + \dot{\mathbf{v}}dt, t + dt) = f(\mathbf{r}, \mathbf{v}, \mathbf{t}) \tag{18.37}$$

$$f(\mathbf{r} + \dot{\mathbf{r}}dt, \mathbf{v} + \dot{\mathbf{v}}dt, t + dt) - f(\mathbf{r}, \mathbf{v}, \mathbf{t}) = 0 \tag{18.38}$$

The above equation can be expressed in terms of partial derivatives and reduces to

$$\left[ \left\{ \frac{\partial f}{\partial x}\dot{x} + \frac{\partial f}{\partial y}\dot{y} + \frac{\partial f}{\partial z}\dot{z} \right\} + \left\{ \frac{\partial f}{\partial v_x}\dot{v}_x + \frac{\partial f}{\partial v_y}\dot{v}_y + \frac{\partial f}{\partial v_z}\dot{v}_z \right\} + \frac{\partial f}{\partial t} \right] dt = 0 \qquad (18.39)$$

Noting that

$$\nabla \equiv \frac{\partial}{\partial \mathbf{r}} = \mathbf{i}\frac{\partial}{\partial x} + \mathbf{j}\frac{\partial}{\partial y} + \mathbf{k}\frac{\partial}{\partial z}$$

$$\frac{\partial}{\partial \mathbf{v}} = \mathbf{i}\frac{\partial}{\partial v_x} + \mathbf{j}\frac{\partial}{\partial v_y} + \mathbf{k}\frac{\partial}{\partial v_z}$$

Equation (18.39) can be put in the form

$$\dot{\mathbf{r}} \cdot \frac{\partial f}{\partial \mathbf{r}} + \dot{\mathbf{v}} \cdot \frac{\partial f}{\partial \mathbf{v}} + \frac{\partial f}{\partial t} = 0 \qquad (18.40)$$

which is known as the *Boltzmann equation without collisions*, or, alternatively can be put in the form

$$\mathbf{v} \cdot \frac{\partial f}{\partial \mathbf{r}} + \frac{\mathbf{F}}{m} \cdot \frac{\partial f}{\partial \mathbf{v}} + \frac{\partial f}{\partial t} = 0 \qquad (18.41)$$

It can be noted that the Boltzmann equation without collisions is a linear partial differential equation satisfied by $f$. It asserts that $f$ remains unchanged if one moves along with the particles in phase space. This is a special case of Liouville's theorem proved in section (16.2). This equation is also encountered in plasma physics where it is sometimes referred to as *Vlasov equation*.

## 18.3   Boltzmann Equation with collisions

In this section we consider the *Boltzmann Equation with collisions*. Consider a particle with velocity $\mathbf{v}$. Let $P(t)$ be the probability that such a particle survives a time $t$ *without* suffering a collision. As time increases, the probability of suferring a collision increases and there must be a collision, and hence $P(t)$ decreases.

$$P(t) \rightarrow \begin{cases} 1 & \text{as } t \rightarrow \\ 0 & \text{as } t \rightarrow \infty \end{cases} \qquad (18.42)$$

$$\begin{aligned} P(t+dt) &= P(t)(1 - \omega dt) \\ P(t) + \frac{dP}{dt}dt &= P(t) - P(t)\omega dt \\ \frac{1}{P}\frac{dP}{dt} &= -\omega \\ \int \frac{dP}{P} &= \int -\omega dt \end{aligned}$$

$$\ln P = -\omega t + C_1 \qquad (18.43)$$

Using boundary conditions that at $t = 0$, $P(0) = 1$, it is obtained that $C_1 = 0$. Hence

$$P(t) = e^{-\omega t} \tag{18.44}$$

Let $\pi(t)dt$ be the probability that the particle, after surviving without collisions in time $t$ suffers a collision in time $t + dt$. $\pi(t)dt$ and $P(t)$ are related by

$$
\begin{aligned}
\pi(t)dt &= P(t)\omega dt \tag{18.45} \\
&= e^{-\omega t}\omega dt \tag{18.46}
\end{aligned}
$$

Using the normalisation condition in the form

$$\int_0^\infty \pi(t)dt = 1 \tag{18.47}$$

we obtain

$$\pi(t)dt = e^{-t/\tau}\frac{dt}{\tau} \tag{18.48}$$

If we wish to relate the didtribution functions at two positions, we have seen that *in the absence of collisions*

$$f(\mathbf{r}, \mathbf{v}, \mathbf{t}) = \mathbf{f}(\mathbf{r_0}, \mathbf{v_0}, \mathbf{t_0})$$

*In the presence of collisions*, the relation becomes

$$f(\mathbf{r}, \mathbf{v}, \mathbf{t}) = \int_0^\infty \mathbf{f^{(0)}}(\mathbf{r_0}, \mathbf{v_0}, \mathbf{t} - \mathbf{t'})e^{-\mathbf{t'}/\tau}\frac{\mathbf{dt'}}{\tau} \tag{18.49}$$

Similarly,

$$f(\mathbf{r} + \dot{\mathbf{r}}dt, \mathbf{v} + \dot{\mathbf{v}}dt, t + dt) = \int_0^\infty f^{(0)}(\mathbf{r_0}, \mathbf{v_0}, \mathbf{t} + \mathbf{dt} - \mathbf{t'})e^{-\mathbf{t'}/\tau}\frac{\mathbf{dt'}}{\tau} \tag{18.50}$$

Subtracting equation (18.49) from (18.50), we obtain

$$f(\mathbf{r} + \dot{\mathbf{r}}dt, \mathbf{v} + \dot{\mathbf{v}}dt, t + dt) - f(\mathbf{r}, \mathbf{v}, \mathbf{t}) = \int_0^\infty \frac{\partial}{\partial \mathbf{t}}\mathbf{f^{(0)}}(\mathbf{r_0}, \mathbf{v_0}, \mathbf{t} - \mathbf{t'})e^{-\mathbf{t'}/\tau}\frac{\mathbf{dt'}}{\tau} \tag{18.51}$$

But

$$\frac{\partial}{\partial t}f^{(0)}(\mathbf{r_0}, \mathbf{v_0}, \mathbf{t} - \mathbf{t'}) = -\frac{\partial}{\partial \mathbf{t'}}\mathbf{f^{(0)}}(\mathbf{r_0}, \mathbf{v_0}, \mathbf{t} - \mathbf{t'}) \tag{18.52}$$

The left hand side of equation (18.51) can be expressed in terms of partial derivatives and the integral on the right hand side can be evaluated by parts to obtain

$$\dot{\mathbf{r}} \cdot \frac{\partial f}{\partial \mathbf{r}} + \dot{\mathbf{v}} \cdot \frac{\partial f}{\partial \mathbf{v}} + \frac{\partial f}{\partial t} = -\frac{1}{\tau} \int_0^\infty \frac{\partial}{\partial t'} f^{(0)}(\mathbf{r}_0, \mathbf{v}_0, t - t') e^{-t'/\tau} dt'$$

$$= -\frac{1}{\tau} \left\{ \left[ f^{(0)}(\mathbf{r}_0, \mathbf{v}_0, t - t') e^{-t'/\tau} \right]_0^\infty + \int_0^\infty f^{(0)}(\mathbf{r}_0, \mathbf{v}_0, t - t') e^{-t'/\tau} \frac{dt'}{\tau} \right\}$$

$$= -\frac{1}{\tau} \left\{ 0 - f^{(0)}(\mathbf{r}_0, \mathbf{v}_0, t) + f(\mathbf{r}, \mathbf{v}, t) \right\}$$

$$\dot{\mathbf{r}} \cdot \frac{\partial f}{\partial \mathbf{r}} + \dot{\mathbf{v}} \cdot \frac{\partial f}{\partial \mathbf{v}} + \frac{\partial f}{\partial t} = -\left( \frac{f - f^{(0)}}{\tau} \right) \tag{18.53}$$

which is known as the *Boltzmann equation with collisions*, or, alternatively can be put in the form

$$\mathbf{v} \cdot \frac{\partial f}{\partial \mathbf{r}} + \frac{\mathbf{F}}{m} \cdot \frac{\partial f}{\partial \mathbf{v}} + \frac{\partial f}{\partial t} = -\left( \frac{f - f^{(0)}}{\tau} \right) \tag{18.54}$$

### 18.3.1   Application to electrical conductivity I

Let us apply the Boltzmann transport equation with collisions to calculate the electrical conductivity, assuming that the distribution function $g(E)$ is given by Maxwell-Boltzmann statistics

$$g(E) = n \left( \frac{m}{2\pi k_B T} \right)^{3/2} e^{-E/k_B T}$$

$$= n \left( \frac{m\beta}{2\pi} \right)^{3/2} e^{-E/k_B T}$$

$$= f^{(0)}$$

where $\beta = 1/(k_B T)$.

When an electric field $\mathcal{E}$ is applied, say in the $z$ direction, the electrons experience a force $F = e\mathcal{E}$. Let $f$ be

$$f = f^{(0)} + f^{(1)} \quad \text{where} \quad f^{(1)} << f^{(0)} \tag{18.55}$$

$$= g + f^{(1)} \quad \text{where} \quad g = g(E) = f^{(0)} \tag{18.56}$$

and $f$ satisfies the Boltzmann equation with collisions

$$\mathbf{v} \cdot \frac{\partial f}{\partial \mathbf{r}} + \frac{e\mathcal{E}}{m} \cdot \frac{\partial f}{\partial \mathbf{v}} + \frac{\partial f}{\partial t} = -\left( \frac{f - f^{(0)}}{\tau} \right) \tag{18.57}$$

from which

$$\frac{\partial f}{\partial \mathbf{r}} = 0$$

$$\frac{\partial f}{\partial t} = 0$$

$$-\left( \frac{f - f^{(0)}}{\tau} \right) = -\frac{f^{(1)}}{\tau}$$

and hence

$$\frac{e\mathcal{E}}{m}\frac{\partial f^{(0)}}{\partial v_z} = -\frac{f^{(1)}}{\tau}$$

$$f^{(1)} = -\frac{e\mathcal{E}\tau}{m}\frac{\partial f^{(0)}}{\partial v_z}$$

$$= -\frac{e\mathcal{E}\tau}{m}\frac{\partial g}{\partial v_z}$$

$$= -\frac{e\mathcal{E}\tau}{m}\left(\frac{\partial g}{\partial E}\right)\left(\frac{\partial E}{\partial v_z}\right) \quad \text{(where} \quad E = \frac{1}{2}mv_z^2\text{)}$$

$$= -e\mathcal{E}\tau v_z\frac{\partial g}{\partial E} \tag{18.58}$$

The current density $j$ is given by

$$j = \int f e v_n d^3\mathbf{v} \tag{18.59}$$

which in the $z$-direction becomes

$$j_z = \int f e v_z d^3 v$$

$$= \int (f^{(0)} + f^{(1)}) e v_z d^3 v$$

$$= \int g e v_z d^3 v + \int f^{(1)} e v_z d^3 v$$

$$= -\int e^2 \mathcal{E}\tau v_z^2 \frac{\partial g}{\partial E} d^3 v \quad \text{(where equation (18.58) has been used)}$$

$$= e^2 \mathcal{E}\beta \int g\tau v_z^2 d^3 v \quad \text{(where} \quad \frac{\partial g}{\partial E} = -\beta g \quad \text{has been used)} \tag{18.60}$$

$$= e^2 \mathcal{E}\beta\bar{\tau}_c n\bar{v}_z^2 \tag{18.61}$$

where from equation (18.58) to (18.61) average values have been introduced by using

$$\int g\tau v_z^2 d^3 v = \bar{\tau}_c n\bar{v}_z^2$$

By using the principle of equipartition of energy $(m\bar{v}_z^2/2 = k_B T/2)$, and noting tha $\beta = 1/(k_B T)$, equation (18.61) can be written in the form

$$j_z = e^2 \mathcal{E}\beta\bar{\tau}_c n\bar{v}_z^2$$

$$= e^2 \mathcal{E}\bar{\tau}_c n\left(\frac{1}{k_B T}\right)\left(\frac{k_B T}{m}\right)$$

$$j_z = \left(\frac{ne^2\bar{\tau}_c}{m}\right)\mathcal{E}$$

$$j_z = \sigma\mathcal{E} \tag{18.62}$$

from which the electrical conductivity $\sigma$ can be identified as

$$\sigma = \frac{ne^2\bar{\tau}_c}{m} \tag{18.63}$$

## 18.3.2   Application to electrical conductivity II

Let us apply the Boltzmann transport equation with collisions to calculate the electrical conductivity, assuming that the distribution function $g(E)$ is given by Fermi-Dirac statitistics

$$g(E) = \frac{1}{e^{(E-\mu)/k_B T} + 1} \tag{18.64}$$

where $\mu$ is the chemical potential.

When an electric field $\mathcal{E}$ is applied, say in the $z$ direction, the electrons experience a force $F = e\mathcal{E}$. Let $f$ be

$$
\begin{aligned}
f &= f^{(0)} + f^{(1)} \quad \text{where} \quad f^{(1)} << f^{(0)} & (18.65)\\
&= g + f^{(1)} \quad \text{where} \quad g = g(E) = f^{(0)} & (18.66)
\end{aligned}
$$

and $f$ satisfies the Boltzmann equation with collisions

$$\mathbf{v} \cdot \frac{\partial f}{\partial \mathbf{r}} + \frac{e\mathcal{E}}{m} \cdot \frac{\partial f}{\partial \mathbf{v}} + \frac{\partial f}{\partial t} = -\left( \frac{f - f^{(0)}}{\tau} \right) \tag{18.67}$$

from which

$$
\begin{aligned}
\frac{\partial f}{\partial \mathbf{r}} &= 0\\
\frac{\partial f}{\partial t} &= 0\\
-\left( \frac{f - f^{(0)}}{\tau} \right) &= -\frac{f^{(1)}}{\tau}
\end{aligned}
$$

and hence

$$
\begin{aligned}
\frac{e\mathcal{E}}{m} \frac{\partial f^{(0)}}{\partial v_z} &= -\frac{f^{(1)}}{\tau}\\
f^{(1)} &= -\frac{e\mathcal{E}\tau}{m} \frac{\partial f^{(0)}}{\partial v_z}\\
&= -\frac{e\mathcal{E}\tau}{m} \frac{\partial g}{\partial v_z}\\
&= -\frac{e\mathcal{E}\tau}{m} \left( \frac{\partial g}{\partial E} \right) \left( \frac{\partial E}{\partial v_z} \right) \quad \left( \text{where} \quad E = \frac{1}{2} m v_z^2 \right)\\
&= c\mathcal{E}\tau v_z \frac{\partial g}{\partial E} & (18.68)
\end{aligned}
$$

The current density $j_z$ is given by

$$j_z = \int f e v_z d^3 v$$

$$\begin{aligned}
&= \int (f^{(0)} + f^{(1)}) e v_z d^3v \\
&= \int g e v_z d^3v + \int f^{(1)} e v_z d^3v \\
&= -e^2 \mathcal{E} \int v_z^2 \tau \frac{\partial g}{\partial E} d^3v \\
&= -e^2 \mathcal{E} \int v_z^2 \tau_F \frac{\partial g}{\partial E} d^3v \quad \text{(where } \tau \text{ has been replaced by } \tau_F \text{ since only values close to} \\
&\qquad\qquad\qquad\qquad\qquad \text{the Fermi surface are finite because of the form of } \frac{\partial g}{\partial E} \text{)}
\end{aligned}$$

$$\begin{aligned}
&= -\frac{e^2 \mathcal{E} \tau_F}{m} \int_\infty^\infty v_z \frac{dg}{dv_z} dv_z \quad \text{(where } \frac{dg}{dE} = \frac{dg}{dv_z} \frac{1}{m v_z} \text{ has been used)} \\
&= -\frac{e^2 \mathcal{E} \tau_F}{m} \left\{ v_z g |_\infty^\infty - \int_\infty^\infty g \, dv_z \right\} \quad \text{where integration by parts has been performed} \\
&= -\frac{e^2 \mathcal{E} \tau_F}{m} \{ 0 - n \} \\
j_z &= \left( \frac{n e^2 \bar{\tau}_F}{m} \right) \mathcal{E} \\
j_z &= \sigma \mathcal{E}
\end{aligned} \qquad (18.69)$$

from which the electrical conductivity $\sigma$ can be identified as

$$\sigma = \frac{n e^2 \bar{\tau}_F}{m} \qquad (18.70)$$

## 18.4   Exercises

**18.1**. Consider $N$ molecules in a volume $V$, each moving with a velocity $v$. Show the pressure $P$ is given by

$$P = \frac{2}{3}\frac{N}{V}\left(\frac{1}{2}m\bar{v^2}\right)$$

**18.2**. Consider a container with $n$ molecules per unit volume, each of mass $m$ and moving with a mean velocity $\bar{v}$, and the mean free path is $l$, show that the coefficient of thermal conductivity $\kappa$ is given by

$$\kappa = \frac{1}{3}n\bar{v}cl \tag{18.71}$$

**18.3**. Consider a container with $n$ molecules per unit volume, each of mass $m$ and moving with a mean velocity $\bar{v}$, show that the coefficient of viscosity $\eta$ is identified as

$$\eta = \frac{1}{3}n\bar{v}ml$$

**18.4**. Consider molecules moving with a mean velocity $\bar{v}$ and the mean free path is $l$, show that the coefficient of self-diffusion $D$ is given by

$$D = \frac{1}{3}\bar{v}l$$

**18.5**. Using Drude's theory, show that electrical conductivity in a metal is given by

$$\sigma = \frac{ne^2\tau}{m}$$

where all symbols are in the usual notation.

**18.6**. An argon atom forms a close-packed structure having a density of 1.65 gm.cm$^{-3}$. The atomic weight of argon is 39.9. Estimate the magnitude of the coefficient of viscosity at 25°C and 1 atmosphere pressure.

**18.7**. (a) Calculate the electrical conductivity, $\sigma$, for aluminium given that, for aluminium: Atomic weight = 26.7 g/mol, Density = 2.7 gm.cm$^{-3}$, Avogadro's number, N=6.022 × 10$^{23}$ mol$^{-1}$, electronic charge, $e = 1.602 \times 10^{-19}$ C, electronic mass, $m = 9.11 \times 10^{-31}$ kg, $\tau = 2.07 \times 10^{-14}$ s.
(b) Calculate the resistivity of aluminium from the data above.

**18.8**. Particles move without collisions in one-dimension under the influence of an acceleration $a$ that is constant, independent of $x$ and $v$.
(a) Find the characteristics of the Boltzmann's equation without collisions for the distribution function, and sketch them in phase space i.e $v$ vs $x$.
(b) Consider the region $x > 0$, for the case when $a$ is negative, Suppose the particles enter the region at $x = 0$ from below $(v > 0)$ with a known velocity distribution

$$f(v) = \frac{f_0}{1 + v^2/v_l^2}$$

where $f_0$ and $v_l$ are constants.

(i) Solve the Boltzmann's equation to find $f(v)$ for $(v < 0)$, at $x = 0$.

(ii) Solve to find $f(v)$ for all $v$ at $x = 1$.

**18.9.** The Boltzmann equation with collisions is given by

$$\frac{\partial f}{\partial t} + \underline{v} \cdot \frac{\partial f}{\partial \underline{r}} + \frac{\underline{F}}{m} \cdot \frac{\partial f}{\partial \underline{v}} = -\frac{f - f^{(0)}}{\tau}$$

Assuming that the equilibrium distribution function is given by Maxwell-Boltzmann statistics in the form

$$f^{(0)} = g(E) = n(\frac{m}{2\pi k_B T})^{3/2} e^{-E/k_B T}$$

where all symbols are in the usual notation, show that the electrical conductivity is given by

$$\sigma = \frac{ne^2 \tau}{m}$$

**18.10.** The Boltzmann equation with collisions is given by

$$\frac{\partial f}{\partial t} + \underline{v} \cdot \frac{\partial f}{\partial \underline{r}} + \frac{\underline{F}}{m} \cdot \frac{\partial f}{\partial \underline{v}} = -\frac{f - f^{(0)}}{\tau}$$

Assuming that the equilibrium distribution function is given by Fermi-Dirac statistics in the form

$$f^{(0)} = g(E) = \frac{1}{e^{(E-\mu)/k_B T} + 1}$$

where all symbols are in the usual notation, show that the electrical conductivity is given by

$$\sigma = \frac{ne^2 \tau_F}{m}$$

# Chapter 19

# Phase Transitions

## 19.1 Introduction

In phase transitions, one observes changes of state from one form to another at some temperatures and/or pressures. Examples of phase transitions include the following:

- Ice $\rightleftharpoons$ Water $\rightleftharpoons$ Vapour

- Ferromagnet $\rightleftharpoons$ Paramagnet

- Antiferromagnet $\rightleftharpoons$ Paramagnet

- Ferroelectric $\rightleftharpoons$ Paraelectric

- Superconductor $\rightleftharpoons$ Normal metal

- Superfluid $\rightleftharpoons$ Normal fluid

- Liquid crystals $\rightleftharpoons$ Normal crysal

## 19.2 Orders of Phase Transitions

At the phase transition, the Gibb's free energy, $G$ of each phase must have the same value. However, the volume, $V$, and entropy, $S$, are not continuous. Orders of phase transitions are defined in terms of the discontinuity of the derivatives of the Gibb's free energy.

*First Order Phase Transitions* are those transitions in which the first order derivatives of the Gibb's free energy $G$ are discontinuous. The volume, $V$, and entropy, $S$, are first order derivatives of the Gibb's free energy, given by

$$V \;=\; \left(\frac{\partial G}{\partial P}\right)_T \tag{19.1}$$

$$S \;=\; -\left(\frac{\partial G}{\partial T}\right)_P \tag{19.2}$$

Graphs of $F$ vs $T$, $S$ vs $T$, $G$ vs $P$ and $V$ vs $P$ are shown in Figure 19.1.

*Higher Order Phase Transitions* are those transitions in which the second or higher order derivatives of the Gibb's free energy $G$ are discontinuous, for example:

$$\left(\frac{\partial^2 G}{\partial P^2}\right)_T, \cdots, \left(\frac{\partial^n G}{\partial P^n}\right)_T \tag{19.3}$$

$$\left(\frac{\partial^2 G}{\partial T^2}\right)_P, \cdots, \left(\frac{\partial^n G}{\partial T^n}\right)_P \tag{19.4}$$

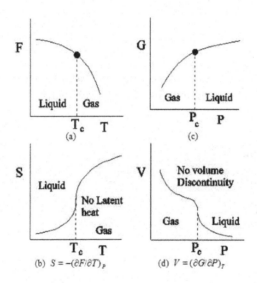

Figure 19.1: Graphs of $F$ vs $T$, $S$ vs $T$, $G$ vs $P$ and $V$ vs $P$. First order phase transitions have first order derivatives of $G$ discontinous.

## 19.3   Response Functions

In our study of phase transitions, it is useful to observe a remarkable analogy of properties of *fluid systems* and *magnetic systems*, as illustrated in Table (19.1).

**Table 19.1:** Analogy of properties of fluid systems and magnetic systems.

| **FLUID SYSTEMS** | **MAGNETIC SYSTEMS** |
|---|---|
| Pressure, $P$ | Magnetic field, $H$ |
| Temperature, $T$ | Temperature, $T$ |
| Volume, $V$ | Magnetization, $-M$ |
| Entropy, $S$ | Entropy, $S$ |
| Internal energy, $U$ | Internal energy, $U$ |
| Enthalpy, $E = U + PV$ | Enthalpy, $E = U - PV$ |
| Gibb's free energy, $G = U - TS + PV$ | Gibb's free energy, $G = U - TS - HM$ |
| Helmholtz free energy, $F = U - TS$ | Helmholtz free enrgy, $F = U - TS$ |

| First Law of Thermodynamics in fluid systems | First Law of Thermodynamics in magnetic systems |
|---|---|
| $dU = TdS - PdV$ | $dU = TdS + HdM$ |
| $dE = TdS + VdP$ | $dE = TdS - MdH$ |
| $dG = -SdT + VdP$ | $dG = -SdT - MdH$ |
| $dF = -SdT - PdV$ | $dU = -SdT + HdM$ |

## 19.3.1  Response Functions for Fluids

*(i) Specific heats, $C_V$ and $C_p$*

$$
\begin{aligned}
C_V &= \left(\frac{dQ}{dT}\right)_V \\
&= T\left(\frac{\partial S}{\partial T}\right)_V \\
&= \left(\frac{\partial U}{\partial T}\right)_V \\
&= -T\left(\frac{\partial^2 F}{\partial T^2}\right)_V \\
C_P &= -T\left(\frac{\partial^2 G}{\partial T^2}\right)_P
\end{aligned}
$$

*(ii) Compressibility, $K_T$ and $K_S$*

Isothermal compressibility, $K_T$

$$
\begin{aligned}
K_T &= -\frac{1}{V}\left(\frac{\partial V}{\partial P}\right)_T \\
&= \frac{1}{\rho}\left(\frac{\partial \rho}{\partial P}\right)_T \\
&= -\frac{1}{V}\left(\frac{\partial^2 G}{\partial P^2}\right)_V
\end{aligned}
$$

Adiabatic compressibility, $K_S$

$$
\begin{aligned}
K_S &= -\frac{1}{V}\left(\frac{\partial V}{\partial P}\right)_S \\
&= -\frac{1}{V}\left(\frac{\partial^2 E}{\partial P^2}\right)_S
\end{aligned}
$$

*(iii) Volume expansion, $\alpha_P$*

$$
\alpha_P = \frac{1}{V}\left(\frac{\partial V}{\partial T}\right)_P \tag{19.5}
$$

The various response functions are related, for example

$$
\begin{aligned}
K_T(C_P - C_V) &= TV\alpha_P^2 \tag{19.6} \\
C_P(K_T - K_S) &= TV\alpha_P^2 \tag{19.7}
\end{aligned}
$$

## 19.3.2  Response Functions for Magnetic Systems

*(i) Specific heats, $C_M$ and $C_H$*

$$
\begin{aligned}
C_M &= T\left(\frac{\partial S}{\partial T}\right)_M \\
&= -T\left(\frac{\partial^2 F}{\partial T^2}\right)_M \\
C_H &= -T\left(\frac{\partial^2 G}{\partial T^2}\right)_H
\end{aligned}
$$

*(ii) Susceptibility, $\chi_T$ and $\chi_S$*

Isothermal susceptibility, $\chi_T$

$$
\begin{aligned}
\chi_T &= \left(\frac{\partial M}{\partial H}\right)_T \\
&= -\left(\frac{\partial^2 G}{\partial H^2}\right)_T
\end{aligned}
$$

Adiabatic susceptibility, $\chi_S$

$$
\begin{aligned}
\chi_S &= \left(\frac{\partial M}{\partial H}\right)_S \\
&= -\left(\frac{\partial^2 E}{\partial H^2}\right)_S
\end{aligned}
$$

*(iii)* $\alpha_H$

$$\alpha_H = \left(\frac{\partial M}{\partial T}\right)_H \qquad (19.8)$$

The various response functions for magnetic systems are related, for example

$$\chi_T(C_H - C_M) = T\alpha_H^2 \qquad (19.9)$$
$$C_H(\chi_T - \chi_S) = T\alpha_H^2 \qquad (19.10)$$

## 19.4 Order Parameters and Critical Point Exponents

In phase transitions, there exists an order parameter, $\mathcal{P}$, which is such that

$$\mathcal{P}\begin{cases} > 0, & \text{for } T < T_C \\ = 0, & \text{for } T > T_C \end{cases} \qquad (19.11)$$

where $T_C$ is the critical temperature.

An example of an order parameter is a function $f(\epsilon)$, where $\epsilon$ is a dimensionless variable. The order parameter $f(\epsilon)$ is related to a critical point exponent $\lambda$ as

$$f(\epsilon) \sim \epsilon^\lambda \qquad (19.12)$$

where

$$\epsilon = \frac{T - T_C}{T_C} \qquad (19.13)$$

It can be noted that

$$\lambda = \lim_{\epsilon \to 0} \frac{\ln f(\epsilon)}{\ln \epsilon} \qquad (19.14)$$

Some examples of the relation between $f(\epsilon)$ and $\lambda$ are discussed below.

*Case (i):$\lambda$ is positive*

*Case (ii):$\lambda$ is negative*

*Case (iii):$\lambda$ is positive*

*Case (iv):$\lambda$ is logarithmic*

### 19.4.1 Order parameters and critical point exponents for fluids and magnetic systems

In this section we study several order parameters and their corresponding critical point exponents.

*(i) $\alpha$-exponent: Specific Heats, $C_V$ and $C_H$*

Fluids can be described by an order parameter known as the specific heat at constant volume, $C_V$, which satifies

$$C_V = \begin{cases} \mathcal{A}\left(-\epsilon\right)^{-\alpha'}, & \text{for } T < T_C \\ \\ \mathcal{A}\left(\epsilon\right)^{-\alpha}, & \text{for } T > T_C \end{cases} \tag{19.15}$$

Magnets can be described by an order parameter known as the specific heat at constant magnetic field, $C_H$, which satifies

$$C_H = \begin{cases} \mathcal{A}\left(-\epsilon\right)^{-\alpha'}, & \text{for } T < T_C \\ \\ \mathcal{A}\left(\epsilon\right)^{-\alpha}, & \text{for } T > T_C \end{cases} \tag{19.16}$$

*(ii) $\beta$-exponent: Density and Magnetization*

Fluids can also be described by an order parameter defined in terms of the liquid-gas density difference, which satifies

$$\frac{\rho_L(T) - \rho_G(T)}{2\rho_c} = \begin{cases} \mathcal{B}\left(-\epsilon\right)^{\beta}, & \text{for } T < T_C \\ \\ 0, & \text{for } T > T_C \end{cases} \tag{19.17}$$

Magnets can be described by an order parameter known as the magnetization, $M(T)$, which satifies

$$\frac{M(T)}{M(0)} = \begin{cases} \mathcal{B}'\left(-\epsilon\right)^{\beta}, & \text{for } T < T_C \\ \\ 0, & \text{for } T > T_C \end{cases} \tag{19.18}$$

*(iii) $\gamma$-exponent: Compressibility and Susceptibility*

Fluids can be described by an order parameter known as the isothermal compressibility, $K_T$, which satifies

$$\frac{K_T}{K_T^0} = \begin{cases} \mathcal{G}\left(-\epsilon\right)^{-\gamma'}, & \text{for } T < T_C \\ \\ \mathcal{G}\left(\epsilon\right)^{-\gamma}, & \text{for } T > T_C \end{cases} \tag{19.19}$$

Magnets can be described by an order parameter known as the isothermal susceptibility, $\chi_T$, which satifies

$$\frac{\chi_T}{\chi_T^0} = \begin{cases} \mathcal{G}\left(-\epsilon\right)^{-\gamma'}, & \text{for } T < T_C \\ \\ \mathcal{G}\left(\epsilon\right)^{-\gamma}, & \text{for } T > T_C \end{cases} \tag{19.20}$$

*(iv) δ-exponent: Pressure and Magnetic Field*

Fluids can be described by an order parameter, $\delta$, which satisfies

$$\frac{P - P_C}{P_C} = \mathcal{D} \mid \frac{\rho}{\rho_C} - 1 \mid^{\delta} \text{Sign}(\rho - \rho_C)$$

Magnets can be described by an order parameter, $\delta$, which satisfies

$$\frac{H}{H_C} = \mathcal{D} \mid \frac{M_H(T_C)}{M_0(0)} \mid^{\delta}$$

*(v) ν-exponent: Correlation length*

Fluids can be described by an order parameter known as the correlation length, $\xi$, which satifies

$$\xi = \begin{cases} \xi_0 \left(-\epsilon\right)^{-\nu'}, & \text{for } T < T_C \\ \\ \xi_0 \left(\epsilon\right)^{-\nu}, & \text{for } T > T_C \end{cases} \tag{19.21}$$

Magnets can be described by an order parameter known as the correlation length, $\xi$, which satifies

$$\xi = \begin{cases} \xi_0 \left(-\epsilon\right)^{-\nu'}, & \text{for } T < T_C \\ \\ \xi_0 \left(\epsilon\right)^{-\nu}, & \text{for } T > T_C \end{cases} \tag{19.22}$$

*(vi) η-exponent: Pair correlation function*

Fluids can be described by an order parameter known as a pair correlation length, which satisfies

$$G(r) = \frac{\omega}{r^{d-2+\eta}} \tag{19.23}$$

where $d$ is dimensionality.

Magnets can be described by an order parameter known as a pair correlation length, which satisfies

$$\Gamma(r) = \frac{\omega'}{r^{d-2+\eta}} \tag{19.24}$$

where $d$ is dimensionality.

## 19.4.2 Molecular field theory of critical point exponents

In this section, molecular field theory is applied to find the critical point exponent of the following order parameters.

- Magnetization, $M$

- Susceptibility, $\chi$

## Magnetization

To obtain magnetization, we start from the *Heisenberg model* of the ferromagnet, where the Hamiltonian of the spin-ordered system is given by

$$H = -J\sum_{i,j}\mathbf{S}_i\cdot\mathbf{S}_j \quad \text{in the } absence \text{ of an external magnetic field} \tag{19.25}$$

$$H = -J\sum_{i,j}\mathbf{S}_i\cdot\mathbf{S}_j - g\mu_B H_0\sum_i S_i^z \quad \text{in the } presence \text{ of an external magnetic field,} \tag{19.26}$$

where $J$ is the exchange integral, $\mathbf{S}_i$ and $\mathbf{S}_j$ are the spins of the $i^{th}$ and $j^{th}$ atoms respectively, $g$ is Lande's $g$ factor, $\mu_B$ is the Bohr magneton, and $S_i^z$ is the $z$-component of the spin of the $i^{th}$ atom.

The exchange integral $J$ satisfies the following

$$J = \begin{cases} > 0 & \text{for parallel spins (ferromagnets)} \\ < 0 & \text{for antiparallel spins (antiferromagnets)} \end{cases} \tag{19.27}$$

According to *molecular field theory*, certain approximations to the Hamiltonian given in equation (19.26) can be done as shown below.

$$\begin{aligned} H &= -J\sum_{i,j}\mathbf{S}_i\cdot\mathbf{S}_j - g\mu_B H_0\sum_i S_i^z \\ &= \{Jz<S^z> - g\mu_B H_0\}\sum_i S_i^z \\ &= -g\mu_B\left\{H_0 + \frac{Jz<S^z>}{g\mu_B}\right\}\sum_i S_i^z \\ &= -g\mu_B H_{eff}\sum_i S_i^z \end{aligned}$$

where $H_{eff}$ can be identified as

$$\begin{aligned} H_{eff} &= H_0 + \frac{Jz<S^z>}{g\mu_B} \\ &= H_0 + \frac{JzNg\mu_B<S^z>}{Ng^2\mu_B^2} \\ &= H_0 + \frac{JzM}{Ng^2\mu_B^2} \tag{19.28} \\ &= H_0 + \lambda M \tag{19.29} \end{aligned}$$

where $M = Ng\mu_B<S^z>$ has been used, and $\lambda = Jz/(Ng^2\mu_B^2)$ has been introduced. In equation (19.29), the first term is the external magnetic field and the second term $\lambda M$ is the *molecular field*.

Let us now calculate the *magnetization*, $M = Ng\mu_B<S^z>$, where the average spin $<S^z>$ is calculated using the Gibb's distribution from statistical mechanics.

$$< S^z > = \frac{\sum_i S_i^z e^{-E/k_B T}}{\sum_i e^{-E/k_B T}} \qquad (19.30)$$

where the energy $E = -g\mu_B H_{eff} S_i^z$. Hence the magnetization is given by

$$\begin{aligned}
M &= N g\mu_B \frac{\sum_i S_i^z e^{-E/k_B T}}{\sum_i e^{-E/k_B T}} \\
&= N g\mu_B \frac{\sum_{i=-\frac{1}{2}}^{+\frac{1}{2}} S_i^z e^{g\mu_B H_{eff} S_i^z / k_B T}}{\sum_{i=-\frac{1}{2}}^{+\frac{1}{2}} e^{g\mu_B H_{eff} S_i^z / k_B T}} \\
&= N g\mu_B \frac{\left\{ \frac{1}{2} e^{a/2} - \frac{1}{2} e^{-a/2} \right\}}{e^{a/2} + e^{-a/2}} \\
&= \frac{N g\mu_B}{2} \tanh \left[ \frac{a}{2} \right] \quad \text{where } g = 2 \\
M &= N\mu_B \tanh \left[ \frac{\mu_B H_{eff}}{k_B T} \right] \qquad (19.31)
\end{aligned}$$

Let us study the expression for magnetization when $H_0 = 0$, and hence $H_{eff} = \lambda M$. For $T = T_C$,

$$M = N\mu_B \tanh \left[ \frac{\mu_B \lambda M}{k_B T_C} \right] \qquad (19.32)$$

Using the approximation $\tanh x \approx x$ in equation (19.32) gives the critical temperature as

$$T_C = \frac{N\mu_B^2 \lambda}{k_B} \qquad (19.33)$$

which is also known as the *Curie temperature*. Using this, the expression for the magnetization can also be written in the form

$$\begin{aligned}
M &= N\mu_B \tanh \left( \frac{\mu_B \lambda M}{k_B T} \right) \\
&= N\mu_B \tanh \left( \frac{M N \mu_B^2 \lambda}{T N \mu_B k_B} \right) \\
&= M_s \tanh \left( \frac{M T_C}{T M_s} \right) \quad \text{where } M_s = N\mu_B \\
\frac{M}{M_s} &= \tanh \left( \frac{T_C M}{T M_s} \right) \qquad (19.34)
\end{aligned}$$

There are three limits of physical interest to equation (19.34).

*(i) Low temperature limit, $T \ll T_C$*

Using the definition of $\tanh x$

$$
\begin{aligned}
\tanh x &= \frac{e^x - e^{-x}}{e^x - e^{-x}} \\
\frac{M}{M_s} &= \tanh\left(\frac{T_C M}{T M_s}\right) \\
&= \frac{e^{\frac{T_C M}{T M_s}} - e^{-\frac{T_C M}{T M_s}}}{e^{\frac{T_C M}{T M_s}} + e^{-\frac{T_C M}{T M_s}}} \\
&\approx 1 - 2e^{-2\frac{T_C M}{T M_s}} \quad \text{as } T \to 0
\end{aligned}
\tag{19.35}
$$

which shows an exponential deviation.

*(ii) Close to $T_C$, $T \leq T_C$*

Using the definition of $\tanh x$

$$
\begin{aligned}
\tanh x &= x - \frac{1}{3}x^3 + \frac{2}{5}x^5 \cdots \\
\frac{M}{M_s} &\approx \frac{T_C M}{T M_s} - \frac{1}{3}\left(\frac{T_C M}{T M_s}\right)^3 \\
1 &= \frac{T_C}{T} - \frac{1}{3}\left(\frac{T_C}{T}\right)^3 \left(\frac{M}{M_s}\right)^2 \\
\left(\frac{M}{M_s}\right)^2 &= 3\left(\frac{T_C}{T} - 1\right)\left(\frac{T}{T_C}\right)^3 \\
\frac{M}{M_s} &= 3^{1/2}\left(\frac{T}{T_C}\right)\left(\frac{T_C - T}{T_C}\right)^{1/2}
\end{aligned}
\tag{19.36}
$$

which when compared with equation (19.18)

$$
\frac{M(T)}{M(0)} = \mathcal{B}'\left(-\epsilon\right)^\beta
$$

implies that

$$
\beta = \frac{1}{2}
\tag{19.37}
$$

*(iii) High temperature limit, $T \gg T_C$*

$$
M = 0 \text{ ie Paramagnetic}
$$

## Magnetic Susceptibility

To obtain the magnetic usceptibility, recall equation (19. ), and consider the case when there is a finite external magnetic field, $H_0 \neq 0$, and hence $H_{eff} = H_0 + \lambda M$. Hence the magnetization is given by

$$
\begin{aligned}
M &= N\mu_B \tanh\left[\frac{\mu_B H_{eff}}{k_B T}\right] \\
&= N\mu_B \tanh\left[\frac{\mu_B(H_0 + \lambda M)}{k_B T}\right] \\
&= \frac{N\mu_B^2(H_0 + \lambda M)}{k_B T} \quad \text{where} \quad \tanh x \approx x \quad \text{has been used} \\
M &= \frac{N\mu_B^2 H_0}{k_B T} + \frac{N\mu_B^2 \lambda M}{k_B T} \\
M\left[1 - \frac{N\mu_B^2 \lambda}{k_B T}\right] &= \frac{N\mu_B^2 H_0}{k_B T} \\
M &= \frac{N\mu_B^2 H_0}{k_B T - N\mu_B^2 \lambda}
\end{aligned}
\tag{19.38}
$$

from which the magnetic susceptibility, $\chi_m$ is given by

$$
\begin{aligned}
\chi_m &= \frac{M}{H_0} \\
&= \frac{N\mu_B^2}{k_B T - N\mu_B^2 \lambda} \\
&= \frac{N\mu_B^2}{k_B(T - T_C)}
\end{aligned}
\tag{19.39}
$$

where $T_C$ is the critical temperature (or Curie temperature) given by equation (19.33). Equation (19.39 is known as *Curie-Weiss law* which shows that as $T \to T_C$, the susceptibility becomes very large.

Comparing equation (19.39) and (19.20),

$$
\frac{\chi_T}{\chi_T^0} = \begin{cases} \mathcal{G}(-\epsilon)^{-\gamma'}, & \text{for } T < T_C \\ \mathcal{G}(\epsilon)^{-\gamma}, & \text{for } T > T_C \end{cases}
$$

we can identify

$$
\gamma = 1
\tag{19.40}
$$

### 19.4.3 Landau's theory of critical point exponents

In this section, Landau's theory is applied to find the critical point exponent of the following order parameters.

- Magnetization, $M$

- Susceptibility, $\chi$

**Magnetization**

In Landau's theory of critical point exponents, it is assumed that the Helmholtz free energy can be expanded as power series in the magnetization

$$F(T, M) = \sum_{j=0} L_j(T) M^j \quad \text{where } j \text{ is even} \tag{19.41}$$

$$= L_0 + L_2 M^2 + L_4 M^4 + \cdots \tag{19.42}$$

where the coefficients $L_j$ are expandable about te critical point.

$$L_j(T) = \sum_{k=0}^{\infty} l_{jk}(T - T_C)^k \tag{19.43}$$

The magnetic field $H$ can be obtained by differentiating the free energy with respect to the magnetization, and solving for $M$ when $H = 0$.

$$
\begin{aligned}
H &= \left( \frac{\partial F}{\partial M} \right)_T \\
&= 2L_2 M + 4L_4 M^3 + \cdots \\
&= 2M(L_2 + 2M^2 L_4 + \cdots) \\
0 &= \{l_{20} + l_{21}(T - T_C) + \cdots\} + 2M^2 \{l_{40} + l_{41}(T - T_C) + \cdots\}
\end{aligned}
$$

Taking $l_{20} = 0$ and lowest order terms,

$$
\begin{aligned}
l_{21}(T - T_C) + 2M^2 l_{40} &= 0 \\
2M^2 l_{40} &= -l_{21}(T - T_C) \\
M &= \left( \frac{l_{21}}{2l_{40}} \right)^{1/2} (T_C - T)^{1/2} \tag{19.44}
\end{aligned}
$$

which when compared with equation (19.18)

$$\frac{M(T)}{M(0)} = \mathcal{B}' \, (-\epsilon)^\beta$$

implies that

$$\beta = \frac{1}{2}$$

**Susceptibility**

The magnetic susceptibility $\chi_T$ can be obtained by differentiating the magnetization with respect to the magnetic field, that is

$$\chi_T = \left( \frac{\partial M}{\partial H} \right)_T \quad \text{or} \quad \chi_T^{-1} = \left( \frac{\partial H}{\partial M} \right)_T$$

and hence

$$\begin{aligned}
\chi_T^{-1} &= \left(\frac{\partial H}{\partial M}\right)_T \\
&= 2L_2 + 12L_4M^2 + \cdots \\
&= 2L_2 \ \text{if} \ M \to 0 \\
&= 2\left\{l_{20} + l_{21}(T - T_C) + \cdots\right\}
\end{aligned} \tag{19.45}$$

Taking $l_{20} = 0$ and lowest order terms,

$$\begin{aligned}
\chi_T^{-1} &= 2l_{21}(T - T_C) \\
\chi_T &= \frac{1}{2l_{21}(T - T_C)}
\end{aligned} \tag{19.46}$$

which when compared with equation (19.20), we can identify

$$\gamma = 1$$

## 19.4.4 Exponential Inequalities

There are several relations between the critical points, and a few will be studied in this section. The critical point exponents also satisfy several relations of interest, for example, if $f(x)$ has a critical point $\lambda$ of the form $x^\lambda$, and $g(x)$ has a critical point $\phi$ of the form $x^\phi$, then if $f(x) \le g(x)$, then $\lambda \ge \phi$. The proof of this is straight forward.

$$\begin{aligned}
f(x) &\le g(x) \\
\ln f(x) &\le \ln g(x) \quad \text{if} \quad x < 1, \ln x < 0 \\
\frac{\ln f(x)}{\ln x} &\ge \frac{\ln g(x)}{\ln x} \\
\lambda &\ge \phi
\end{aligned}$$

which completes the proof.

**Rushbrooke inequality**

The Rushbrooke inequality relates critical point exponents of a magnetic system which exhibits a first-order phase transition in the thermodynamic limit for non-zero temperature.

$$\begin{aligned}
\text{Recall that,} \quad C_H - C_M &= \frac{T}{\chi_T}\left(\frac{\partial M}{\partial T}\right)_H^2 \\
C_H &\ge \frac{T}{\chi_T}\left(\frac{\partial M}{\partial T}\right)_H^2
\end{aligned}$$

$$C_H \sim (-\epsilon)^{-\alpha'}$$
$$\chi_T \sim (-\epsilon)^{-\gamma'}$$
$$\text{But} \quad M \sim (-\epsilon)^{\beta} \quad \text{and} \quad T \sim (-\epsilon)^1$$
$$\left(\frac{\partial M}{\partial T}\right)_H \sim (-\epsilon)^{\beta-1}$$
$$\text{Combining, we obtain} \quad -\alpha' \leq \gamma' + 2(\beta - 1)$$
$$\alpha' + 2\beta + \gamma' \geq 2 \tag{19.47}$$

which is known as the Rushbrooke inequality.

## Griffiths inequality

The Griffiths inequality is given by

$$\alpha' + \beta(1 + \delta) \geq 2 \tag{19.48}$$

## Fisher inequality

The Fisher inequality is given by

$$\gamma \leq (2 - \eta)\nu \tag{19.49}$$

## Josephson inequality

The Josephson inequality is given by

$$d\nu \geq 2 - \alpha \tag{19.50}$$

## 19.5 Exercises

**19.1**. (a) Give expressions for the following response functions:
(i) Specific heat at constant pressure, $C_p$
(ii) Specific heat at constant volume, $C_v$
(iii) Isothermal compressibility, $k_T$
(iv) Adiabatic compressibility, $k_S$
(v) Volume expansion, $\alpha_p = \frac{1}{V}\left(\frac{\partial V}{\partial P}\right)_P$
(b) Prove that

$$k_T(C_p - C_v) = TV\alpha_p^2$$

(c) Prove that

$$C_p(k_T - k_s) = TV\alpha_p^2$$

**19.2**. (a) The critical point for a gas/liquid is defined a point where $(\partial p/\partial V)_T = 0$ and $(\partial^2 p/\partial V^2)_T = 0$. Assuming the van der Waals' equation of state,

$$\left[p + a\left(\frac{N}{V}\right)^2\right](V - Nb) = NkT$$

where $a$ and $b$ are constants specific for a substance. Show that the critical point is determined by

$$V_c = 3Nb, \quad p_c = \frac{1}{27}\frac{a}{b^2}, \quad kT_c = \frac{8}{27}\frac{a}{b}$$

(b) Water has a critical point at $T = 374°C$ and $p = 218$ atm. Calculate the parameters $a$ and $b$ and comment on the results.

**19.3**. (a) Give expressions for the following response functions:
(i) Specific heat at constant magnetisation, $C_M$
(ii) Specific heat at constant magnetic field strength, $C_H$
(iii) Isothermal susceptibility, $\chi_T$
(iv) Adiabatic compressibility, $\chi_S$
(b) Prove that

$$\chi_T(C_H - C_M) = TV\alpha_H^2$$

(c) Prove that

$$C_H(\chi_T - \chi_s) = T\alpha_H^2$$

**19.4**. (a) Explain what is meant by a *critical point exponent*, and give four examples of such quantities.
(b) If $f(x)$ has a critical point $\lambda$ of the form $x^\lambda$, and $g(x)$ has a critical point $\phi$ of the form $x^\phi$, then if $f(x) \leq g(x)$, show that $\lambda \geq \phi$.

**19.5**. (a) Explain what is meant by a *phase transition*, and give three examples of this phenomenon.

(b) According to molecular field theory, the magnetization $M$ of some ferromagnetic system is given by

$$M = N\mu_B \tanh[\frac{\mu_B H_{eff}}{k_B T}]$$

where all symbols are in the usual notation.
(i) Plot a schematic graph for magnetization with temperature and discuss the spin orientations for $T < T_c$ and $T > T_c$.
(ii) Show that close to the critical temperature $T_c$, the magnetization is given by

$$\frac{M}{M_s} = \frac{3^{1/2} T (T_c - T)^{1/2}}{T_c^{3/2}}$$

where $M_s = N\mu_B$, $H_{eff} = \lambda M$ and $T_c = N\mu_B^2 \lambda / k_B$,                     (10 marks)
You may assume that

$$\tanh x \approx x - \frac{1}{3}x^3.$$

(iii) From above, identify the critical point exponent for this phase transition.

**19.6.** According to *Ginzburg-Landau's theory of superconductivity*, the free energy of a superconductor can be expanded as a power series, and in presence of a magnetic field, is given by

$$F_{sc} = F_{no} + \sum_{j=2,even} \frac{2}{j} g_j(T)|\psi|^j + \frac{1}{2m^*}|(-i\hbar\nabla + \frac{e^*A}{c})\psi|^2 + \frac{B^2}{8\pi}$$

where $g_2(T) = a, g_4(T) = b, ...$, and all symbols are in the usual notation.
(a) Write down the expansion of the free energy up to the $j = 4$ term, in presence of the magnetic field.
(b) From the expression you have written in (a) above, minimise the free energy with respect to the order parameter $\psi$, and show that for a 1-d case when the vector potential $\vec{A} \to 0$, $\psi$ satisfies

$$-\frac{\hbar^2}{2m^*}\frac{d^2}{dz^2} + a\psi + b|\psi^2|\psi = 0$$

where all symbols are in the usual notation.
(c) (i) Hence show that if a function $f(z) = \frac{\psi(z)}{|\psi_\infty|}$ is introduced, it will satisfy the following differential equation

$$\frac{d^2 f}{dz^2} + \frac{f}{\xi^2}(1 - f^2) = 0$$

where

$$\xi = \frac{\hbar}{\sqrt{2m^*|a|}}$$

(ii) Hence show that the solution of the differential equation in c(i) is

$$f(z) = \tanh(\frac{z}{\sqrt{2}\xi})$$

**19.7**. According to *Landau's theory of phase transitions*, the free energy of a ferroelectric can be expanded as a power series in the form

$$F(P, T, E) = -EP + \sum_{j=2, even} [g_0 + \frac{g_j(T)}{j} P^j]$$

where all symbols are in the usual notation.
(a) Write down the expansion of the free energy up to the $j = 6$ term.
(b) From the expression you have written in (a) above, give an expression for $\frac{\partial F}{\partial P}$.
(c) If the differential in (b) above is zero in a zero electric field, with $g_2 = \gamma(T - T_c)$, and expansion only up to the $g_4$ term is considered, show that, one of the solutions for $P_s$ is given by

$$|P_s| = |\frac{\gamma}{g_4}|^{1/2}(T_c - T)^{1/2}$$

(d) If the differential in (b) above is zero in a zero electric field, with $g_2 = \gamma(T - T_c)$, $g_4 < 0$ and expansion only up to the $g_6$ term is considered, show that, one of the solutions for $P_s$ satisfies

$$|P_s|^2 = \frac{|g_4| \pm \sqrt{|g_4|^2 - 4g_6\gamma(T - T_c)}}{2g_6}$$

**19.8**. According to *Landau's theory of phase transitions* as applied to the study of magnetization $M$, the Helmholtz free energy $F(T, M)$ can be expanded as a power series

$$F(T, M) = \sum_{j=0} L_j(T)M^j$$

where

$$L_j(T) = \sum_{k=0} l_{jk}(T - T_c)^k$$

and all symbols are in the usual notation.
(a) What are the conditions for the values of (i) $j$? (ii)$k$ ?
(b) Using the condition for $j$, write down the expansion of the Helmholtz free energy up to the $j = 4$ term.
(c) (i) Outline a proof, using lowest order terms, to show that the magnetization satisfies

$$M = \{\frac{l_{21}}{2l_{40}}\}^{1/2}(T_c - T)^{1/2}$$

where you may assume the thermodynamic relation $dF = -SdT + HdM$ and that $l_{20} = 0$.
Hence, what is the critical point exponent for magnetization?
(ii) Outline a proof, using lowest order terms, to show that the magnetic susceptibility satisfies

$$\chi = \{\frac{1}{2l_{21}}\}(T_c - T)^{-1}$$

Hence, what is the critical point exponent for he magnetic susceptibility?

**19.9**. The Rushbrooke inequality is given by

$$\alpha' + 2\beta + \gamma' \geq 2$$

where all symbols are in the usual notation.

(i) Prove the Rushbrooke inequality.

(ii) Verify that the Rushbrooke inequality is satified by magnetic systems by assuming values of $\alpha'$, $\beta$ and $\gamma'$.

**19.10**. Below are given a number of inequalities. Prove and verify each of them by an example.

(i) The Griffiths inequality is given by

$$\alpha' + \beta(1 + \delta) \geq 2$$

(ii) The Fisher inequality is given by

$$\gamma \leq (2 - \eta)\nu$$

(iii) The Josephson inequality is given by

$$d\nu \geq 2 - \alpha$$

# Chapter 20

# Relativity

## 20.1 Galilean Relativity

Consider two coordinate systems (also known as frames of reference) $S_1$ and $S_2$ as illustrated in Figure 20.1. The coordinates in $S_1$ are $(x_1, y_1)$ at time $t_1$, while in $S_2$ are $(x_2, y_2)$ at time $t_2$.

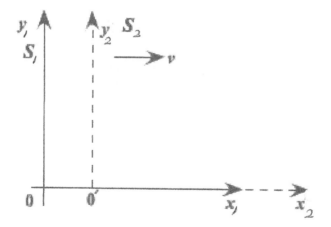

Figure 20.1: Two coordinate systems $S_1$ and $S_2$, initially together and $S_2$ moving with velocity $v$.

Initially (at time $t_1 = t_2 = 0$), the two coordinate systems are together, and hence

$$x_1 = x_2 \tag{20.1}$$
$$y_1 = y_2 \tag{20.2}$$
$$z_1 = z_2 \tag{20.3}$$
$$t_1 = t_2 = 0 \tag{20.4}$$

## 20.1.1   The Galilean Coordinate Transformations

Now, suppose the coordinate system $S_2$ moves with velocity $\mathbf{v}$, then the coordinates in $S_2$ and $S_1$ are related by

$$x_2 = x_1 - vt_1 \tag{20.5}$$
$$y_2 = y_1 \tag{20.6}$$
$$z_2 = z_1 \tag{20.7}$$
$$t_2 = t_1 \tag{20.8}$$

and the Inverse Galilean Coordinate Transformations are given by

$$x_1 = x_2 + vt_2 \tag{20.9}$$
$$y_1 = y_2 \tag{20.10}$$
$$z_1 = z_2 \tag{20.11}$$
$$t_1 = t_2 \tag{20.12}$$

## 20.1.2   The Galilean Velocity Transformations

Now, suppose the coordinate system $S_2$ moves with velocity $\mathbf{v}$, then the velocities in $S_2$ and $S_1$ are related by

$$v_{2x} = v_{1x} - v \tag{20.13}$$
$$v_{2y} = v_{1y} \tag{20.14}$$
$$v_{2z} = v_{1z} \tag{20.15}$$
$$\mathbf{v_2} = \mathbf{v_1} - \mathbf{v} \tag{20.16}$$

and the Inverse Galilean Velocity Transformations are given by

$$v_{1x} = v_{2x} + v \tag{20.17}$$
$$v_{1y} = v_{2y} \tag{20.18}$$
$$v_{1z} = v_{2z} \tag{20.19}$$
$$\mathbf{v_1} = \mathbf{v_2} + \mathbf{v} \tag{20.20}$$

## 20.1.3   The Galilean Acceleration Transformations

Now, suppose the coordinate system $S_2$ moves with velocity $\mathbf{v}$, then the velocities in $S_2$ and $S_1$ are related by

$$a_{2x} = a_{1x} \tag{20.21}$$
$$a_{2y} = a_{1y} \tag{20.22}$$
$$a_{2z} = a_{1z} \tag{20.23}$$
$$\mathbf{a_2} = \mathbf{a_1} \tag{20.24}$$

## 20.2 Special Theory of Relativity

### 20.2.1 Assumptions of the Special Theory of Relativity

In the Special Theory of Relativity, the following assumptions are made:

1. Principle of relativity: All laws of physics have the same mathematical form in all frames of reference.
The above assumption has two implications:
(i) There is no absolute frame of reference.
(ii) Measured values are relative to a frame of reference.

2. The speed of light in vacuum has the same value in all inertial frames of reference.
The above assumption has he following implication:
(i) Electromagnetic waves move with the same speed and do not need a material medium to go through, as was verified by the Michelson-Morley experiment in 1887, and this is illustrated in figure 20.2.
The purpose of the Michelson-Morley experiment is to find out whether electromagnetic waves need a material medium to propagate through. If such a medium did exit, referred to as ether, then it would have a velocity, $v$. Let us find the differerence, $\Delta T$, between time $T_\parallel$ and $T_\perp$.

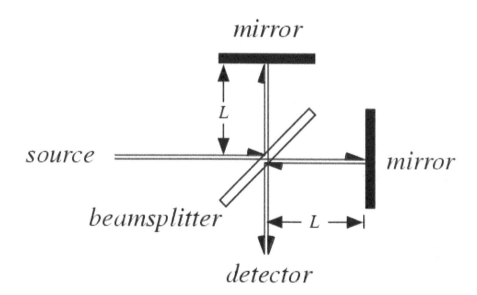

Figure 20.2: Michelson-Morley Experiment.

$$
\begin{aligned}
T_\parallel &= \frac{L}{c+v} + \frac{L}{c-v} \\
&= \frac{L(c-v+c+v)}{(c+v)(c-v)}
\end{aligned}
$$

$$= \frac{2Lc}{c^2 - v^2}$$

$$= \frac{2L/c}{(1 - v^2/c^2)} \tag{20.25}$$

$$T_\perp = \frac{L}{\sqrt{c^2 - v^2}} + \frac{L}{\sqrt{c^2 - v^2}}$$

$$= \frac{2L}{\sqrt{c^2 - v^2}}$$

$$= \frac{2L/c}{(1 - v^2/c^2)^{1/2}} \tag{20.26}$$

Hence,     $$\Delta T = T_\parallel - T_\perp \tag{20.27}$$

$$= \frac{2L/c}{(1 - v^2/c^2)} - \frac{2L/c}{(1 - v^2/c^2)^{1/2}}$$

$$= \frac{2L}{c} \left\{ \left(1 - \frac{v^2}{c^2}\right)^{-1} - \left(1 - \frac{v^2}{c^2}\right)^{-1/2} \right\}$$

$$= \frac{2L}{c} \left\{ \left[1 + (-1)(-\frac{v^2}{c^2})\right] - \left[1 + (-)(\frac{1}{2})(-\frac{v^2}{c^2})\right] \right\}$$

$$= \frac{2L}{c} \left\{ \left[1 + \frac{v^2}{c^2} - 1 - \frac{1}{2}\frac{v^2}{c^2}\right] \right\}$$

$$= \frac{2L}{c} \frac{1}{2} \frac{v^2}{c^2}$$

$$\Delta T = \frac{Lv^2}{c^3} \tag{20.28}$$

The result of the Michelson-Morley experiment is that $\Delta T = 0$ since $T_\parallel = 2L/c$ and $T_\perp = 2L/c$, and thus $v = 0$ meaning that *ether does not exist*.

### 20.2.2   The Lorentz Coordinate Transformations

Consider two frames of reference $S_1$ and $S_2$ are initially together at $t_1 = t_2 = 0$, when a spherical pulse of light moving with the velocity $c$ is sent out at their common origin, as illustrated in figure 20.3. Subsequently, $S_2$ moves with a constant velocity $v$. Since $c$ is the same for all observers in both $S_1$ and $S_2$ must detect a spherical wavefront of radius $r$ satisfying the equation of a sphere.

$$x^2 + y^2 + z^2 = r^2$$

For $S_1$,     $$x_1^2 + y_1^2 + z_1^2 - c^2 t_1^2 = 0$$

For $S_2$,     $$x_2^2 + y_2^2 + z_2^2 - c^2 t_2^2 = 0$$

Let     $$x_2 = \gamma(x_1 - vt_1)$$

$$y_2 = y_1$$

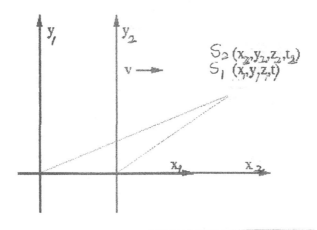

Figure 20.3: Two frames of reference $S_1$ and $S_2$ are initially together at $t_1 = t_2 = 0$, when a spherical pulse of light moving with the velocity $c$ is sent out at their common origin

$$z_2 = z_1$$

We obtain

$$x_1^2 + y_1^2 + z_1^2 - c^2 t_1^2 = \gamma^2 (x_1 - vt_1)^2 + y_2^2 + z_2^2 - c^2 t_2^2$$
$$x_1^2 - c^2 t_1^2 = \gamma^2 (x_1 - vt_1)^2 - c^2 t_2^2$$
$$= \gamma^2 (x_1^2 - 2vt_1 x_1 + v^2 t_1^2) - c^2 t_2^2$$
$$x_1^2(1 - \gamma^2) + x_1[\gamma^2 2vt_1] - t_1^2[c^2 + \gamma^2 v^2] + c^2 t_2^2 = 0$$

$$(20.29)$$

Equation (20.29) can only have a solution if and only if $t_2$ depends on both $t_1$ and $x_1$. Try $t_2$ of the form

$$t_2 = At_1 - Bx_1 \qquad (20.30)$$

Hence, we obtain

$$x_1^2(1 - \gamma^2) + x_1[\gamma^2 2vt_1] - t_1^2[c^2 + \gamma^2 v^2] + c^2[At_1 - Bx_1]^2 = 0$$
$$x_1^2(1 - \gamma^2) + x_1[\gamma^2 2vt_1] - t_1^2[c^2 + \gamma^2 v^2] + c^2[A^2 t_1^2 - 2ABt_1 x_1 + B^2 x_1^2] = 0$$
$$x_1^2[1 - \gamma^2 + c^2 B^2] + 2x_1 t_1[\gamma^2 v - c^2 AB] + t_1^2[-c^2 - \gamma^2 v^2 + c^2 A^2] = 0$$

If all the coefficients are zero, then

$$1 - \gamma^2 + c^2 B^2 = 0$$
$$\gamma^2 v - c^2 AB = 0$$
$$c^2 A^2 - \gamma^2 v^2 - c^2 = 0$$

or

$$c^2 B^2 = \gamma^2 - 1$$
$$\gamma^4 = \frac{c^4 A^2 B^2}{v^2} = \frac{c^2 A^2 B^2 c^2}{v^2}$$
$$c^2 A^2 = \gamma^2 v^2 + c^2$$

Hence

$$\gamma^4 = \frac{[\gamma^2 v^2 + c^2][\gamma^2 - 1]}{v^2}$$

$$= \frac{1}{v^2} \left[ \gamma^4 v^2 - \gamma^2 v^2 + \gamma^2 c^2 - c^2 \right]$$

$$= \gamma^4 + \frac{\gamma^2}{v^2} [c^2 - v^2] - \frac{c^2}{v^2}$$

$$\gamma^4 = \gamma^4 + \gamma^2 [\frac{c^2}{v^2} - 1] - \frac{c^2}{v^2}$$

$$-\gamma^2 [\frac{c^2}{v^2} - 1] = -\frac{c^2}{v^2}$$

$$\gamma^2 = \frac{c^2/v^2}{\frac{c^2}{v^2} - 1} = \frac{1}{[1 - \frac{v^2}{c^2}]}$$

$$\gamma = \frac{1}{\sqrt{1 - \frac{v^2}{c^2}}} = \frac{1}{\sqrt{1 - \beta^2}} \qquad (20.31)$$

where $\beta = \frac{v}{c}$.
Let us find $A$ and $B$ in equation (20.30) in terms of $\gamma$.

$$A^2 = \gamma^2 \frac{v^2}{c^2} + \frac{c^2}{c^2}$$

$$= \frac{v^2/c^2}{1 - \frac{v^2}{c^2}} + 1$$

$$= \frac{\frac{v^2}{c^2} + 1 - \frac{v^2}{c^2}}{1 - \frac{v^2}{c^2}}$$

$$= \frac{1}{\sqrt{1 - \frac{v^2}{c^2}}}$$

$$A^2 = \gamma^2$$

$$A = \gamma \qquad (20.32)$$

which means in equation (20.30) $A = \gamma$. How about $B$?

$$B^2 = \frac{\gamma^2 - 1}{c^2} = \frac{\gamma^2 v^2/c^2}{c^2}$$

$$B^2 = \frac{\gamma^2 \beta^2}{c^2}$$

$$B = \frac{\gamma \beta}{c} \qquad (20.33)$$

Hence $t_2$ from equation (20.30) is

$$
\begin{aligned}
t_2 &= At_1 - Bx_1 \\
&= \gamma t_1 - \frac{\gamma\beta}{c}x_1 \\
&= \gamma[t_1 - \frac{\beta}{c}x_1]
\end{aligned}
\tag{20.34}
$$

Hence the *Lorentz Coordinate Transformations* are given by

$$
x_2 = \gamma(x_1 - vt_1) = \frac{x_1 - vt_1}{\sqrt{1 - \frac{v^2}{c^2}}}
\tag{20.35}
$$

$$
y_2 = y_1
\tag{20.36}
$$

$$
z_2 = z_1
\tag{20.37}
$$

$$
t_2 = \gamma[t_1 - \frac{\beta}{c}x_1] = \frac{[t_1 - \frac{v}{c^2}x_1]}{\sqrt{1 - \frac{v^2}{c^2}}}
\tag{20.38}
$$

and the *Inverse Lorentz Coordinate Transformations* are given by

$$
x_1 = \gamma(x_2 + vt_2) = \frac{x_2 + vt_2}{\sqrt{1 - \frac{v^2}{c^2}}}
\tag{20.39}
$$

$$
y_1 = y_2
\tag{20.40}
$$

$$
z_1 = z_2
\tag{20.41}
$$

$$
t_1 = \gamma(t_2 + \frac{\beta}{c}x_2) = \frac{[t_2 + \frac{v}{c^2}x_2]}{\sqrt{1 - \frac{v^2}{c^2}}}
\tag{20.42}
$$

The *Lorentz Coordinate Transformations* can also be written in the form

$$
t_2 = \gamma[t_1 - \frac{\beta}{c}x_1] \quad \text{or} \quad ct_2 = \gamma[ct_1 - \beta x_1]
\tag{20.43}
$$

$$
x_2 = \gamma(x_1 - c\beta t_1) = \gamma(x_1 - \beta ct_1)
\tag{20.44}
$$

$$
y_2 = y_1
\tag{20.45}
$$

$$
z_2 = z_1
\tag{20.46}
$$

Putting in a matrix form, we obtain

$$
\begin{bmatrix} ct_2 \\ x_2 \\ y_2 \\ z_2 \end{bmatrix} =
\begin{bmatrix} \gamma & -\gamma\beta & 0 & 0 \\ -\gamma\beta & \gamma & 0 & 0 \\ 0 & 0 & 1 & 0 \\ 0 & 0 & 0 & 1 \end{bmatrix}
\begin{bmatrix} ct_1 \\ x_1 \\ y_1 \\ z_1 \end{bmatrix}
\tag{20.47}
$$

where

$$
\gamma = \frac{1}{\sqrt{1 - \beta^2}}
$$

$$
\beta = \frac{v}{c}
$$

The matrix in equation (20.47) can be written as a type of rotation matrix through an angle $\phi$ which satisfies

$$\tanh\phi = \frac{\sinh\phi}{\cosh\phi} = \frac{v}{c}$$

$$\sinh^2\phi = (\frac{v}{c})^2 \cosh^2\phi$$

Noting that

$$\cosh^2\phi - \sinh^2\phi = 1$$

$$\cosh^2\phi - (\frac{v}{c})^2\cosh^2\phi = 1$$

$$\cosh^2\phi\left(1 - \frac{v^2}{c^2}\right) = 1$$

$$\cosh\phi = \frac{1}{\sqrt{1 - \frac{v^2}{c^2}}} = \gamma$$

$$\text{and}\quad \sinh\phi = \frac{v}{c}\gamma = \beta\gamma, \quad\text{where}\quad \beta = \frac{v}{c}$$

Hence, the matrix in equation (20.47) becomes

$$\begin{bmatrix} \cosh\phi & -\sinh\phi & 0 & 0 \\ -\sinh\phi & \cosh\phi & 0 & 0 \\ 0 & 0 & 1 & 0 \\ 0 & 0 & 0 & 1 \end{bmatrix} \qquad (20.48)$$

### 20.2.3   Length Contraction

Consider a fixed frame $S_1$ and a frame moving with velocity $v$. Suppose there is a rod of length $L$ whose ends have coordinates $x_1$ and $x_1'$ in $S_1$ and have coordinates $x_2$ and $x_2'$ in $S_2$. These are related as

$$x_2' = \gamma(x_1' - vt_1)$$

$$x_2 = \gamma(x_1 - vt_1)$$

$$\text{where}\quad \gamma = \frac{1}{\sqrt{1 - v^2/c^2}}$$

Hence

$$x_2' - x_2 = \gamma(x_1' - vt_1) - \gamma(x_1 - vt_1)$$

$$= \gamma(x_1' - x_1)$$

$$\text{Noting}\quad L = x_1' - x_1)$$

$$L_0 = x_2' - x_2$$

Hence

$$L_0 = \gamma L$$

$$L = \frac{L_0}{\gamma} = L_0\sqrt{1 - v^2/c^2} \qquad (20.49)$$

which implies that since $v < c$, always $L < L_0$ which is *Length contraction*.

### 20.2.4 Time Dilation

Consider a fixed frame $S_1$ and a frame moving with velocity $v$. Suppose there is an event timed during a duration $T$ between $t_1$ and $t_1'$ in $S_1$ and is between $t_2$ and $t_2'$ in $S_2$. These are related as

$$t_2' = \gamma(t_1' - \frac{\beta}{c}x_1)$$

$$t_2 = \gamma(t_1 - \frac{\beta}{c}x_1)$$

$$t_1' = \gamma(t_2' + \frac{\beta}{c}x_2')$$

$$t_1 = \gamma(t_2 + \frac{\beta}{c}x_2)$$

where $\quad \beta = \frac{v}{c}$

Hence

$$t_1' - t_1 = \gamma(t_2' + \frac{\beta}{c}x_2') - \gamma(t_2 + \frac{\beta}{c}x_2)$$

$$= \gamma(t_2' - t_2)$$

Noting $\quad T = t_1' - t_1)$

$$T_0 = t_2' - t_2$$

Hence

$$T = \gamma T_0 = \frac{T_0}{\sqrt{1 - v^2/c^2}} \tag{20.50}$$

which implies that since $v < c$, always $T > T_0$ which is *Time dilation*.

### 20.2.5 The Lorentz Velocity Transformations

If we differentiate the Lorentz coordinate transformations given in equations 20.35 to 20.38, we obtain

$$dx_2 = \gamma(dx_1 - vdt_1) = \frac{x_1 - vt_1}{\sqrt{1 - \frac{v^2}{c^2}}}$$

$$dy_2 = dy_1$$

$$dz_2 = dz_1$$

$$dt_2 = \gamma[dt_1 - \frac{\beta}{c}dx_1]$$

Dividing $dx_2, dy_2$ and $dz_2$ by $dt_2$, we obtain the *Lorentz Velocity Transformations* given by

$$v_{2x} = \frac{v_{1x} - v}{1 - (\beta/c)v_{1x}} = \frac{v_{1x} - v}{1 - v \cdot v_{1x}/c^2} \tag{20.51}$$

$$v_{2y} = \frac{v_{1y}}{\gamma\left[1-(\beta/c)v_{1x}\right]} = \frac{v_{1y}\sqrt{1-v^2/c^2}}{1-v\cdot v_{1x}/c^2} \qquad (20.52)$$

$$v_{2z} = \frac{v_{1z}}{\gamma\left[1-(\beta/c)v_{1x}\right]} = \frac{v_{1z}\sqrt{1-v^2/c^2}}{1-v\cdot v_{1x}/c^2} \qquad (20.53)$$

and the *Inverse Lorentz Velocity Transformations* given by

$$v_{1x} = \frac{v_{2x}+v}{1+(\beta/c)v_{2x}} = \frac{v_{2x}+v}{1+v\cdot v_{2x}/c^2} \qquad (20.54)$$

$$v_{1y} = \frac{v_{2y}}{\gamma\left[1+(\beta/c)v_{2x}\right]} = \frac{v_{2y}\sqrt{1-v^2/c^2}}{1+v\cdot v_{2x}/c^2} \qquad (20.55)$$

$$v_{1z} = \frac{v_{2z}}{\gamma\left[1+(\beta/c)v_{2x}\right]} = \frac{v_{2z}\sqrt{1-v^2/c^2}}{1+v\cdot v_{2x}/c^2} \qquad (20.56)$$

## 20.3   General Theory of Relativity

In the General Theory of Relativity, the following assumption is made:

*Principle of Equivalence*: No experiment can distinguish between a uniform gravitational field and an equivalent uniform acceleration.

### 20.3.1   Einstein's Field Equation for General Relativity

Einstein developed the field equations for General Relativity in 1915 given by a deceptively simple equation which hides a lot of complexity

$$G_{\mu\nu} = \frac{8\pi G}{c^4}T_{\mu\nu} \qquad (20.57)$$

where

$$
\begin{array}{ll}
G_{\mu\nu} & \text{is the Einstein Tensor} \\
T_{\mu\nu} & \text{is the Stress-Energy-Momentum Tensor} \\
G & \text{is the Newton's Gravitational constant} \\
c & \text{is the velocity of light}
\end{array}
$$

The Einstein Tensor $G_{\mu\nu}$ is given by

$$G_{\mu\nu} = R_{\mu\nu} - \frac{1}{2}g_{\mu\nu}R \qquad (20.58)$$

which means that the Einstein Field Equation can also be written as

$$R_{\mu\nu} - \frac{1}{2}g_{\mu\nu}R = \frac{8\pi G}{c^4}T_{\mu\nu} \qquad (20.59)$$

where

$$R_{\mu\nu} \qquad \text{is the Ricci Curvature Tensor}$$
$$g_{\mu\nu} \qquad \text{is the Metric Tensor}$$
$$R \qquad \text{is the Curvature scalar}$$

The Ricci Tensor, $R_{\mu\nu}$, is defined as

$$R_{\mu\nu} = \frac{\partial \Gamma^{\lambda}_{\mu\lambda}}{\partial x^{\nu}} - \frac{\partial \Gamma^{\lambda}_{\mu\nu}}{\partial x^{\lambda}} + \Gamma^{\beta}_{\mu\lambda}\Gamma^{\lambda}_{\nu\beta} - \Gamma^{\beta}_{\mu\nu}\Gamma^{\lambda}_{\beta\lambda} \tag{20.60}$$

where $\Gamma^{\lambda}_{\mu\nu}$ is the Christoffel symbol given by

$$\Gamma^{\lambda}_{\mu\nu} = \frac{1}{2}g^{\lambda\beta}\left(\frac{\partial g_{\mu\beta}}{\partial x^{\nu}} + \frac{\partial g_{\nu\beta}}{\partial x^{\mu}} - \frac{\partial g_{\mu\nu}}{\partial x^{\beta}}\right) \tag{20.61}$$

The Einstein Field Equation were further generalized to include the cosmological constant $\Lambda$ to be in the form

$$R_{\mu\nu} - \frac{1}{2}g_{\mu\nu}R + g_{\mu\nu}\Lambda = \frac{8\pi G}{c^4}T_{\mu\nu} \tag{20.62}$$

The value of the cosmological constant is small, and is important when one studies cosmology. For this reason, we shall not consider this further since cosmology is beyond the scope of this book. The reader interested in cosmology is referred to texts on cosmology which deal with this subject.

The complexity of the Einstein field equations is simplified by an observation of John Wheeler who said;

"Spacetime tells matter how to move,

Matter tells spacetime how to curve"

Note that since each of $\mu$ and $\nu$ can take four values, there are $4 \times 4 = 16$ equations arising from equation (20.59), with $R_{\mu\nu}, g_{\mu\nu}$ and $T_{\mu\nu}$ of the form:

$$R_{\mu\nu} = \begin{bmatrix} R_{00} & R_{01} & R_{02} & R_{03} \\ R_{10} & R_{11} & R_{12} & R_{13} \\ R_{20} & R_{21} & R_{22} & R_{23} \\ R_{30} & R_{31} & R_{32} & R_{33} \end{bmatrix} \tag{20.63}$$

$$g_{\mu\nu} = \begin{bmatrix} g_{00} & g_{01} & g_{02} & g_{03} \\ g_{10} & g_{11} & g_{12} & g_{13} \\ g_{20} & g_{21} & g_{22} & g_{23} \\ g_{30} & g_{31} & g_{32} & g_{33} \end{bmatrix} \tag{20.64}$$

$$T_{\mu\nu} = \begin{bmatrix} T_{00} & T_{01} & T_{02} & T_{03} \\ T_{10} & T_{11} & T_{12} & T_{13} \\ T_{20} & T_{21} & T_{22} & T_{23} \\ T_{30} & T_{31} & T_{32} & T_{33} \end{bmatrix} \tag{20.65}$$

where $T_{00}$ is related to energy density, $T_{01}, T_{02}, T_{03}$ are related to energy flux, $T_{10}, T_{20}, T_{30}$ are related to momentum density, $T_{11}, T_{22}, T_{33}$ are related to pressure, $T_{12}, T_{13}, T_{23}, T_{21}, T_{31}, T_{32}$ are related to shear stress.

The complexity of the tensors is, however, slightly reduced because the matrices are symmetrical, that is, for $\mu = i$ and $\nu = j$ where $i = 1, 2, 3$, elements with $\mu\nu = \nu\mu$ are equal, and there are six of these, hence the total number of equations reduce to 10.

$$R_{00} - \frac{1}{2}g_{00}R = \frac{8\pi G}{c^4}T_{00} \tag{20.66}$$

$$R_{01} - \frac{1}{2}g_{01}R = \frac{8\pi G}{c^4}T_{01} \tag{20.67}$$

$$R_{02} - \frac{1}{2}g_{02}R = \frac{8\pi G}{c^4}T_{02} \tag{20.68}$$

$$R_{03} - \frac{1}{2}g_{03}R = \frac{8\pi G}{c^4}T_{03} \tag{20.69}$$

$$R_{11} - \frac{1}{2}g_{11}R = \frac{8\pi G}{c^4}T_{11} \tag{20.70}$$

$$R_{12} - \frac{1}{2}g_{12}R = \frac{8\pi G}{c^4}T_{12} \tag{20.71}$$

$$R_{13} - \frac{1}{2}g_{13}R = \frac{8\pi G}{c^4}T_{13} \tag{20.72}$$

$$R_{22} - \frac{1}{2}g_{22}R = \frac{8\pi G}{c^4}T_{22} \tag{20.73}$$

$$R_{23} - \frac{1}{2}g_{23}R = \frac{8\pi G}{c^4}T_{23} \tag{20.74}$$

$$R_{33} - \frac{1}{2}g_{00}R = \frac{8\pi G}{c^4}T_{33} \tag{20.75}$$

It can be noted that the Einstein's equations are essentially non linear second order differential equations for the metric tensor, as can be seen from equations (20.60) and (20.61), where the Ricci tensor is a derivative of the Christoffel symbol which in turn is a derivative of the metric tensor, and the other terms in the Ricci tensor enter as second order in the Christoffel symbol. These differential equations are extremely complicated, and generally it is necessary to make some approximations. This is beyond the level for which this text is intended and the reader will have to refer to specialized texts to solve these equations. We shall, however, study few solutions of special interest.

### 20.3.2   Schwarzschild solution of the Einstein's Field Equations

In the Schwarzschild solution of the Einstein's Field Equations, one considers the gravitational field due to an isolated sphere of mass $M$, and take $T_{\mu\nu} = 0$. It is convenient to use spherical coordinates, with relations between $(x, y, z)$ and $(\rho, \phi, \theta)$ given by

$$x = \rho \sin\phi \cos\theta$$

$$y = \rho \sin\phi \sin\theta$$

$$z = \rho \cos\theta$$

Hence we have

$$
\begin{aligned}
ds^2 &= -dt^2 + dx^2 + dy^2 + dz^2 \\
&= -d\tau^2 \\
d\tau^2 &= dt^2 - dx^2 - dy^2 - dz^2
\end{aligned}
$$

After straight forward algebra, we obtain

$$
ds^2 = -dt^2 + d\rho^2 + \rho^2 d\phi^2 + \rho^2 \sin^2 \phi d\theta^2 \tag{20.76}
$$

If $\rho$ and $t$ are constant, we have

$$
\begin{aligned}
dl^2 &= \rho^2 (d\phi^2 + \sin^2 \phi d\theta^2) \\
&= f(\rho, l)[d\phi^2 + \sin^2 \phi d\theta^2]
\end{aligned}
$$

Let $f(\rho, t) = r$ and noting that

$$
\begin{aligned}
(x^0, x^1, x^2, x^3) &\rightarrow (t, r, \phi, \theta) \\
ds^2 &= g_{\mu\nu} dx^\mu dx^\nu
\end{aligned}
$$

where

$$
g_{\mu\nu} =
\begin{bmatrix}
g_{00} & g_{01} & g_{02} & g_{03} \\
g_{10} & g_{11} & g_{12} & g_{13} \\
g_{20} & g_{21} & g_{22} & g_{23} \\
g_{30} & g_{31} & g_{32} & g_{33}
\end{bmatrix}
=
\begin{bmatrix}
g_{tt} & g_{tr} & g_{t\phi} & g_{t\theta} \\
g_{rt} & g_{rr} & g_{r\phi} & g_{r\theta} \\
g_{\phi t} & g_{\phi r} & g_{\phi\phi} & g_{\phi\theta} \\
g_{\theta t} & g_{\theta r} & g_{\theta\phi} & g_{\theta\theta}
\end{bmatrix}
\tag{20.77}
$$

and $g_{\mu\nu}$ reduces to

$$
g_{\mu\nu} =
\begin{bmatrix}
g_{tt} & 0 & 0 & 0 \\
0 & g_{rr} & 0 & 0 \\
0 & 0 & g_{\phi\phi} & 0 \\
0 & 0 & 0 & g_{\theta\theta}
\end{bmatrix}
\tag{20.78}
$$

$$
ds^2 = g_{tt} dt^2 + g_{rr} dr^2 + g_{\phi\phi} d\phi^2 + g_{\theta\theta} d\theta^2 \tag{20.79}
$$

Define so that

$$
ds^2 = -dt^2 + d\rho^2 + \rho^2 d\phi^2 + \rho^2 \sin^2 \phi d\theta^2
$$

becomes

$$
ds^2 = -A(r) dt^2 + B(r) dr^2 + r^2 d\phi^2 + r^2 \sin^2 \phi d\theta^2 \tag{20.80}
$$

so that

$$
\begin{aligned}
A(r) &= e^{2m(r)} = e^{2m} \\
B(r) &= e^{2n(r)} = e^{2n}
\end{aligned}
$$

and hence we have

$$
ds^2 = -e^{2m} dt^2 + e^{2n} dr^2 + r^2 d\phi^2 + r^2 \sin^2 \phi d\theta^2 \tag{20.81}
$$

which implies

$$g_{\mu\nu} = \begin{bmatrix} -e^{2m} & 0 & 0 & 0 \\ 0 & -e^{2n} & 0 & 0 \\ 0 & 0 & r^2 & 0 \\ 0 & 0 & 0 & r^2\sin^2\phi \end{bmatrix} \qquad (20.82)$$

$$det(g_{ij} = -e^{2m+2n}r^4\sin^2\phi \qquad (20.83)$$

After long but tedious algebra, using definitions for the Ricci Tensor, $R_{\mu\nu}$,

$$R_{\mu\nu} = \frac{\partial\Gamma^\lambda_{\mu\lambda}}{\partial x^\nu} - \frac{\partial\Gamma^\lambda_{\mu\nu}}{\partial x^\lambda} + \Gamma^\beta_{\mu\lambda}\Gamma^\lambda_{\nu\beta} - \Gamma^\beta_{\mu\nu}\Gamma^\lambda_{\beta\lambda}$$

where $\Gamma^\lambda_{\mu\nu}$ is the Christoffel symbol

$$\Gamma^\lambda_{\mu\nu} = \frac{1}{2}g^{\lambda\beta}\left(\frac{\partial g_{\mu\beta}}{\partial x^\nu} + \frac{\partial g_{\nu\beta}}{\partial x^\mu} - \frac{\partial g_{\mu\nu}}{\partial x^\beta}\right)$$

one obtains the following relations

$$\begin{aligned} m &= -n \\ e^{2m} &= 1 + \frac{\alpha}{r} \\ e^{2n} &= e^{-2n} = \frac{1}{e^{2m}} = \frac{1}{\left(1 + \frac{\alpha}{r}\right)} \\ \text{where} \quad \alpha &= -2M \\ e^{2m} &= 1 - \frac{2M}{r} \\ e^{2n} &= \frac{1}{1 - \frac{2M}{r}} \end{aligned}$$

and hence the solution is

$$ds^2 = -\left(1 - \frac{2M}{r}\right)dt^2 + \frac{dr^2}{\left(1 - \frac{2M}{r}\right)} + r^2d\phi^2 + r^2\sin^2\phi d\theta^2 \qquad (20.84)$$

which is known as *Schwarzchild metric*, first published by Karl Schwartzchild in 1916. If we define

$$r^2d\phi^2 + r^2\sin^2\phi d\theta^2 = r*2(d\phi^2 + \sin^2\phi d\theta^2) = r^2d\Omega^2$$

and introduce units for which $G \neq 1$ and $c \neq 1$, the Schwarzchild metric can be written as

$$ds^2 = -\left(1 - \frac{2MG}{c^2r}\right)dt^2 + \frac{dr^2}{\left(1 - \frac{2MG}{c^2r}\right)} + \frac{r^2d\Omega^2}{c^2} \qquad (20.85)$$

$$d\tau^2 = \left(1 - \frac{2MG}{c^2r}\right)dt^2 - \frac{dr^2}{\left(1 - \frac{2MG}{c^2r}\right)} - \frac{r^2d\Omega^2}{c^2} \qquad (20.86)$$

### 20.3.3 Do Einstein's field equations reduce to Newton's theory of gravitation in certain limits?

In this section, we wish to answer the question: Do Einstein's field equations reduce to Newton's theory of gravitation in certain limits?

Let us review some of the well known results of Newton's theory of gravitation.

1. *First*, Recall Newton's law of gravitation between masses $M$ and $m$ at a distance $r$ apart is given by

$$F = \frac{GMm}{r^2} \tag{20.87}$$

where $G = 6.67 \times 10^{-11} \ Nm^2 kg^{-2}$ is the Newton's Gravitational constant.

2. *Secondly*, The gravitational field strength due to a mass $M$ at a distance $r$ is the force per unit mass, given by

$$\frac{\mid F \mid}{m} = \frac{GM}{r} \tag{20.88}$$

3. *Thirdly*, The gravitational potential energy, $U$, due to masses $M$ and $m$ separated by a distance $r$ is given by

$$U = -\frac{GMm}{r} \tag{20.89}$$

4. *Fourthly*, The gravitational potential, $\phi(r)$, due to masses $M$ at a point distance $r$ away is given by

$$\phi(r) = -\frac{GM}{r} \tag{20.90}$$

5. *Fifthly*, The force $F$ and gravitational potential, $\phi(r)$ are related by

$$\vec{F} = -\nabla\phi(r) \tag{20.91}$$

where

$$\nabla = \mathbf{i}\frac{\partial}{\partial \mathbf{x}} + \mathbf{j}\frac{\partial}{\partial \mathbf{y}} + \mathbf{k}\frac{\partial}{\partial \mathbf{z}} \tag{20.92}$$

Below, we derive a relation between the gravitational potential, $\phi$, and mass density $\rho$. Consider a sphere, and pick a spherical element of radius $s$ with an element $ds$ such that the spherical element has mass, $dM$. The mass density is given by

$$\rho = \frac{dM}{4\pi s^2 ds} \tag{20.93}$$

The potential at P, $\phi_P$ is given by

$$\phi_P = -\int_0^r \frac{GdM}{s} ds$$

$$= -\int_0^r \frac{G4\pi\rho s^2}{s} ds$$

$$= -4\pi G\rho \int_0^r s\, ds$$

$$= -4\pi G\rho \frac{s^2}{2} \Big|_0^r$$

$$= -2\pi G\rho(r^2 - 0)$$

$$\phi_P = -2\pi G\rho r^2 \qquad (20.94)$$

We can also find the potential at $q$ given by

$$\phi_q = -\frac{GM}{r}$$

$$= -\frac{G\left(\frac{4}{3}\pi r^3 \rho\right)}{r}$$

$$\phi_q = -\frac{4\pi G\rho r^2}{3} \qquad (20.95)$$

$$\text{Hence,} \quad \phi(r) = \phi_q - \phi_P$$

$$= -\frac{4\pi G\rho r^2}{3} - (-2\pi G\rho r^2)$$

$$= \pi G\rho r^2\left(-\frac{4}{3} + 2\right)$$

$$\phi(r) = \frac{2}{3}\pi G\rho r^2 = \frac{2}{3}\pi G\rho(x^2 + y^2 + z^2) \qquad (20.96)$$

$$\frac{\partial\phi}{\partial x} = \frac{\partial}{\partial x}\left\{\frac{2}{3}\pi G\rho(x^2 + y^2 + z^2)\right\}$$

$$= \frac{2}{3}\pi G\rho 2x$$

$$\frac{\partial\phi}{\partial x} = \frac{4\pi G\rho x}{3}$$

$$\text{Hence,} \quad \frac{\partial^2\phi}{\partial x^2} = \frac{\partial}{\partial x}\left\{\frac{4\pi G\rho x}{3}\right\} = \frac{4\pi G\rho}{3} \qquad (20.97)$$

Similarly

$$\frac{\partial^2\phi}{\partial y^2} = \frac{4\pi G\rho}{3}$$

$$\frac{\partial^2\phi}{\partial z^2} = \frac{4\pi G\rho}{3}$$

Hence,

$$\frac{\partial^2\phi}{\partial x^2} + \frac{\partial^2\phi}{\partial y^2} + \frac{\partial^2\phi}{\partial z^2} = \frac{4\pi G\rho}{3} + \frac{4\pi G\rho}{3} + \frac{4\pi G\rho}{3}$$

$$\nabla^2\phi = 4\pi G\rho \qquad (20.98)$$

which is known as Poisson's equation.

In Einstein's general relativity, we have

$$
\begin{aligned}
g_{00} &= 1 - \frac{2MG}{r} \\
&= 1 + 2\phi \\
\text{where} \quad \phi &= -\frac{2MG}{r} \\
\text{Hence} \quad \nabla^2 g_{00} &= 8\pi G\rho \\
\nabla^2 g_{00} &= 8\pi G T_{00}
\end{aligned}
\tag{20.99}
$$

which is a tensor form from Einstein's field equations of general relativity corresponding to Poisson's equation (20.98) in Newton's gravitational theory.

### 20.3.4 Gravitational bending of light

The Sun's gravity bends light towards it so that light from a star as viewed from the earth appears to be deflected giving rise to the effect of gravitational lensing as illustrated in Figure 20.4.

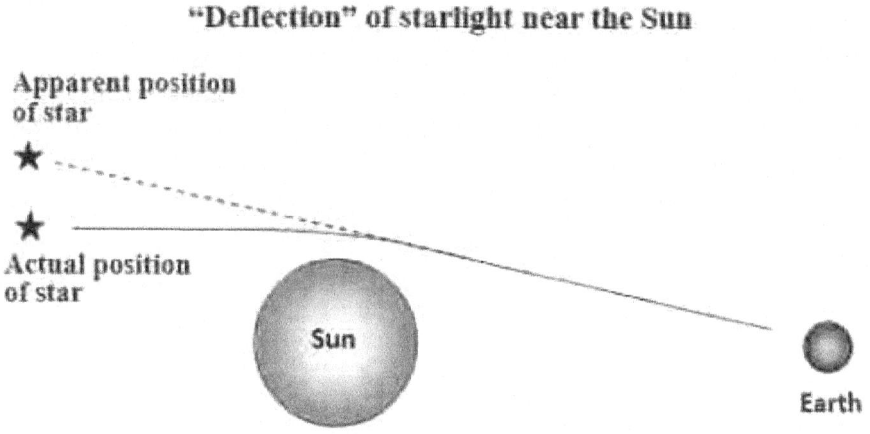

Figure 20.4: Gravitational bending of light

### 20.3.5 Gravitational waves

Gravitational waves are ripples in the curvature of spacetime propagating as waves at the speed of light.Experiments for the detection of gravitational waves are performed using a network of LIGO (Laser Interferometer Gravitational-Wave Observatory) detectors. , infrastructure that has been established in Hanford and Livingston in the USA, and in India. The basic structure is illustrated Figure 20.5.

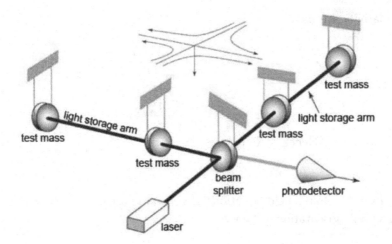

Figure 20.5: LIGO set up for observation of Gravitational waves (Source: www.ligo.org).

## 20.3.6   Global Positioning System (GPS)

The Global Positioning System (GPS) consists of 24 satellites which circle the earth at an altitude of approximately 20,000km (or 12,550 miles) above the earth's surface, spaced such that at any location four satellites are in the range. User devices such as cellphones receive signals from the satellites so that locations can be obtained. GPS woulf fail if corrections due to corrections arising from the Special Theory and General theory were not included. The GPS consists of three major segments: the Space segment (SS), the Control segment (CS) and the User segment US), as illustrated in Figure 20.6.

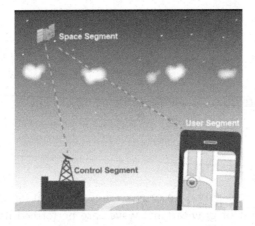

Figure 20.6: Segments of GPS: SS, CS and US.

### 20.3.7 Black holes

A black hole is a region of spacetime in which the gravitational efects are so strong that nothing can escape from it, including particles and light. Black holes have been studied for many years as long back as two hundred years ago, long before the term *"black hole"* was popularized by John Wheeler around 1969. There is still a lot to be learnt about them. In this book, we can only discuss briefly about this subject. Actually, the idea is not that complicated, starting from the concept of "escape velocity". It is well known that the escape velocity, $v_e$, is a minimum velocity required to leave a sphere of mass $M$ and radius $R$, given by

$$v_e = \sqrt{\frac{2GM}{R}} \qquad (20.100)$$

Values for the escape velocity for the Earth can be easily calculated and is $11.2 \ km/s$, while that of the Moon is $2.4 \ km/s$. Now, obviously, the Earth is not a black hole. However, imagine if the mass $M$ of the Earth is compressed into a small volume of the size of a marble of diameter 1.0 cm or radius 0.5 cm. It is clear that the escape velocity would be $4.0 \times 10^8 \ ms^{-1}$, which is greater than the velocity of light, and hence nothing could escape from the marble, in which case it would be a black hole!

The same result is obtained from the the Schwarzschild solution of the Einstein's field equations of general relativity which can be written in the form:

$$ds^2 = -\left(1 - \frac{r_s}{r}\right)dt^2 + \frac{dr^2}{\left(1 - \frac{r_s}{r}\right)c^2} + \frac{r^2 d\Omega^2}{c^2} \qquad (20.101)$$

$$d\tau^2 = \left(1 - \frac{r_s}{r}\right)dt^2 - \frac{dr^2}{\left(1 - \frac{r_s}{r}\right)c^2} - \frac{r^2 d\Omega^2}{c^2} \qquad (20.102)$$

$$\text{where} \qquad r_s = \frac{2MG}{c^2} \quad \text{is known as the } \textit{Schwarzschild Radius} \qquad (20.103)$$

A simple illustration of a black hole is shown in Figure 20.7. At the centre of he black hole, there is a *Singularity*. There is a region where nothing can escape known as the *Event Horizon* with a radius known as the *Schwarzschild Radius*, and within this region the escape velocity is equal to the velocity of light. The*Accretion Disk* is a disk consisting of material spiralling towards the black hole.

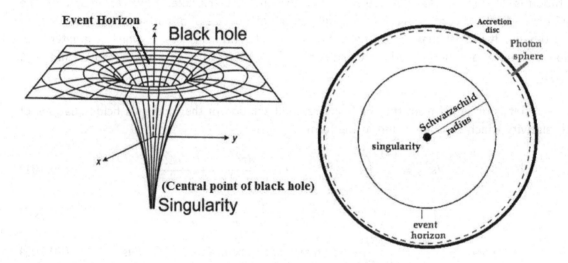

Figure 20.7: An illustration of the structure of a black hole.

## 20.4  Exercises

**20.1**. (a)State the Galilean coordinate transformation equations between $(x_2, y_2, z_2)$ at time $t_2$, and $(x_1, y_1, z_1)$ at time $t_1$.

(b) Show that the Galilean coordinate transformations can be written in a matrix form as

$$
\begin{bmatrix} x_2 \\ y_2 \\ z_2 \\ ict_2 \end{bmatrix} = \begin{bmatrix} 1 & 0 & 0 & i\beta \\ 0 & 1 & 0 & 0 \\ 0 & 0 & 1 & 0 \\ 0 & 0 & 0 & 1 \end{bmatrix} \begin{bmatrix} x_1 \\ y_1 \\ z_1 \\ ict_1 \end{bmatrix}
$$

where $\beta = v/c$.

**20.2**. (a) Find the $4 \times 4$ matrix for the inverse Galilean transformation to the one in problem 20.1.

(b) Show that the product of the two $4 \times 4$ matrices from problem 20.1 and Part (a) of this problem gives the identity matrix,

**20.3**. (a) State the Lorentz coordinate transformation equations between $(t_2, x_2, y_2, z_2)$ and $(t_1, x_1, y_1, z_1)$.

(b) Show that the Lorentz coordinate transformations can be written in a matrix form as

$$
\begin{bmatrix} ct_2 \\ x_2 \\ y_2 \\ z_2 \end{bmatrix} = \begin{bmatrix} \gamma & -\gamma\beta & 0 & 0 \\ -\gamma\beta & \gamma & 0 & 0 \\ 0 & 0 & 1 & 0 \\ 0 & 0 & 0 & 1 \end{bmatrix} \begin{bmatrix} ct_1 \\ x_1 \\ y_1 \\ z_1 \end{bmatrix}
$$

where

$$
\gamma = \frac{1}{\sqrt{1 - \beta^2}}
$$
$$
\beta = \frac{v}{c}
$$

**20.4**. (a) Find the $4 \times 4$ matrix for the inverse Lorentz transformation to the one in problem 20.3.

(b) Show that the product of the two $4 \times 4$ matrices from problem 20.3 and Part (a) of this problem gives the identity matrix,

**20.5** A pole-vaulter runs at the rate of $\frac{\sqrt{3}}{2}$ m/s and carries a pole that is 20 m long in his reference frame, given that the length of the pole in the rest frame is 20 m. He approaches a barn that is open at both ends and is 10 m long, as measured by a ground observer.
(i) To the ground oberver, will the pole fit inside the barn?
(ii) What is the length of the barn to the runner? What is the pole-vaulter's conclusion?

**20.6** Consider the following twin paradox. One of the twins embarks on a journey into outer space, taking one year earth time to accelerate to $\frac{3}{4}c$, then spends the next 20 years cruising to reach a galaxy 15 light years away. An additional year is spent decelerating in order to to explore one of its solar systems. After one year of exploration ($\beta = 0$), the twin returns to earth by the same schedule - one year of acceleration, 20 years cruising, and one year of deceleration.
(i) How ols is the twin who went to space?
(ii) How old is the twin who remained on earth?
(iii) What is the difference in the ages of the twins after the journey has ended?

**20.7.** Given Maxwell's equations in the form: $\nabla \cdot \mathbf{B} = 0$, $\nabla \wedge \mathbf{E} = -\frac{\partial \mathbf{B}}{\partial t}$ and $\nabla \cdot \mathbf{E} = 0$, $\nabla \wedge \mathbf{B} = \frac{1}{c^2} \frac{\partial \mathbf{E}}{\partial t}$, show that Maxwell's equations lead to two matrices, $\mathbf{F}$ and $\tilde{\mathbf{F}}$ given by

$$
\mathbf{F} = \begin{bmatrix}
0 & -B_x & -B_y & -B_z \\
B_x & 0 & \frac{E_z}{c} & -\frac{E_y}{c} \\
B_y & -\frac{E_z}{c} & 0 & \frac{E_x}{c} \\
B_z & \frac{E_y}{c} & -\frac{E_x}{c} & 0
\end{bmatrix}
$$

$$
\tilde{\mathbf{F}} = \begin{bmatrix}
0 & \frac{E_x}{c} & \frac{E_y}{c} & \frac{E_z}{c} \\
-\frac{E_x}{c} & 0 & B_z & -B_y \\
-\frac{E_y}{c} & -B_z & 0 & B_x \\
-\frac{E_z}{c} & B_y & -B_x & 0
\end{bmatrix}
$$

**20.8.** The Einstein field equations for General Relativity are given as

$$
G_{\mu\nu} = \frac{8\pi G}{c^4} T_{\mu\nu}
$$

$$
\text{where } G_{\mu\nu} = R_{\mu\nu} - \frac{1}{2} g_{\mu\nu} R
$$

(a) Explain all the terms: $G_{\mu\nu}, T_{\mu\nu}, G, c, R_{\mu\nu}, g_{\mu\nu}$ and $R$.
(b) Do Einstein's field equations reduce to Newton's theory of gravitation in certain limits?

**20.9.** Show that the Schwarzschild solution of the Einstein's field equations of general relativity is given as

$$
ds^2 = -\left(1 - \frac{2MG}{c^2 r}\right) dt^2 + \frac{dr^2}{\left(1 - \frac{2MG}{c^2 r}\right)} + \frac{r^2 d\Omega^2}{c^2}
$$

$$
d\tau^2 = \left(1 - \frac{2MG}{c^2 r}\right) dt^2 - \frac{dr^2}{\left(1 - \frac{2MG}{c^2 r}\right)} - \frac{r^2 d\Omega^2}{c^2}
$$

**20.10**. Explain the following with respect to general relativity.

(a) Bending of light by gravity.

(b) Gravitational waves, and describe the set up of LIGO (Laser Interferometer Gravitational-Wave Observatory) detector.

(c) Black hole.

# Appendix A

# Some Useful Constants etc

*Table A.1: Some Fundamental Physical Constants*

| Name | Symbol | Value | Units |
|------|--------|-------|-------|
| Acceleration due to gravity | $g$ | $9.807^\dagger$ | $ms^{-2}$ |
| Universal gravitational constant | $G$ | $6.673 \times 10^{-11}$ | $Nm^2 kg^{-2}$ |
| | | | |
| Electronic charge | $e$ | $1.602 \times 10^{-19}$ | $C$ |
| Rest mass of electron | $m_e$ | $9.109 \times 10^{-31}$ | $kg$ |
| Rest mass of proton | $m_p$ | $1.673 \times 10^{-27}$ | $kg$ |
| Rest mass of neutron | $m_n$ | $1.675 \times 10^{-27}$ | $kg$ |
| | | | |
| Mass of hydrogen atom | $m_H$ | $1.673 \times 10^{-27}$ | $kg$ |
| Bohr radius | $a_o$ | $0.529 \times 10^{-10}$ | $m$ |
| Binding energy of hydrogen | $E_o$ | $13.606$ | $eV$ |
| | | | |
| Speed of light in vacuum | $c$ | $2.998 \times 10^8$ | $ms^{-1}$ |
| Permittivity of free space | $\epsilon_o = \frac{1}{\mu_o c^2}$ | $8.854 \times 10^{-12}$ | $C^2 (Nm^2)^{-1}$ |
| Permeability of free space | $\mu_o$ | $4\pi \times 10^{-7}$ | $NA^{-2}$ |
| Coulomb's constant for free space | $k = \frac{1}{4\pi\epsilon_o}$ | $8.988 \times 10^9$ | $Nm^2 C^{-2}$ |
| | | | |
| Planck's constant | $h$ | $6.626 \times 10^{-34}$ | $Js$ |
| | $\hbar = \frac{h}{2\pi}$ | $1.055 \times 10^{-34}$ | $Js$ |
| Stefan-Boltzmann constant | $\sigma$ | $5.67 \times 10^{-8}$ | $Wm^{-2} K^{-4}$ |
| Boltzmann's constant | $k = \frac{R}{N_A}$ | $1.381 \times 10^{-23}$ | $JK^{-1}$ |
| Avogadro's number | $N_A$ | $6.022 \times 10^{26}$ | $(kmol)^{-1}$ |
| Universal gas constant | $R$ | $8.315$ | $J(mol-K)^{-1}$ |
| | | | |
| Standard temperature | $T$ | $273.15$ | $K$ |
| Standard pressure | $P$ | $1$ | $atmosphere$ |

$\dagger$: Average value at sea level; decreases to $9.598\ ms^{-2}$ at an altitude of $100,000\ m$. Varies with latitude from $9.780\ ms^{-2}$ at the equator to $9.832\ ms^{-2}$ at the poles.

*Table A.2: Some common physical properties and parameters*

| Name | Symbol | Value | Units |
|------|--------|-------|-------|
| Volume of ideal gas at STP | $V$ | $22.415 \times 10^3$ | $l\,(kmol)^{-1}$ |
| Density of air at STP | $\rho_{air}$ | 1.293 | $kgm^{-3}$ |
| Average molecular weight of air | $m_{air}$ | 28.970 | $kg(kmol)^{-1}$ |
| Density of water at $4^oC$ | $\rho_w$ | 1000 | $kgm^{-3}$ |
| Specific heat capacity of water | $C_w$ | 4200 | $Jkg^{-1}$ |
| Latent heat of fusion of ice | $L_f$ | $3.360 \times 10^5$ | $Jkg^{-1}$ |
| Latent heat of vaporization of water | $L_v$ | $2.260 \times 10^6$ | $Jkg^{-1}$ |
| Speed of sound in air at STP | $v_{s,air}$ | 331 | $ms^{-1}$ |
| Absolute zero (0 K) | $K$ | -273.15 | $^oC$ |

*Table A.3: Earth, Sun and Moon data*

| | | |
|---|---|---|
| Earth | Mass | $5.98 \times 10^{24} kg$ |
| | Mean radius | $6.378 \times 10^6 m$ |
| | Period of revolution (sun) | $1y = 365d\ 5h\ 48m\ 45.97s$ |
| | Period of rotation (axial) | $86164.1s = 23h\ 56m\ 4.1s$ |
| | Mean Earth - sun distance | $1.496 \times 10^{11} m$ |
| | Mean Earth - moon distance | $3.844 \times 10^8 m$ |
| Moon | Mass | $7.35 \times 10^{22} kg$ |
| | Mean radius | $1.738 \times 10^6 m$ |
| | Period of revolution (earth) | $27.32\ days$ |
| Sun | Mass | $1.99 \times 10^{30} kg$ |
| | Mean radius | $6.96 \times 10^8 m$ |
| | Period of rotation (axial) | $25 - 35\ days$ |

Table A.4: Some conversion factors

| Fundamental units: | | | |
|---|---|---|---|
| | 1 $slug$ | = | 14.59 $kg$ |
| | 1 $slug$ | = | 32.17 $lb$ mass where $g = 9.81\ ms^{-2}$ |
| | 1 $kg$ | = | 2.20 $lb$ mass where $g = 9.81\ ms^{-2}$ |
| | 1 $year\ (y)$ | = | 365.242(19878) $days = 3.156 \times 10^7 s$ |
| | 1 $in$ | = | 2.54 $cm$ |
| | 1 $mi$ | = | 5280 $ft = 1.61\ km$ |
| | 1 $nautical\ mile\ (U.S.)$ | = | 1.15 $mi = 6074\ ft = 1.852\ km$ |

Secondary units:

| | | | |
|---|---|---|---|
| Speed | 1 $knot$ | = | 1.852 $kmh^{-1} = 1.151\ mih^{-1}$ |
| Volume | 1 $gallon\ (= 4qt)\ (U.S.)$ | = | 3.78 $l = 0.83\ gal\ (Imperial)$ |

| | | | |
|---|---|---|---|
| Atomic mass unit ($u$) | 1 $u$ | = | $1.661 \times 10^{-27} kg = 931.434\ MeV\ c^{-2}$ |
| Electron volt ($eV$) | 1 $eV$ | = | $1.602 \times 10^{-19} J$ |
| Mechanical equivalent of heat (J) | J | = | 4.185 $J\ (cal)^{-1} = 3.97 \times 10^{-3}\ Btu\ (cal)^{-1}$ |
| Angstrom ($\mathring{A}$) | 1 $\mathring{A}$ | = | $1 \times 10^{-10} m$ |
| Fermi ($fm$) | 1 $fm$ | = | $1 \times 10^{-15} m$ |
| Light-year ($ly$) | 1 $ly$ | = | $9.46 \times 10^{15} m$ |
| Atmospheric pressure ($atm$) | 1 atm | = | $1.013 \times 10^5 Nm^{-2} = 1.013\ bar = 760\ torr$ |
| | 1 $radian\ (rad)$ | = | $57.30^o = 57^o\ 18'$ |

Table A.5: The Greek Alphabet

| | | | | | | |
|---|---|---|---|---|---|---|
| Alpha | A | $\alpha$ | | Nu | N | $\nu$ |
| Beta | B | $\beta$ | | Xi | $\Xi$ | $\xi$ |
| Gamma | $\Gamma$ | $\gamma$ | | Omicorn | O | $o$ |
| Delta | $\Delta$ | $\delta$ | | Pi | $\Pi$ | $\pi,\ \varpi$ |
| Epsilon | E | $\epsilon,\ \varepsilon$ | | Rho | P | $\rho,\ \varrho$ |
| Zeta | Z | $\zeta$ | | Sigma | $\Sigma$ | $\sigma,\ \varsigma$ |
| Eta | H | $\eta$ | | Tau | T | $\tau$ |
| Theta | $\Theta$ | $\theta,\ \vartheta$ | | Upsilon | $\Upsilon$ | $\upsilon$ |
| Iota | I | $\iota$ | | Phi | $\Phi$ | $\Phi,\ \varphi$ |
| Kappa | K | $\kappa$ | | Chi | X | $\chi$ |
| Lambda | $\Lambda$ | $\lambda$ | | Psi | $\Psi$ | $\psi$ |
| Mu | M | $\mu$ | | Omega | $\Omega$ | $\omega$ |

# Appendix B

# Mathematical Relations

## B2.1. Mathematical Symbols and Constants

*Table B.2.1. Mathemitcal symbols and numerical constants*

| Expression | Explanation |
|---|---|
| $a \propto b$ | $a$ is proportional to $b$ |
| $a \approx b$ | $a$ is approximately equal to $b$ |
| $a \sim b$ | $a$ is same order of magnitude (*i.e.* same power of 10) as $b$ |
| $a \neq b$ | $a$ is not equal to $b$ |
| $a \perp b$ | $a$ is perpendicular to $b$ |
| $a \parallel b$ | $a$ is parallel to $b$ |
| $a > b$ | $a$ is greater than $b$ |
| $a < b$ | $a$ is less than $b$ |
| $a \geq b$ | $a$ is greater than or equal to $b$ OR $a$ is not less than $b$ |
| $a \leq b$ | $a$ is less than or equal to $b$ OR $a$ is not greater than $b$ |
| $a \gg b$ | $a$ is much larger than $b$ |
| $a \ll b$ | $a$ is much less (smaller) than $b$ |
| $\sum_i a_i$ | Summation: $a_1 + a_2 + a_3 + ...$ |
| $\prod_i a_i$ | Product: $a_1 \times a_2 \times a_3 \times ...$ |
| n! | Factorial $n$(integer): $n \times (n-1) \times (n-2)... \times 3 \times 2 \times 1.$ |
| $\Delta x$ | Small change in $x$ |
| $\Delta x \to 0$ | $\Delta x$ approaches zero |
| $\bar{x}$ | Average value of $x$ |
| $\infty$ | Infinity |

| Constant | Value |
|---|---|
| $\pi$(rad) | $3.14159\ 26535\ 89793...\ \text{rad} = 180^o$ |
| 1 rad | $57.2957795^o = 57^o17'44.8''$ |
| $e$ | $2.7182\ 818...$ |
| ln 2 | $0.6931472...$ |
| ln 10 | $2.3025851...$ |
| $\log_{10} e$ | $0.4342945...$ |

## B.2.2. Algebra

NOTE: In this section some of the common algebraic mistakes made by students are also include as a cautionary measure.

### Quadratic equation

The quadratic equation of the form: $ax^2 + bx + c = 0$ has two roots given by:

$$x = \frac{(-)b \pm \sqrt{b^2 - 4ac}}{2a}$$

### Ratios, fractions, and inversion

$$\text{If } \frac{a}{b} = \frac{c}{d}, \text{ then}$$
$$\frac{a}{c} = \frac{b}{d},$$
$$\frac{b}{a} = \frac{d}{c}, \text{ and}$$
$$\frac{c}{a} = \frac{d}{b}$$

$$\text{If } a = (b + c + d), \text{ then}$$
$$\frac{1}{a} = \frac{1}{b + c + d)}, \text{ and}$$
$$\frac{1}{a} \neq \frac{1}{a} + \frac{1}{b} + \frac{1}{c}$$

$$\text{If } \frac{a}{b} = \frac{c}{d} + \frac{e}{f}, \text{ then}$$
$$\frac{b}{a} = \frac{1}{\{\frac{c}{d} + \frac{e}{f}\}}, \text{ and}$$
$$\frac{b}{a} \neq \frac{d}{c} + \frac{f}{e}$$

### Algebraic expansions and factorization

$$(x \pm a)^2 = x^2 \pm 2ax + a^2$$
$$(x \pm a)^3 = x^3 \pm 3ax^2 + 3a^2x \pm a^3$$
$$(x \pm a)^n = x^n + C_1 an^{n-1} + C_2 a^2 x^{n-2} + \ldots + C_j a^j x^{(n-j)} + \ldots C_n a^n$$
$$\text{where} C_j = (\pm 1)^j \frac{\{n(n-1)(n-2)\ldots(n-j+1)\}}{j!}$$

$$
\begin{aligned}
(x^2 - a^2) &= (x+a)(x-a) \\
(x^3 \pm a^3) &= (x \pm a)(x^2 \mp ax + a^2) \\
(x^4 - a^4) &= (x^2 + a^2)(x^2 - a^2) = (x^2 + a^2)(x+a)(x-a)
\end{aligned}
$$

## Manipulation of the powers and roots

$$
\begin{aligned}
x^0 &= 1 \\
x^1 &= x \\
\frac{1}{x^n} &= x^{-n} \\
x^{\frac{1}{n}} &= \sqrt[n]{x} \\
x^n . x^m &= x^{n+m} \\
(x^n)^m &= x^{nm} \\
\sqrt{(a^2 + b^2 + c^2)} &\neq (a+b+c)
\end{aligned}
$$

## Logrithmic and exponential functions

$$
\begin{aligned}
\ln x &= \ln_e x \\
\text{If } \ln x &= y, \text{ then} \\
x &= e^y \\
\ln e &= 1 \\
\ln 1 &= 0 \\
\ln 10 &= 2.302585...
\end{aligned}
$$

$$
\begin{aligned}
\ln (ab) &= \ln a + \ln b \\
\ln \frac{a}{b} &= \ln a - \ln b \\
\ln a^b &= b \ln a \\
\text{If } z &= ae^{bx}, \text{ then} \\
\ln z &= bx + \ln a
\end{aligned}
$$

$$
\begin{aligned}
\log x &= \log_{10} x \\
\text{If } \log x &= y, \text{ then} \\
y &= 10^x \\
\log 1 &= 0 \\
\log 10 &= 1 \\
\log 100 &= 2
\end{aligned}
$$

$$\log 10^n = n$$
$$\ln f(x) = \ln 10 \times \log f(x)$$
$$= (2.302585) \times \log f(x)$$

$$e^{\pm ax} = 1 \pm ax + \frac{(ax)^2}{2!} \pm \frac{(ax)^3}{3!} + \ldots (\pm 1)^j \frac{(ax)^j}{j!} \ldots, \qquad (j = 0, 1, 2, 3, \ldots \infty)$$

For $(-)1 < x < 1$,

$$\ln(1+x) = x - \frac{1}{2}x^2 + \frac{1}{3}x^3 - \frac{1}{4}x^4 + \frac{1}{5}x^5 - \ldots$$
$$\ln(1-x) = -x - \frac{1}{2}x^2 - \frac{1}{3}x^3 - \frac{1}{4}x^4 - \frac{1}{5}x^5 - \ldots$$

## Approximations

For $x \ll 1$:

$$e^{\pm x} \cong (1 \pm x)$$
$$\ln(1 \pm x) \cong \pm x$$
$$(1 \pm x)^n \cong (1 \pm nx)$$
$$(1 \pm x)^{\frac{1}{2}} \cong 1 \pm \frac{1}{2}x$$
$$(1 \pm x)^{-1} \cong 1 \mp x$$
$$(1 \pm x)^{-\frac{1}{2}} \cong 1 \mp \frac{1}{2}x$$

## B.2.3.  Plane Geometry

### Triangles

For a general triangle of sides *a*, *b*, and *c* and opposite angles *A*, *B*, and *C*:

$$A + B + C = 180^o = \pi \text{ radians}$$
$$c^2 = a^2 + b^2 - 2\,a\,b\,\cos C$$

$$\text{Area of the triangle} = \frac{1}{2}a\,b\,\sin C = \frac{1}{2}b\,c\,\sin A = \frac{1}{2}c\,a\,\sin B$$
$$= \sqrt{s \times (s-a) \times (s-b) \times (s-c)}$$
$$\text{where } s = \frac{1}{2}(a+b+c)$$

$$\frac{a}{\sin A} = \frac{b}{\sin B} = \frac{c}{\sin C}$$

For a right angled trangle in which $\angle C = 90^o$, the opposite side *c* is called the hypotenous, and $c^2 = a^2 + b^2$. This relationship between the sides of a right angle triangle is called the *Pythagora's Theorem*.

## Straight line:

General equation of a straight line is: $y = mx + c$.
where $c$ is the $y-$ intercept of the line, *i.e.* the value of $y$ when $x = 0$, and
$m$ is the slope of the line given as: $m = \frac{y_2 - y_1}{x_2 - x_1} = \Delta y / \Delta x = \tan\theta$. The angle $\theta$ is measured from the positive
$x-$ axis and is positive anticlockwise and negative clockwise.

Angle between two straight lines of slope $m_1$ and $m_2$ is given by:

$$\tan\alpha = \frac{m_2 - m_1}{1 + m_1 m_2}$$
$$\text{For parallel lines: } m_1 = m_2, \text{ and}$$
$$\text{for perpendicular lines: } m_1 \times m_2 = (-)1.$$

Equation of a straight line passing through points $(x_1, y_1)$ and $(x_2, y_2)$ is:
$y = m(x - x_1) + y_1 = m(x - x_2) + y_2$ where $m = \frac{(y_2 - y_1)}{(x_2 - x_1)}$

Distance $d$ between two points $(x_1, y_1)$ and $(x_2, y_2)$ is given by: $[(x_2 - x_1)^2 + (y_2 - y_1)^2]^{\frac{1}{2}}$

## Circle

The general equation of a circle is: $x^2 + y^2 + 2bx + 2cy + d = 0$
with center at $(-b, -c)$ and radius, $r = \sqrt{b^2 + c^2 - d}$

Equation of a circle with center at $(x_c, y_c)$ and radius $R$ is: $(x - x_c)^2 + (y - y_c)^2 = R^2$

Area of a circle of radius $R$ is: $\pi R^2$, and its circumfrence is: $2\pi R$ where $\pi = 3.1415...$ radians.

Area of a sector of angle $\theta$ (radians) of a circle of radius $R$ is: $\frac{1}{2}\theta R^2$ and the arc of the sector is: $\theta R$

## Ellipse

The general equation of an ellipse is: $Ax^2 + Bxy + Cy^2 + Dx + Ey + F = 0$ where $(B^2 - 4AC) < 0$.

Equation of an ellipse with center at the origin, and the axes of the ellipse of lengths $2a$ and $2b, (a \neq b)$, along
the $x-$ and $y-$ axes respectively is: $\frac{x^2}{a^2} + \frac{y^2}{b^2} = 1$.
Area of the ellipse is: $\pi a b$ and its circumfrence is $\approx 2\pi \left[\frac{1}{2}(a^2 + b^2)\right]^{\frac{1}{2}}$

Equation of an ellipse with center at $(x_o, y_o)$ and axes of the ellipse parallel to the coordinate axes is:
$\frac{(x - x_o)^2}{a^2} + \frac{(y - y_o)^2}{b^2} = 1$

Longer of the two axis is know as the major axis, and the shorter is the minor axis. Let $2a$ be the major axis
and $2b$ be the minor axis, $\{(a/b) > 1\}$, then $a$ and $b$ are the semi-major and semi-minor axes respectively.

The eccentricity $(e)$ of the ellipse is given by: $e = \frac{\sqrt{(a^2 - b^2)}}{a} < 1$

Circle is a special case of an ellipse for which $a = b, (a/b) = 1$, ecentricity $e = 0$, and the radius of the circle
is $R = a = b$

## Parabola

Eccentricity of a parabola: $e = 1$.

The general equation of a parabola is: $Ax^2 + Bxy + Cy^2 + Dx + Ey + F = 0$ where $(B^2 - 4AC) = 0$, and the axis of the parabola is oblique to the coordinate axes.

The equation of a parabola with its axis parallel to the $x-$ axis is: $x = ay^2 + by + c$.

The equation of a parabola with its axis parallel to the $y-$ axis is: $y = ax^2 + bx + c$.

The equation of a parabola with its axis along the $x-$ axis, vertex at the origin, and focus at $(p, 0)$ is: $y^2 = 4px$.

## Hyperbola

Let $2a$ be the transverse axis, and $2b$ be the conjugate axis of the hyperbola. The eccentricity of a hyperbola is given by: $e = \frac{\sqrt{(a^2 + b^2)}}{a} > 1$

For a rectangular hyperbola $a = b$ and $e = \sqrt{2}$, and the asymptotes are perpendicular.

General equation of a hyperbola with axes oblique to the coordinate axes is: $Ax^2 + Bxy + Cy^2 + Dx + Ey + F = 0$ where $(B^2 - 4AC) > 1$

Equation of a hyperbola with center at the origin, and the transverse axis of length $2a$ and and the conjugate axis of length $2b$ along the $x-$ and $y-$ axes respectively is: $\frac{x^2}{a^2} - \frac{y^2}{b^2} = 1$.

Equation of an ellipse with center at $(x_o, y_o)$, and the transverse axis $2a$ and the conjugate axis $2b$ parallel to the $x-$ and $y-$ axes respectively is: $\frac{(x - x_o)^2}{a^2} - \frac{(y - y_o)^2}{b^2} = 1$.

## Transformation of coordinates

Let the coordinates of a point $P$ in a 3-D space in cartesian, spherical and cylinderical coordinates system be: $P = (x, y, z) = (r, \theta, \phi) = (\rho, \phi, z)$ respectively. Then the spherical and the cylindrical coordinates are related to the cartesian coordinates by the following expressions.

**Spherical and cartesian coordinates:**

$$
\begin{aligned}
x &= r \sin \theta \cos \phi \\
y &= r \sin \theta \sin \phi \\
z &= r \cos\theta
\end{aligned}
$$

$$
\begin{aligned}
r &= \sqrt{x^2 + y^2 + z^2} \\
\theta &= \cos^{-1}\left(\frac{z}{\sqrt{x^2 + y^2 + z^2}}\right) = \cos^{-1}\left(\frac{z}{r}\right) \\
\phi &= \tan^{-1}\left(\frac{y}{x}\right)
\end{aligned}
$$

**Cylindrical and cartesian coordinates:**

$$x = \rho \cos \phi$$
$$y = \rho \sin \phi$$
$$z = z$$

$$\rho = \sqrt{x^2 + y^2}$$
$$\phi = \tan^{-1}\left(\frac{y}{x}\right)$$
$$z = z$$

**Volume element in the three coordinate systems**:
Cartesian coordinates: $dv = dx\, dy\, dz$
Spherical coordinates: $dv = r^2\, dr\, \sin\theta\, d\theta\, d\phi$
Cylindrical coordinates: $dv = \rho d\rho d\phi dz$

**Surface element in the three coordinate systems**:
Cartesian coordinates: $ds = dx\, dy$
Spherical coordinates: $ds = R^2 \sin\theta\, d\theta\, d\phi$
Cylindrical coordinates: $ds = R\, d\phi\, dz$
where $R$ is the radius of the spherical (cylinderical) surface.

# B.2.4. Solid (3-D) geometry

The following symbols are used for expressions in this section:
$V$ = volume,
$T$ = total surface area,
$S$ = lateral surface area, and
$s$ = surface area of one of the face where all faces are equal.

## Parallelepiped

Let the three edges of the parallelepiped are given by vectors **a, b,** and **c**, and the dihedral angles are $\alpha \neq \beta \neq \gamma$.

$V = \mathbf{a}.(\mathbf{b} \times \mathbf{c}) = \mathbf{b}.(\mathbf{c} \times \mathbf{a}) = \mathbf{c}.(\mathbf{a} \times \mathbf{b})$
$T = 2 \times |(\mathbf{a} \times \mathbf{b}) + (\mathbf{b} \times \mathbf{c}) + (\mathbf{c} \times \mathbf{a})| = 2(a\, b\ \sin\gamma + b\, c\ \sin\alpha + c\, a\ \sin\beta)$

For a rectangular parallelepiped:
$\alpha = \beta = \gamma = 90^o$,
$V = a \times b \times c$,
$T = 2(ab + bc + ca)$, and
body diagonal: $D = (a^2 + b^2 + c^2)^{\frac{1}{2}}$.

## Tetrahedron

A solid bounded by four equalateral triangles of side $a$ each.

Height: $h = a\sqrt{\frac{2}{3}}$, $s = \frac{\sqrt{3}}{4}a^2$, and $V = \frac{1}{6\sqrt{2}}a^3$.

## Pyramid

$V = \frac{1}{3}$(area of base) × (altitude)

$T = $ (area of slant surfaces) + (area of base) $= \frac{1}{2}$(perimeter of base) × (slant height) + (area of base).

## Cone

Cone is a regular pyramid with a circular base of radius $R$, and height $h$.

$V = \frac{1}{3}\pi R^2 h$

Stant height $h_s = (R^2 + h^2)^{\frac{1}{2}}$

$S = \pi R h_s = \pi R (R^2 + h^2)^{\frac{1}{2}}$.

$T = S + \pi R^2$.

## Cylinder

For a right circular cylinder[†] of base radius $R$, and height $h$.

$V = \pi R^2 h$, cylinder surface area: $S = 2\pi R h$, and $T = \pi R \times (R + 2h)$

† Same relations apply to a disk of radius $R$ and thickness $t$ whereby $h$ is replaced with $t$.

## Sphere

$V = \frac{4}{3}\pi R^3$, and $T = 4\pi R^2$, where $R$ is the radius of the sphere.

## Ellipsoid

$V = \frac{4}{3}\pi\, a\, b\, c$, where $a$, $b$, and $c$ are the semi axes of the ellipsoid.

## Spheroid

Formed by rotating an ellipse of major and minor semiaxes $a$ and $b$ respectively and ecentricity $e$ about one of the axes.

**Oblate spheroid: Rotation about the minor axis:** $V = \frac{4}{3}\pi a^2 b$, and $T = 2\pi a^2 + \pi\, \frac{b^2}{e}\, \ln \frac{(1+e)}{(1-e)}$

**Prolate spheroid: Rotation about the major axis:** $V = \frac{4}{3}\pi a b^2$, and $T = 2\pi b^2 + 2\pi\, \frac{ab}{e}\, \sin^{-1} e$

## B.2.5.  Trigonometry

Consider a right angle triangle $ABC$ with sides $a, b$ and $c$.  $\angle C = 90^o$, $\angle A = \theta$ and $\angle B = (90 - \theta)^o$. The side $c$ of the triangle is the hypotenuse, and with respect to $\angle A = \theta$, $a$ is designated as the opposite side, and $b$ is designated as the adjacent side. The three sides of the triangle are related by the Pythagora's theorem: $c^2 = (a^2 + b^2)$. Using this nomenclature, we define the following trigonometric functions for $\angle A = \theta$.

$$\text{sine } \theta \quad = \quad \sin \theta = \frac{a}{c} = \frac{\text{opposite}}{\text{hypotenuse}}$$

$$\text{cosine } \theta \quad = \quad \cos \theta = \frac{b}{c} = \frac{\text{adjacent}}{\text{hypotenuse}}$$

$$\text{tangent } \theta \quad = \quad \tan \theta = \frac{a}{b} = \frac{\text{opposite}}{\text{adjacent}}$$

and

$$\text{cosecant } \theta \quad = \quad \text{cosec } \theta = \frac{1}{\sin \theta}$$

$$\text{secent } \theta \quad = \quad \sec \theta = \frac{1}{\cos \theta}$$

$$\text{cotangent } \theta \quad = \quad \cot \theta = \frac{1}{\tan \theta}$$

## Table of trigonometric functions

| $\theta^o$ | $\sin \theta$ | $\cos \theta$ | $\tan \theta$ |
|---|---|---|---|
| 0 | 0 | 1 | 0 |
| 30 | $\frac{1}{2}$ | $\frac{\sqrt{3}}{2}$ | $\frac{1}{\sqrt{3}}$ |
| 45 | $\frac{1}{\sqrt{2}}$ | $\frac{1}{\sqrt{2}}$ | 1 |
| 60 | $\frac{\sqrt{3}}{2}$ | $\frac{1}{2}$ | $\sqrt{3}$ |
| 90 | 1 | 0 | $\infty$ |
| In the $I^{st}$ quadrant | $(+)$ | $(+)$ | $(+)$ |
| In the $II^{nd}$ quadrant | $(+)$ | $(-)$ | $(-)$ |
| In the $III^{rd}$ quadrant | $(-)$ | $(-)$ | $(+)$ |
| In the $IV^{th}$ quadrant | $(-)$ | $(+)$ | $(-)$ |
| $(-)\alpha$ | $(-) \sin \alpha$ | $(+) \cos \alpha$ | $(-) \tan \alpha$ |
| $90 \pm \alpha$ | $(+) \cos \alpha$ | $(\mp) \sin \alpha$ | $(\mp) \cot \alpha$ |
| $180 \pm \alpha$ | $(\mp) \sin \alpha$ | $(-) \cos \alpha$ | $(\pm) \tan \alpha$ |
| $270 \pm \alpha$ | $(-) \cos \alpha$ | $(\pm) \sin \alpha$ | $(\mp) \cot \alpha$ |
| $360 \pm \alpha$ | $(\pm) \sin \alpha$ | $(+) \cos \alpha$ | $(\pm) \tan \alpha$ |

## Trigonometric identities

$$\sin^2 \theta + \cos^2 \theta \quad = \quad 1$$

$$\sec^2 \theta \quad = \quad 1 + \tan^2 \theta$$

$$\text{cosec}^2 \theta \quad = \quad 1 + \cot^2 \theta$$

$$\sin (\alpha \pm \beta) \quad = \quad \sin \alpha \cos \beta \pm \cos \alpha \sin \beta$$

$$\cos (\alpha \pm \beta) \quad = \quad \cos \alpha \cos \beta \mp \sin \alpha \sin \beta$$

$$\tan (\alpha \pm \beta) \quad = \quad \frac{(\tan \alpha \pm \tan \beta)}{(1 \mp \tan \alpha \tan \beta)}$$

$$\sin 2\alpha = 2 \sin \alpha \cos \alpha$$

$$\cos 2\alpha = \cos^2 \alpha - \sin^2 \alpha = 2 \cos^2 \alpha - 1 = 1 - 2 \sin^2 \alpha$$

$$\tan 2\alpha = \frac{2 \tan \alpha}{(1 - \tan^2 \alpha)}$$

$$\sin 3\alpha = 3 \sin \alpha - 4 \sin^2 \alpha$$

$$\cos 3\alpha = 4 \cos^2 \alpha - 3 \cos \alpha$$

$$\tan 3\alpha = \frac{3 \tan \alpha - \tan^3 \alpha}{(1 - 3 \tan^2 \alpha)}$$

$$\sin \alpha + \sin \beta = 2 \sin \left( \frac{\alpha + \beta}{2} \right) \cos \left( \frac{\alpha - \beta}{2} \right)$$

$$\sin \alpha - \sin \beta = 2 \cos \left( \frac{\alpha + \beta}{2} \right) \sin \left( \frac{\alpha - \beta}{2} \right)$$

$$\cos \alpha + \cos \beta = 2 \cos \left( \frac{\alpha + \beta}{2} \right) \cos \left( \frac{\alpha - \beta}{2} \right)$$

$$\cos \alpha - \cos \beta = (-)2 \sin \left( \frac{\alpha + \beta}{2} \right) \sin \left( \frac{\alpha - \beta}{2} \right)$$

$$\tan \alpha + \tan \beta = \frac{\sin (\alpha + \beta)}{\cos \alpha \cos \beta}$$

$$\tan \alpha - \tan \beta = \frac{\sin (\alpha - \beta)}{\cos \alpha \cos \beta}$$

$$\sin \alpha \sin \beta = \frac{1}{2} \cos (\alpha - \beta) - \frac{1}{2} \cos (\alpha + \beta)$$

$$\cos \alpha \cos \beta = \frac{1}{2} \cos (\alpha - \beta) + \frac{1}{2} \cos (\alpha + \beta)$$

$$\sin \alpha \cos \beta = \frac{1}{2} \sin (\alpha + \beta) + \frac{1}{2} \sin (\alpha - \beta)$$

$$\cos \alpha \sin \beta = \frac{1}{2} \sin (\alpha + \beta) - \frac{1}{2} \sin (\alpha - \beta)$$

**For angle $\alpha$ in radians:**

$$\sin \alpha = \frac{e^{i\alpha} - e^{-i\alpha}}{2i}; \quad \text{where } i = \sqrt{-1}$$

$$\cos \alpha = \frac{e^{i\alpha} + e^{-i\alpha}}{2}$$

$$e^{\pm i\alpha} = \cos \alpha \pm i \sin \alpha$$

$$\sin \alpha = \alpha - \frac{\alpha^3}{3!} + \frac{\alpha^5}{5!} - \frac{\alpha^7}{7!} + ... + (-1)^{2j+1} \frac{\alpha^{2j+1}}{(2j+1)!} + ..., \quad \text{where } j = 0, 1, 2, 3, ...$$

$$\cos \alpha = 1 - \frac{\alpha^2}{2!} + \frac{\alpha^4}{4!} - \frac{\alpha^6}{6!} + ... + (-1)^j \frac{\alpha^{2j}}{(2j)} + ...$$

For $\alpha$ (radians) $\ll 1$: $\sin \alpha = \alpha$; $\cos \alpha = 1$; $\tan \alpha = \alpha$

## Hyperbolic functions

For a real argument $u$:

$$\text{hyperbolic sine of u} = \sinh u = \frac{e^u - e^{-u}}{2}$$

$$\text{hyperbolic cosine of u} = \cosh u = \frac{e^u + e^{-u}}{2}$$

$$\text{hyperbolic tangent of u} = \tanh u = \frac{\sinh u}{\cosh u}$$

$$\text{hyperbolic cosecant of u} = \operatorname{cosech} u = \frac{1}{\sinh u}$$

$$\text{hyperbolic secant of u} = \operatorname{sech} u = \frac{1}{\cosh u}$$

$$\text{hyperbolic cotangent of u} = \coth u = \frac{1}{\tanh u}$$

### Relationship between the squares of functions

$$\cosh^2 u - \sinh^2 u = 1$$
$$\tanh^2 u + \operatorname{sech}^2 u = 1$$
$$\coth^2 u - \operatorname{cosech}^2 u = 1$$
$$\operatorname{cosech}^2 u - \operatorname{sech}^2 u = \operatorname{cosech}^2 u \operatorname{sech}^2 u$$

### Symmetry and periodicity

$$\sinh (-u) = (-)\sinh u$$
$$\cosh (-u) = \cosh u$$
$$\tanh (-u) = (-)\tanh u$$
$$\operatorname{cosech} (-u) = (-)\operatorname{cosech} u$$
$$\operatorname{sech} (-u) = \operatorname{sech} u$$
$$\coth (-u) = (-)\coth u$$

**Range of functions for real argument** $u$

| Function | Range of u | Range of function |
|----------|------------|-------------------|
| sinh u | $(-\infty, +\infty)$ | $(-\infty, +\infty)$ |
| cosh u | $(-\infty, +\infty)$ | $(1, +\infty)$ |
| tanh u | $(-\infty, +\infty)$ | $(-1, +1)$ |
| cosech u | $(-\infty, 0)$ | $(0, -\infty)$ |
|  | $(0, +\infty)$ | $(+\infty, 0)$ |
| sech u | $(-\infty, +\infty)$ | $(0, 1)$ |
| coth u | $(-\infty, 0)$ | $(1, -\infty)$ |
|  | $(0, +\infty)$ | $(+\infty, 1)$ |

**Special values of hyperbolic functions**

| x | 0 | $\frac{\pi}{2}i$ | $\pi i$ | $\frac{3\pi}{2}i$ | $\infty$ |
|---|---|---|---|---|---|
| sinh x | 0 | $i$ | 0 | $-i$ | $\infty$ |
| cosh x | 1 | 0 | -1 | 0 | $\infty$ |
| tanh x | 0 | $\infty i$ | 0 | $-\infty i$ | 1 |
| cosech x | $\infty$ | $-i$ | $\infty$ | $i$ | 0 |
| sech x | 1 | $\infty$ | -1 | $\infty$ | 0 |
| coth x | $\infty$ | 0 | $\infty$ | 0 | 1 |

# B.2.6. Differential Calculus

**NOTE:** In the differential expressions given below, and in the integral expressions in the next section $A, B$ and $C$ are the functions of $x$, $f(A)$ is a function of $A$, and $a, b, c, n$, and $m$ are real constants. The arguments of trigonometric functions are expressed in radians.

**Definition:**

$$\frac{d}{dx}A = Lim_{\Delta x \to 0} \frac{\Delta A}{\Delta x}$$

$$\frac{d}{dx}(A \pm B \mp C) = \frac{d}{dx}A \pm \frac{d}{dx}B \mp \frac{d}{dx}C$$

$$\frac{d}{dx}(ABC) = BC\frac{d}{dx}A + CA\frac{d}{dx}B + AB\frac{d}{dx}C$$

$$\frac{d}{dx}\left(\frac{1}{ABC}\right) = \frac{1}{BC}\frac{d}{dx}\frac{1}{A} + \frac{1}{CA}\frac{d}{dx}\frac{1}{B} + \frac{1}{AB}\frac{d}{dx}\frac{1}{C}$$

$$\frac{d}{dx}\left(\frac{A}{BC}\right) = \frac{1}{BC}\frac{d}{dx}A + A\frac{d}{dx}\frac{1}{BC}$$

$$\frac{d}{dx}f(A) = \frac{d}{dA}f(A)\frac{d}{dx}A$$

$$\frac{d}{dx}A^B = B\,A^{B-1}\frac{d}{dx}A + A^B \ln A\frac{d}{dx}B$$

**Table of some common differential expressions**

| A(x) | $\frac{d}{dx}$ A |
|------|------------------|
| $a\ x^n$ | $a\ nx^{n-1}$ |
| $a^x$ | $a^x \ln a$ |
| $e^x$ | $e^x$ |
| $\ln x$ | $\frac{1}{x}$ |
| $\sin x$ | $\cos x$ |
| $\cos x$ | $(-)\sin x$ |
| $\tan x$ | $\sec^2 x$ |
| $\text{cosec } x$ | $(-)\cot x \text{ cosec } x$ |
| $\sec x$ | $\tan x \sec x$ |
| $\cot x$ | $(-)\text{cosec}^2 x$ |

| f(A) | $\frac{d}{dx}$ f(A) |
|------|---------------------|
| $a\ A^n$ | $a\ nA^{(n-1)} \frac{d}{dx}A$ |
| $a^A$ | $a^A \ln a \frac{d}{dx}A$ |
| $e^A$ | $e^A \frac{d}{dx}A$ |
| $\ln A$ | $\frac{1}{A} \frac{d}{dx}A$ |
| $\sin A$ | $\cos A \frac{d}{dx}A$ |
| $\cos A$ | $(-)\sin A \frac{d}{dx}A$ |
| $\tan A$ | $\sec^2 A \frac{d}{dx}A$ |

$d(ABC) = BC\ dA + CA\ dB + AB\ dC$

If $y = f(x)$, then dy $= \frac{d}{dx}f(x)\ dx$

# B.2.7. Integral Calculus

**Definition:**

$$\sum_{Lim\ \Delta x_i \to 0} (A(x_i)\ \Delta x_i) = \int A(x)\ dx$$

Integral of the product of two functions $AB$ is obtained by**integration by parts** following the expression given below:

$$\int (AB)\ dx = A \int B dx - \int \left(\frac{d}{dx} A \int B\ dx\right) dx$$

## Table of some common integral expressions

| $A(x)$ | $\int A(x)\,dx$ |
|---|---|
| $a\,x^n$ | $\frac{a}{n+1}\,x^{n+1}$ |
| $e^x$ | $e^x$ |
| $\ln ax$ | $x\ln ax - x$ |
| $\sin ax$ | $-\frac{1}{a}\cos ax$ |
| $\cos ax$ | $\frac{1}{a}\sin ax$ |
| $\tan ax$ | $-\frac{1}{a}\ln(\cos ax)$ |
| $\operatorname{cosec} ax$ | $\frac{1}{a}\ln(\operatorname{cosec} ax - \cot ax)$ |
| $\sec ax$ | $\frac{1}{a}\ln(\sec ax + \tan ax)$ |
| $\cot ax$ | $\frac{1}{a}\ln(\sin ax)$ |

# Bibliography

Arya, A. P., (1979). *Introductory College Physics*, McMillan Publishing Co., Inc., New York, USA.

Beiser, A. (1973) *Concepts of Modern Physics*, Mc-graw Hill, Kogakusha, Ltd., Japan

Benson, H., (1996). *University Physics*, revised edition, John Wiley & Sons, Inc., New York, USA.

Born, M. and E. Wolf (1975) *Principles of optics* Pergamon Press, Oxford

Born, M. and E. Wolf (1980) *Principles of Optics: Electromagnetic theory of propagation, interference and diffraction of light*, Cambridge University Press, UK

Brinkmanship, W.F, D. E. Haggan and W. W. Troutman (1997), IEEE J. Solid State Circuits, **12**, No.22, 1858 "A History of the Invention of the Transistor and where it will lead us".

Bueche, F. J., (1986). *Introduction to Physics for Scientists and Engineers*, $4^{th}$ Edition, McGraw-Hill Book Company, New York, USA.

Crease, R. P. (September 2002) *"The most beautiful experiment"*, Physics World, **15**, Number 8, 19, and also *The Editorial Comment: "The double-slit experiment"* page 15

Ditchburn, R.W. (1976) *Light*, Academic Press, London

Fishbane, P. M. S., S. Gasiorowitz and S. T. Thorton, (1996). *Physics for Scientists and Engineers*, $2^{nd}$ Edition, Prentics Hall, New Jersey, USA.

Giancoli, D. C., (2000). *Physics for Scientists and Engineers*, $3^{rd}$ Edition, Prentice Hall, New Jersey, USA.

Griffiths, D. J. (1989) *Introduction to Electrodynamics*, Prentice Hall International, Inc., New jersey

Ingram, D.J.E. (1973) *Radiation and quantum physics*, Clarendon Press, Oxford

Jain, P. K. and J S Nkoma,(2004), *Introduction to Mechanics: Kinematics, Newtonian and Lagrangian*, Bay Publishers, Gaborone.

Jain, P. K. and L. K. Sharma (1988), J Appl. Sc. in Southern Afr., **4**, 80 - 101, *"The Physics of Blackbody Radiation: A Review"*

Jenkins, F. A. and H. E. White, H. E. (1981) *Fundamentals of Optics*, Mc-Graw Hill Book Company, London

Kasap, S. O. (2013), *Optoelectronics and Photonics*, 2nd Edition), Pearson.

Reif, F. (1984) *Fundamentals of Statistical and Thermal Physics*, McGraw-Hill

Kittel, C. (2004), *Introduction to Solid State physics*, 8th Edition, John Wiley& Sons

Landau, L. D. and E. M. Liftshitz, (1960). *Electrodynamics of Continous Media*, Pergamon Press, Oxford, UK.

Lipson, H.S. (ed.) (1972) *Optical transforms*, Academic Press, New York

Longhurst, R.S. (1973) *Geometrical and physical optics*, Longmans, London

Loudon, R. (1973) *Quantum theory of light*, Clarendon Press, Oxford

Midwinter, J.E. (1979) *Optical fibres for transmission*, John Wiley, New York

Miller, D. A. B (2008), *Quantum Mechanics for Scientists and Engineers*, Cambridge.

Mishra, U. K and J. Singh., (2008), *Semiconductor Device Physics and Design*, Springer.

Mynbaev, D. K. and Scheiner, L. L. (2001), *Fiber optic communication technology*, Prentice Hall Intl., New York

Nkoma, J. S., R. Loudon and D. R. Tilley (1974), J.Phys. **C7**, 3547-3559, *"Elementary properties of surface polaritons"*.

Nkoma, J. S. and R. Loudon (1975), J.Phys. **C8**, 1950-1968, *"Theory of Raman scattering by surface polaritons"*.

Nkoma, J. S. (1975), J.Phys. **C8**, 3919-3936, *"Theory of Raman scattering by surface polaritons in a thin film"*.

Nkoma, J. S. (1999), Physica B: Condensed Matter **262** 273 - 283, *"Theory of Raman scattering by bulk polaritons of a composite medium"*

Nkoma, J. S. (1999), J. Phys. Condens. Matter **11** 4093 - 4107, *"Generalized theory of Raman scattering by bulk and surface polaritons"*

Nkoma, J. S. and G. Ekosse (1999), J. Phys. Condens. Matter **11** 121 - 128, *"X-ray Diffraction Study of Chalcopyrite CuFeS$_2$, Pentlandite (Fe,Ni)$_9$S$_8$ and Pyrrhotite Fe$_{1-x}$S obtained from Cu-Ni orebodies"*,

Nkoma, J. S. and P K Jain (2003), *Introduction to Optics: Geometrical, Physical and Quantum*, Bay Publishers, Gaborone.

Ohanian, H. C. (1985) *Physics*, W W Norton & Company

Robinson, F.N.H. (1973) *Electromagnetism*, Clarendon Press, Oxford

Sears, F. W. and Salinger G.L., *Thermodynamics, Kinetic Theory and Statistical Thermo-dynamics*, Blackwel

Sears, F. W., M. W. Zemansky and H. D. Young (1985) *College Physics*, Addison-Wesley

Senior, J. M. and Jamro, M. Y. (2009), *Optical fibre communication: Principles and practice*, $3^{rd}$ Edition), Prentice Hall Int, New York.

Serway, R. A. and J. W. Jewett Jr., (2004). *Physics for Scientists and Engineers* , $6^{th}$ Edition, Thomson Brooks Cole, UK.

Stanley, H. Eugene. (1971) *Introduction to Phase Transitions and Crirical Phenomena*, (Oxford University Press)

Stratton, J.A. (1941) *Electromagnetic theory*, (New York: McGraw-Hill)

Svelto, O. (1976) *Principles of lasers*, (trans. D.C. Hanna). Heyden, London

Tipler, P. A., (1999). *Physics for Scientists and Engineers*, $4^{th}$ Edition, W. H. Freefan and Company, Worth Publishers, New York, USA.

The Tanzania Communication Regulatory Authority (TCRA) Act, (2003), Tanzania Government Printer

The Electronic and Postal Communications Act (EPOCA), (2010), Tanzania Government Printer

The Electronic and Postal Communications Act (EPOCA) Regulations, (2011), Tanzania Government Printer

Tsai, M. R. et.al., (2013), Biomedical Optics **18** (2), 026012

Urone, P. P. , (2001). *College Physics*, $2^{nd}$ Edition, Thomson Brooks/ Cole, UK.

Welford, W.T. (1962) *Geometrical optics; optical instrumentation*, North-Holland, Amsterdam

Welford, W.T. (1974) *Aberrations of the symmetrical optical system*, Academic Press, London

Welford, W.T. (1981) *Optics*, Oxford University Press

Wilson, J. D. and A. J. Bufa, (2003). *College Physics*, $5^{th}$ Edition, Prentica Hall, New Jersey, USA.

Wolfson, R and J. M. Pasachoff (1999) *Physics*, Addison-Wesley

Yariv, A. (1989), *Quantum Electronics*, 3 Edition, John Wiley & Sons

Young, H D and R. A. Freedman (2000) *University Physics*, Addison-Wesley

Ziman, J. (1972), *Principles of the theory of solids*, Cambridge University Press

**Internet Resources**

https://www.andajun.wordpress.com

https://www.britannica.com

https://www.circuitstoday.com

https://www.commons.wikimedia.org

https://www.commsbusiness.co.uk

https://www.ecse.rpc.edu/Schubert, EF

https://www.intechopen.com/A.J.S. Machado et.al.

https://www.ligo.org

https://www.physics.smu.edu/Randall J. Scaliness

https://www.plot.ly/javario

https://www.wikipedia.org

# Index

NOTES

NOTES

Printed in the United States
By Bookmasters